Optimal Control of Induction Heating Processes

MECHANICAL ENGINEERING
A Series of Textbooks and Reference Books

Founding Editor

L. L. Faulkner

Columbus Division, Battelle Memorial Institute
and Department of Mechanical Engineering
The Ohio State University
Columbus, Ohio

1. *Spring Designer's Handbook*, Harold Carlson
2. *Computer-Aided Graphics and Design*, Daniel L. Ryan
3. *Lubrication Fundamentals*, J. George Wills
4. *Solar Engineering for Domestic Buildings*, William A. Himmelman
5. *Applied Engineering Mechanics: Statics and Dynamics*, G. Boothroyd and C. Poli
6. *Centrifugal Pump Clinic*, Igor J. Karassik
7. *Computer-Aided Kinetics for Machine Design*, Daniel L. Ryan
8. *Plastics Products Design Handbook, Part A: Materials and Components; Part B: Processes and Design for Processes*, edited by Edward Miller
9. *Turbomachinery: Basic Theory and Applications*, Earl Logan, Jr.
10. *Vibrations of Shells and Plates*, Werner Soedel
11. *Flat and Corrugated Diaphragm Design Handbook*, Mario Di Giovanni
12. *Practical Stress Analysis in Engineering Design*, Alexander Blake
13. *An Introduction to the Design and Behavior of Bolted Joints*, John H. Bickford
14. *Optimal Engineering Design: Principles and Applications*, James N. Siddall
15. *Spring Manufacturing Handbook*, Harold Carlson
16. *Industrial Noise Control: Fundamentals and Applications*, edited by Lewis H. Bell
17. *Gears and Their Vibration: A Basic Approach to Understanding Gear Noise*, J. Derek Smith
18. *Chains for Power Transmission and Material Handling: Design and Applications Handbook*, American Chain Association
19. *Corrosion and Corrosion Protection Handbook*, edited by Philip A. Schweitzer
20. *Gear Drive Systems: Design and Application*, Peter Lynwander
21. *Controlling In-Plant Airborne Contaminants: Systems Design and Calculations*, John D. Constance
22. *CAD/CAM Systems Planning and Implementation*, Charles S. Knox
23. *Probabilistic Engineering Design: Principles and Applications*, James N. Siddall

24. *Traction Drives: Selection and Application*, Frederick W. Heilich III and Eugene E. Shube
25. *Finite Element Methods: An Introduction*, Ronald L. Huston and Chris E. Passerello
26. *Mechanical Fastening of Plastics: An Engineering Handbook*, Brayton Lincoln, Kenneth J. Gomes, and James F. Braden
27. *Lubrication in Practice: Second Edition*, edited by W. S. Robertson
28. *Principles of Automated Drafting*, Daniel L. Ryan
29. *Practical Seal Design*, edited by Leonard J. Martini
30. *Engineering Documentation for CAD/CAM Applications*, Charles S. Knox
31. *Design Dimensioning with Computer Graphics Applications*, Jerome C. Lange
32. *Mechanism Analysis: Simplified Graphical and Analytical Techniques*, Lyndon O. Barton
33. *CAD/CAM Systems: Justification, Implementation, Productivity Measurement*, Edward J. Preston, George W. Crawford, and Mark E. Coticchia
34. *Steam Plant Calculations Manual*, V. Ganapathy
35. *Design Assurance for Engineers and Managers*, John A. Burgess
36. *Heat Transfer Fluids and Systems for Process and Energy Applications*, Jasbir Singh
37. *Potential Flows: Computer Graphic Solutions*, Robert H. Kirchhoff
38. *Computer-Aided Graphics and Design: Second Edition*, Daniel L. Ryan
39. *Electronically Controlled Proportional Valves: Selection and Application*, Michael J. Tonyan, edited by Tobi Goldoftas
40. *Pressure Gauge Handbook*, AMETEK, U.S. Gauge Division, edited by Philip W. Harland
41. *Fabric Filtration for Combustion Sources: Fundamentals and Basic Technology*, R. P. Donovan
42. *Design of Mechanical Joints*, Alexander Blake
43. *CAD/CAM Dictionary*, Edward J. Preston, George W. Crawford, and Mark E. Coticchia
44. *Machinery Adhesives for Locking, Retaining, and Sealing*, Girard S. Haviland
45. *Couplings and Joints: Design, Selection, and Application*, Jon R. Mancuso
46. *Shaft Alignment Handbook*, John Piotrowski
47. *BASIC Programs for Steam Plant Engineers: Boilers, Combustion, Fluid Flow, and Heat Transfer*, V. Ganapathy
48. *Solving Mechanical Design Problems with Computer Graphics*, Jerome C. Lange
49. *Plastics Gearing: Selection and Application*, Clifford E. Adams
50. *Clutches and Brakes: Design and Selection*, William C. Orthwein
51. *Transducers in Mechanical and Electronic Design*, Harry L. Trietley
52. *Metallurgical Applications of Shock-Wave and High-Strain-Rate Phenomena*, edited by Lawrence E. Murr, Karl P. Staudhammer, and Marc A. Meyers
53. *Magnesium Products Design*, Robert S. Busk
54. *How to Integrate CAD/CAM Systems: Management and Technology*, William D. Engelke

55. *Cam Design and Manufacture: Second Edition; with cam design software for the IBM PC and compatibles*, disk included, Preben W. Jensen
56. *Solid-State AC Motor Controls: Selection and Application*, Sylvester Campbell
57. *Fundamentals of Robotics*, David D. Ardayfio
58. *Belt Selection and Application for Engineers*, edited by Wallace D. Erickson
59. *Developing Three-Dimensional CAD Software with the IBM PC*, C. Stan Wei
60. *Organizing Data for CIM Applications*, Charles S. Knox, with contributions by Thomas C. Boos, Ross S. Culverhouse, and Paul F. Muchnicki
61. *Computer-Aided Simulation in Railway Dynamics*, by Rao V. Dukkipati and Joseph R. Amyot
62. *Fiber-Reinforced Composites: Materials, Manufacturing, and Design*, P. K. Mallick
63. *Photoelectric Sensors and Controls: Selection and Application*, Scott M. Juds
64. *Finite Element Analysis with Personal Computers*, Edward R. Champion, Jr. and J. Michael Ensminger
65. *Ultrasonics: Fundamentals, Technology, Applications: Second Edition, Revised and Expanded*, Dale Ensminger
66. *Applied Finite Element Modeling: Practical Problem Solving for Engineers*, Jeffrey M. Steele
67. *Measurement and Instrumentation in Engineering: Principles and Basic Laboratory Experiments*, Francis S. Tse and Ivan E. Morse
68. *Centrifugal Pump Clinic: Second Edition, Revised and Expanded*, Igor J. Karassik
69. *Practical Stress Analysis in Engineering Design: Second Edition, Revised and Expanded*, Alexander Blake
70. *An Introduction to the Design and Behavior of Bolted Joints: Second Edition, Revised and Expanded*, John H. Bickford
71. *High Vacuum Technology: A Practical Guide*, Marsbed H. Hablanian
72. *Pressure Sensors: Selection and Application*, Duane Tandeske
73. *Zinc Handbook: Properties, Processing, and Use in Design*, Frank Porter
74. *Thermal Fatigue of Metals*, Andrzej Weronski and Tadeusz Hejwowski
75. *Classical and Modern Mechanisms for Engineers and Inventors*, Preben W. Jensen
76. *Handbook of Electronic Package Design*, edited by Michael Pecht
77. *Shock-Wave and High-Strain-Rate Phenomena in Materials*, edited by Marc A. Meyers, Lawrence E. Murr, and Karl P. Staudhammer
78. *Industrial Refrigeration: Principles, Design and Applications*, P. C. Koelet
79. *Applied Combustion*, Eugene L. Keating
80. *Engine Oils and Automotive Lubrication*, edited by Wilfried J. Bartz
81. *Mechanism Analysis: Simplified and Graphical Techniques, Second Edition, Revised and Expanded*, Lyndon O. Barton
82. *Fundamental Fluid Mechanics for the Practicing Engineer*, James W. Murdock
83. *Fiber-Reinforced Composites: Materials, Manufacturing, and Design, Second Edition, Revised and Expanded*, P. K. Mallick

84. *Numerical Methods for Engineering Applications*, Edward R. Champion, Jr.
85. *Turbomachinery: Basic Theory and Applications, Second Edition, Revised and Expanded*, Earl Logan, Jr.
86. *Vibrations of Shells and Plates: Second Edition, Revised and Expanded*, Werner Soedel
87. *Steam Plant Calculations Manual: Second Edition, Revised and Expanded*, V. Ganapathy
88. *Industrial Noise Control: Fundamentals and Applications, Second Edition, Revised and Expanded*, Lewis H. Bell and Douglas H. Bell
89. *Finite Elements: Their Design and Performance*, Richard H. MacNeal
90. *Mechanical Properties of Polymers and Composites: Second Edition, Revised and Expanded*, Lawrence E. Nielsen and Robert F. Landel
91. *Mechanical Wear Prediction and Prevention*, Raymond G. Bayer
92. *Mechanical Power Transmission Components*, edited by David W. South and Jon R. Mancuso
93. *Handbook of Turbomachinery*, edited by Earl Logan, Jr.
94. *Engineering Documentation Control Practices and Procedures*, Ray E. Monahan
95. *Refractory Linings Thermomechanical Design and Applications*, Charles A. Schacht
96. *Geometric Dimensioning and Tolerancing: Applications and Techniques for Use in Design, Manufacturing, and Inspection*, James D. Meadows
97. *An Introduction to the Design and Behavior of Bolted Joints: Third Edition, Revised and Expanded*, John H. Bickford
98. *Shaft Alignment Handbook: Second Edition, Revised and Expanded*, John Piotrowski
99. *Computer-Aided Design of Polymer-Matrix Composite Structures*, edited by Suong Van Hoa
100. *Friction Science and Technology*, Peter J. Blau
101. *Introduction to Plastics and Composites: Mechanical Properties and Engineering Applications*, Edward Miller
102. *Practical Fracture Mechanics in Design*, Alexander Blake
103. *Pump Characteristics and Applications*, Michael W. Volk
104. *Optical Principles and Technology for Engineers*, James E. Stewart
105. *Optimizing the Shape of Mechanical Elements and Structures*, A. A. Seireg and Jorge Rodriguez
106. *Kinematics and Dynamics of Machinery*, Vladimír Stejskal and Michael Valásek
107. *Shaft Seals for Dynamic Applications*, Les Horve
108. *Reliability-Based Mechanical Design*, edited by Thomas A. Cruse
109. *Mechanical Fastening, Joining, and Assembly*, James A. Speck
110. *Turbomachinery Fluid Dynamics and Heat Transfer*, edited by Chunill Hah
111. *High-Vacuum Technology: A Practical Guide, Second Edition, Revised and Expanded*, Marsbed H. Hablanian
112. *Geometric Dimensioning and Tolerancing: Workbook and Answerbook*, James D. Meadows

113. *Handbook of Materials Selection for Engineering Applications*, edited by G. T. Murray
114. *Handbook of Thermoplastic Piping System Design*, Thomas Sixsmith and Reinhard Hanselka
115. *Practical Guide to Finite Elements: A Solid Mechanics* Approach, Steven M. Lepi
116. *Applied Computational Fluid Dynamics*, edited by Vijay K. Garg
117. *Fluid Sealing Technology*, Heinz K. Muller and Bernard S. Nau
118. *Friction and Lubrication in Mechanical Design*, A. A. Seireg
119. *Influence Functions and Matrices*, Yuri A. Melnikov
120. *Mechanical Analysis of Electronic Packaging Systems*, Stephen A. McKeown
121. *Couplings and Joints: Design, Selection, and Application, Second Edition, Revised and Expanded*, Jon R. Mancuso
122. *Thermodynamics: Processes and Applications*, Earl Logan, Jr.
123. *Gear Noise and Vibration*, J. Derek Smith
124. *Practical Fluid Mechanics for Engineering Applications*, John J. Bloomer
125. *Handbook of Hydraulic Fluid Technology*, edited by George E. Totten
126. *Heat Exchanger Design Handbook*, T. Kuppan
127. *Designing for Product Sound Quality*, Richard H. Lyon
128. *Probability Applications in Mechanical Design*, Franklin E. Fisher and Joy R. Fisher
129. *Nickel Alloys*, edited by Ulrich Heubner
130. *Rotating Machinery Vibration: Problem Analysis and Troubleshooting*, Maurice L. Adams, Jr.
131. *Formulas for Dynamic Analysis*, Ronald L. Huston and C. Q. Liu
132. *Handbook of Machinery Dynamics*, Lynn L. Faulkner and Earl Logan, Jr.
133. *Rapid Prototyping Technology: Selection and Application*, Kenneth G. Cooper
134. *Reciprocating Machinery Dynamics: Design and Analysis*, Abdulla S. Rangwala
135. *Maintenance Excellence: Optimizing Equipment Life-Cycle Decisions*, edited by John D. Campbell and Andrew K. S. Jardine
136. *Practical Guide to Industrial Boiler Systems*, Ralph L. Vandagriff
137. *Lubrication Fundamentals: Second Edition, Revised and Expanded*, D. M. Pirro and A. A. Wessol
138. *Mechanical Life Cycle Handbook: Good Environmental Design and Manufacturing*, edited by Mahendra S. Hundal
139. *Micromachining of Engineering Materials*, edited by Joseph McGeough
140. *Control Strategies for Dynamic Systems: Design and Implementation*, John H. Lumkes, Jr.
141. *Practical Guide to Pressure Vessel Manufacturing*, Sunil Pullarcot
142. *Nondestructive Evaluation: Theory, Techniques, and Applications*, edited by Peter J. Shull
143. *Diesel Engine Engineering: Thermodynamics, Dynamics, Design, and Control*, Andrei Makartchouk
144. *Handbook of Machine Tool Analysis*, Ioan D. Marinescu, Constantin Ispas, and Dan Boboc

145. *Implementing Concurrent Engineering in Small Companies*, Susan Carlson Skalak
146. *Practical Guide to the Packaging of Electronics: Thermal and Mechanical Design and Analysis*, Ali Jamnia
147. *Bearing Design in Machinery: Engineering Tribology and Lubrication*, Avraham Harnoy
148. *Mechanical Reliability Improvement: Probability and Statistics for Experimental Testing*, R. E. Little
149. *Industrial Boilers and Heat Recovery Steam Generators: Design, Applications, and Calculations*, V. Ganapathy
150. *The CAD Guidebook: A Basic Manual for Understanding and Improving Computer-Aided Design*, Stephen J. Schoonmaker
151. *Industrial Noise Control and Acoustics*, Randall F. Barron
152. *Mechanical Properties of Engineered Materials*, Wolé Soboyejo
153. *Reliability Verification, Testing, and Analysis in Engineering Design*, Gary S. Wasserman
154. *Fundamental Mechanics of Fluids: Third Edition*, I. G. Currie
155. *Intermediate Heat Transfer*, Kau-Fui Vincent Wong
156. *HVAC Water Chillers and Cooling Towers: Fundamentals, Application, and Operation*, Herbert W. Stanford III
157. *Gear Noise and Vibration: Second Edition, Revised and Expanded*, J. Derek Smith
158. *Handbook of Turbomachinery: Second Edition, Revised and Expanded*, edited by Earl Logan, Jr. and Ramendra Roy
159. *Piping and Pipeline Engineering: Design, Construction, Maintenance, Integrity, and Repair*, George A. Antaki
160. *Turbomachinery: Design and Theory*, Rama S. R. Gorla and Aijaz Ahmed Khan
161. *Target Costing: Market-Driven Product Design*, M. Bradford Clifton, Henry M. B. Bird, Robert E. Albano, and Wesley P. Townsend
162. *Fluidized Bed Combustion*, Simeon N. Oka
163. *Theory of Dimensioning: An Introduction to Parameterizing Geometric Models*, Vijay Srinivasan
164. *Handbook of Mechanical Alloy Design*, edited by George E. Totten, Lin Xie, and Kiyoshi Funatani
165. *Structural Analysis of Polymeric Composite Materials*, Mark E. Tuttle
166. *Modeling and Simulation for Material Selection and Mechanical Design*, edited by George E. Totten, Lin Xie, and Kiyoshi Funatani
167. *Handbook of Pneumatic Conveying Engineering*, David Mills, Mark G. Jones, and Vijay K. Agarwal
168. *Clutches and Brakes: Design and Selection, Second Edition*, William C. Orthwein
169. *Fundamentals of Fluid Film Lubrication: Second Edition*, Bernard J. Hamrock, Steven R. Schmid, and Bo O. Jacobson
170. *Handbook of Lead-Free Solder Technology for Microelectronic Assemblies*, edited by Karl J. Puttlitz and Kathleen A. Stalter
171. *Vehicle Stability*, Dean Karnopp
172. *Mechanical Wear Fundamentals and Testing: Second Edition, Revised and Expanded*, Raymond G. Bayer
173. *Liquid Pipeline Hydraulics*, E. Shashi Menon

174. *Solid Fuels Combustion and Gasification*, Marcio L. de Souza-Santos
175. *Mechanical Tolerance Stackup and Analysis*, Bryan R. Fischer
176. *Engineering Design for Wear*, Raymond G. Bayer
177. *Vibrations of Shells and Plates: Third Edition, Revised and Expanded*, Werner Soedel
178. *Refractories Handbook*, edited by Charles A. Schacht
179. *Practical Engineering Failure Analysis*, Hani M. Tawancy, Anwar Ul-Hamid, and Nureddin M. Abbas
180. *Mechanical Alloying and Milling*, C. Suryanarayana
181. *Mechanical Vibration: Analysis, Uncertainties, and Control, Second Edition, Revised and Expanded*, Haym Benaroya
182. *Design of Automatic Machinery*, Stephen J. Derby
183. *Practical Fracture Mechanics in Design: Second Edition, Revised and Expanded*, Arun Shukla
184. *Practical Guide to Designed Experiments*, Paul D. Funkenbusch
185. *Gigacycle Fatigue in Mechanical Practive*, Claude Bathias and Paul C. Paris
186. *Selection of Engineering Materials and Adhesives*, Lawrence W. Fisher
187. *Boundary Methods: Elements, Contours, and Nodes*, Subrata Mukherjee and Yu Xie Mukherjee
188. *Rotordynamics*, Agnieszka (Agnes) Muszńyska
189. *Pump Characteristics and Applications: Second Edition*, Michael W. Volk
190. *Reliability Engineering: Probability Models and Maintenance Methods*, Joel A. Nachlas
191. *Industrial Heating: Principles, Techniques, Materials, Applications, and Design*, Yeshvant V. Deshmukh
192. *Micro Electro Mechanical System Design*, James J. Allen
193. *Probability Models in Engineering and Science*, Haym Benaroya and Seon Han
194. *Damage Mechanics*, George Z. Voyiadjis and Peter I. Kattan
195. *Standard Handbook of Chains: Chains for Power Transmission and Material Handling, Second Edition*, American Chain Association and John L. Wright, Technical Consultant
196. *Standards for Engineering Design and Manufacturing*, Wasim Ahmed Khan and Abdul Raouf S.I.
197. *Maintenance, Replacement, and Reliability: Theory and Applications*, Andrew K. S. Jardine and Albert H. C. Tsang
198. *Finite Element Method: Applications in Solids, Structures, and Heat Transfer*, Michael R. Gosz
199. *Microengineering, MEMS, and Interfacing: A Practical Guide*, Danny Banks
200. *Fundamentals of Natural Gas Processing*, Arthur J. Kidnay and William Parrish
201. *Optimal Control of Induction Heating Processes*, Edgar Rapoport and Yulia Pleshivtseva

Optimal Control of Induction Heating Processes

Edgar Rapoport
Yulia Pleshivtseva

Taylor & Francis
Taylor & Francis Group
Boca Raton London New York

CRC is an imprint of the Taylor & Francis Group,
an informa business

FIRST INDIAN REPRINT, 2010

CRC Press
Taylor & Francis Group
6000 Broken Sound Parkway NW, Suite 300
Boca Raton, FL 33487-2742

Printed and bound in India by Nutech Photolithographers

International Standard Book Number-10: 0-8493-3754-2
International Standard Book Number-13: 978-0-8493-3754-3

Library of Congress Cataloging-in-Publication Data

Optimal control of induction heating processes / Edgar Rapoport and Yulia Pleshivtseva.
p. cm.
Includes bibliographical references and index.
ISBN-10: 0-8493-3754-2
ISBN-13: 978-0-8493-3754-3
1. Induction heating--Industrial applications. I. Rapoport, Edgar. II. Pleshivtseva, Yulia.

TK4601.O67 2006
621.402--dc22 2006002588

Visit the Taylor & Francis Web site at
http://www.taylorandfrancis.com

and the CRC Press Web site at
http://www.crcpress.com

FOR SALE IN SOUTH ASIA ONLY.

Preface

This book introduces new approaches to the solution of optimal control problems in induction heating process applications. The objective is to demonstrate how to apply and use new optimization techniques for different types of induction heating installations. We describe the processes and the practically accepted solutions for their optimization. The book focuses on practical methods that can be used to solve real problems. This text features a variety of specific optimization examples for induction heater modes and designs, focusing on the most typical and widely used industrial applications.

With a clear and accessible approach, detailed systematic descriptions of basic theory and practical applications of new methods are provided for solving engineering optimization problems. This book describes basic physical phenomena of induction heating processes (IHPs), IHP optimization problems, and IHP mathematical models that could be put to practical use. It explains the fundamentals of a new, highly effective method and the advantages that it offers over other well-known methods.

The book will be a valuable source for engineers, designers, scientists, lecturers, students, the academic community, production managers, and users of induction heating machinery. It could be interesting to all specialists and experts who would like to study, design, and improve processes of induction mass heating. Knowledge of the basics of heat transfer theory, mathematics, and optimal control theory are the only requirements for understanding the material. This textbook can be considered as a core primary or secondary supplemental textbook to standard courses for advanced students and a new tool for design and control of practical, cost-effective modern induction heating processes. It can be used as an introduction to the broad theory of optimal control that enables widening experience and improving erudition for all specialists and experts in this area.

We wish to express our sincere appreciation to all who have made this book possible. We would like to give special thanks to Professor Alfred Mühlbauer whose guidance, constant encouragement, and support were crucial to bringing this work to reality. We particularly want to thank Professors Bernard Nacke and Nikolay Diligenskiy for their generous help and cooperation. It is a great pleasure to acknowledge the assistance of the Samara State Technical University (Russia) in the preparation of the manuscript. Finally, we heartily appreciate the contribution and support of our families, colleagues, and friends.

Edgar Rapoport

Yulia Pleshivtseva

The Authors

Edgar Rapoport is head of the Department of Automatics and Control in Technical Systems at Samara State Technical University, Russia. He received his electrical engineering degree and Ph.D. from Kuibyshev Industrial Institute. Professor Rapoport has 47 years of experience in automatic control and optimization of technological processes and technical systems. His current scientific and technical interests include optimal control of a variety of industrial processes, with concentration on optimization of induction heating processes; optimization methods; and control and synthesis of systems with distributed parameters. Professor Rapoport has been involved in the development of a number of automatic control systems of induction heating installations used in industry. His credits include 250 scientific and engineering publications (among them more than 20 papers in publications of the Russian Academy of Science), 75 patents and inventor's certificates, and 5 monographs.

Dr. Rapoport is an Honored Worker of Science and Engineering of the Russian Federation. He is a scientific expert to the Russian Ministry of Education in the field of automation and control and head of one of the leading scientific educational groups (Samara State Technical University) in the field of mathematical modeling and optimization of thermoelectric processes and systems with distributed parameters. Dr. Rapoport belongs to many professional organizations, including the New York Academy of Sciences and Russian Academy of Nonlinear Sciences.

Yulia Pleshivtseva, an assistant professor, teaches graduate and postgraduate courses in the Heat and Power Engineering Department of Samara State Technical University, Russia. She received an engineering degree and a Ph.D. from Samara State Technical University in the field of control of heat-mass transfer of technological processes. Her current research interests include optimal control of induction heating processes and modeling and developing control systems with distributed parameters.

Dr. Pleshivtseva has 18 years of experience in automatic control and optimization of different technological processes, including induction heating and chemical water purification. Her work appears in more than 50 scientific and engineering publications.

Introduction

Induction heating is widely used in many advanced technologies. In the past three decades, heating by induction has become the preferred technique in metal-working applications. This is one of the most powerful and promising methods in modern electromagnetic processing of materials because heating by induction provides reliable, repeatable, noncontact, and energy-efficient heat in a minimal amount of time. Induction heaters of different types offer certain advantages over similar equipment.

At the same time, modern society would benefit appreciably from optimization of this energy-consuming technology. It is imperative to note here that, due to their flexibility and controllability, electromagnetic processing technologies are very suitable objects for automation and optimization. In time, optimal control technique has emerged as one of the most important and useful methods to improve different technological processes. The problem of optimal control of induction heating processes can be solved using modern optimal control theory and techniques. Some fundamental investigations primarily concerning the application to combustion furnaces, have been successfully implemented in this area. Results show that optimal control methods offer advantages over other well-known optimization techniques.

Work in the field of optimal control of induction heating processes has been primarily devoted to solving particular application optimization problems. Unfortunately, global system optimization has not been developed and general regularities and optimum characteristics of induction processes have not been established. Until recently, it was impossible to develop and implement the optimum design of an induction metal heating system and operating modes on the basis of highly effective and universal methodology.

This text describes a system approach to solve optimization problems for nonstationary heat conductivity processes. An attempt has been made to discuss complex optimal control problems using simple terms. The material is presented in a form applicable to any kind of static and progressive heaters at steady-state operational conditions, as well as to manufacturing-line "heater–hot forming equipment." Novel and universal optimization algorithms are introduced. Practical examples provide illustrations of the theoretical concepts.

The reference list provided is far from complete; it contains only sources that we used when we wrote this book. Feedback, comments, remarks, and suggestions from readers will be appreciated.

Table of Contents

1 Introduction to Theory and Industrial Application of Induction Heating Processes

1.1 SHORT DESCRIPTION OF OPERATING PRINCIPLES OF INDUCTION HEATERS ON THE LEVEL OF BASIC PHYSICAL LAWS

Induction heating is a complex combination of electromagnetic, heat transfer, and metallurgical phenomena involving many factors. The main components of an induction heating system are an induction coil, power supply, load-matching station, quenching system (for heat treating applications), and the workpiece. Induction coils or inductors are usually designed for specific applications and are therefore found in a wide variety of shapes and sizes.

Heat transfer and electromagnetics are tightly interrelated because the physical properties of heated materials depend strongly on magnetic field intensity and temperature. This section is devoted to a discussion of the electromagnetic and heat transfer phenomena and some other aspects relating to them.[1]

1.1.1 Basic Electromagnetic Phenomena in Induction Heating

The basic electromagnetic phenomena in induction heating are quite simple and discussed in several textbooks. An alternating voltage applied to an induction coil (e.g., solenoid coil) will result in an alternating current in the coil circuit. An alternating coil current will produce in its surroundings a time-variable magnetic field that has the same frequency as the coil current. That magnetic field strength depends on the current flowing in the induction coil, the coil geometry, and the distance from the coil. The changing magnetic field induces eddy currents in the workpiece located inside the coil. These induced currents have the same frequency as the coil current; however, their direction is opposite to the coil current.

Alternating eddy currents induced in the workpiece produce their own magnetic fields, which have opposite directions to the direction of the main magnetic

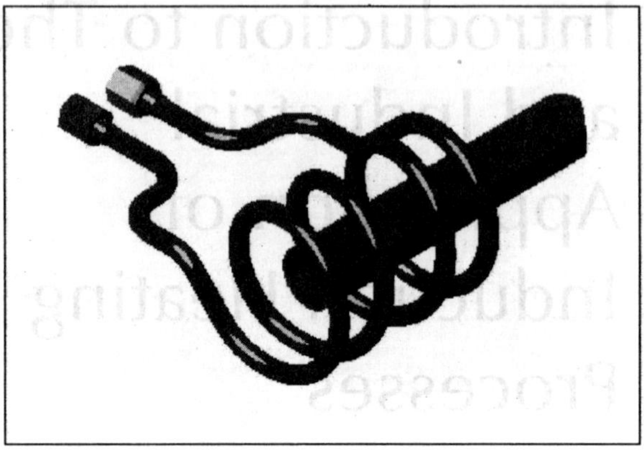

FIGURE 1.1 A conventional induction heating system consisting of a cylindrical load surrounded by a multiturn induction coil.

field of the coil. Therefore, the total magnetic field of the induction coil is a result of the source magnetic field and induced magnetic fields. Alternating eddy currents produce heat by the Joule effect (I^2R). A conventional induction heating system that consists of a cylindrical load surrounded by a multiturn induction coil is shown in Figure 1.1.

The ability of material to conduct electrical current easily is specified by electric conductivity, σ. The reciprocal to the conductivity, σ, is electrical resistivity, ρ. The units for ρ and σ are (Ohm·m) and $(\text{Ohm}\cdot\text{m})^{-1}$, respectively.

Metals and alloys are considered to be good conductors and have much less electrical resistance compared to other materials. Although metals having low electrical resistance are known to be good electrical conductors, they are, in turn, also divided based on their electrical resistivity. We consider some metals to be low-resistive metals (e.g., silver, cooper, gold, aluminum) and others to be high-resistive metals (e.g., stainless steel, titanium, carbon steel). Electrical resistivity of a particular metal varies with temperature, chemical composition, metal microstructure, and grain size. For most metals, ρ rises with temperature.

Electrical resistivity is an imperative physical property. It affects practically all important parameters of an induction heating system, including depth of heating, heat uniformity, coil electrical efficiency, coil impedance, and others. Relative magnetic permeability, μ_r, and relative permittivity, ε, are nondimensional parameters and have similar meanings. Relative magnetic permeability μ_r indicates the ability of a metal to conduct the magnetic flux better than a vacuum or air. Relative permittivity (or dielectric constant) ε indicates the ability of a material to conduct the electric field better than a vacuum or air.

Relative magnetic permeability has a marked effect on all basic induction heating phenomena, coil calculation, and computation of electromagnetic field distribution. Relative permittivity is not widely used in induction heating, but it plays a major role in dielectric heating applications.

The constant $\mu_0 = 4\pi \times 10^{-7}$ *H/m* is called the permeability of free space (the vacuum) and, similarly, the constant $\varepsilon_0 = 8.854 \times 10^{-12}$ *F/m* is called the permittivity of free space. The product of relative magnetic permeability and permeability of free space is called permeability μ and corresponds to the ratio of the magnetic flux density (*B*) to magnetic field intensity (*H*):

$$\frac{B}{H} = \mu_r \mu_0 \text{ or } B = \mu_r \mu_0 H \ . \tag{1.1}$$

Based on their magnetization ability, all materials can be divided into paramagnetic, diamagnetic, and ferromagnetic categories. Relative magnetic permeability of paramagnetic materials is slightly greater than 1 ($\mu_r > 1$). The value of μ_r for diamagnetic materials is slightly less than 1 ($\mu_r < 1$). Due to insignificant differences of μ_r for paramagnetic and diamagnetic materials, those materials are simply called nonmagnetic materials in induction heating practice. Typical nonmagnetic metals are aluminum, copper, titanium, tungsten, etc.

In contrast to paramagnetic and diamagnetic materials, ferromagnetic materials exhibit the high value of relative magnetic permeability ($\mu_r >> 1$). Only a few elements reveal the ferromagnetic properties at room temperature. These include iron, cobalt, and nickel. The ferromagnetic property of the material is a complex function of structure, chemical composition, prior treatment, grain size, frequency, magnetic field intensity, and temperature. The temperature at which a ferromagnetic body becomes nonmagnetic is called the Curie temperature (also known as Curie point).

Because of several electromagnetic phenomena, the current distribution within an inductor and workpiece is not uniform. This heat source nonuniformity causes a nonuniform temperature profile in the workpiece. A nonuniform current distribution can be caused by several electromagnetic phenomena, including skin effect, proximity effect, ring effect, and end and edge effects. These effects play an important role in understanding the induction heating phenomena.[1]

Skin effect. The phenomenon of nonuniform current distribution within the conductor cross-section is called the skin effect, which always occurs when there is an alternating current. As one may know from the basics of electricity, when a direct current flows through a conductor that stands alone, the current distribution is uniform. However, when an alternating current flows through the same conductor, the current distribution is not uniform. The maximum value of the current density will be located on the surface of the conductor; the current density will decrease from the surface of the conductor toward its center. Therefore, the skin effect will also be found in a workpiece located inside an induction coil (Figure 1.2).

This is one of the major factors that cause the concentration of eddy current in the surface layer ("skin") of the workpiece. Due to the circumferential nature of the eddy current induced in the workpiece, there is no current flow at the center of the workpiece.

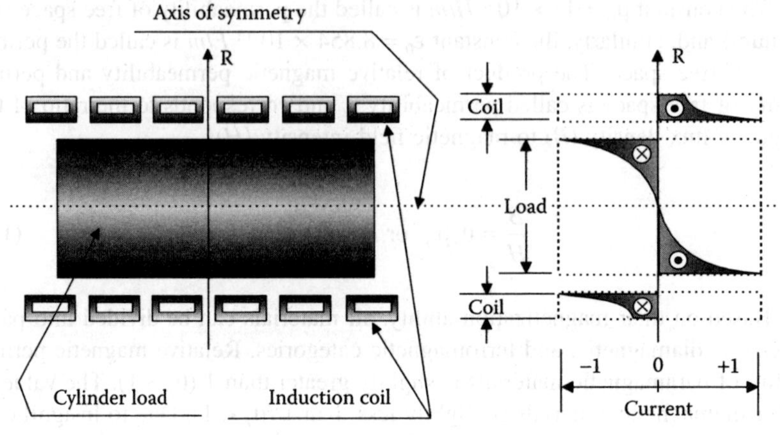

FIGURE 1.2 Current distribution in "coil–workpiece" induction system.

The skin effect is of great practical importance in electrical applications using alternative current. Because of this effect, approximately 86% of the power will be concentrated in the surface layer of the conductor. This layer is called reference (or penetration) depth.

Proximity effect. When we discussed the skin effect, we assumed that the conductor stands alone and no other current-carrying conductors are in the surrounding area. In most practical applications, this is not the case. Most often, other electrically conductive parts are located in close proximity to a current-carrying conductor. These parts have their own magnetic fields, which interact with nearby fields; as a result, the current and power density distribution will be distorted.

An understanding of the physics of the electromagnetic proximity and skin effects is important not only in induction heating but also in power supply and bus design. The proper design of a bus network will significantly decrease its impedance-minimizing voltage drop and power losses.

Ring effect. Up to now, we have discussed current density distribution in straight conductors. If a current-carrying bar is bent to shape it into a ring, then its current will be redistributed. Magnetic flux lines will be concentrated inside the ring and therefore the density of the magnetic field will be higher inside the ring. Outside the ring, the magnetic flux lines will be disseminated. As a result, most of the current will flow within the thin inside surface layer of the ring. The ring effect takes place not only in single-turn inductors but also in multiturn coils.

The appearance of the ring effect can have a positive or negative effect on the process. For example, in conventional induction heating of cylinders, when the workpiece is located inside the induction coil this effect plays a positive role because, in combination with the skin and proximity effects, it will lead to a concentration of the coil current on the inside diameter of the coil. As a result, there will be close coil–workpiece coupling, which leads to good coil efficiency.

The ring effect plays a negative role in the induction heating of internal surfaces when the induction coil is located inside the workpiece. In this case, this effect leads to a coil current concentration on the inside diameter of the coil. This makes the coil–workpiece coupling poor and therefore decreases coil efficiency. Thus, it is necessary to take the ring effect into account when designing the induction coils, power supplies, and cooling circuit for the bus bar.

End and edge effects. To guarantee the required uniform induction heating of the workpiece, it is necessary to accurately predict the electromagnetic field distribution produced by the induction coil under different operating conditions. Among other factors, the temperature profiles along the workpiece's length and width are affected by a distortion of electromagnetic field (emf) in its end and edge areas. Those field distortions and corresponding distributions of induced currents and power densities are referred to as end and edge effects. These effects and the field distortion caused by them are primarily responsible for nonuniform temperature profiles in cylindrical, rectangular, and trapezoidal shaped workpieces. Due to the great importance of these effects in the induction heating applications, much effort has been devoted to their study.

Suppose a slab is placed in an initially uniform magnetic field. If the slab's length and width are much larger than its thickness, the emf in the slab can be viewed as an area consisting of three zones: central part, transverse edge effect area, and longitudinal end effect area (Figure 1.3). In the central part, the emf distribution corresponds to the field in the infinite plate. Basically, end and edge effects have a two-dimensional space distribution, excluding only the zone of three-edge corners where the field is three dimensional and the corresponding field distribution is the result of mixture of the end and edge effects. For many practical applications, the separate study of end and edge effects is of great engineering interest. Depending upon applications, these effects can act differently.[1]

1.1.2 Basic Thermal Phenomena in Induction Heating

In induction heating, all three modes of heat transfer — conduction, convection, and radiation — are present. Heat is transferred by conduction from the high-temperature regions of the workpiece toward the low-temperature regions. The basic law that describes heat transfer by conduction is Fourier's law:

$$q_{cond} = -\lambda(t) \cdot grad\ t \,, \tag{1.2}$$

where

q_{cond} is heat flux by conduction
$\lambda(t)$ is thermal conductivity, W/(m · °C)
t is temperature.

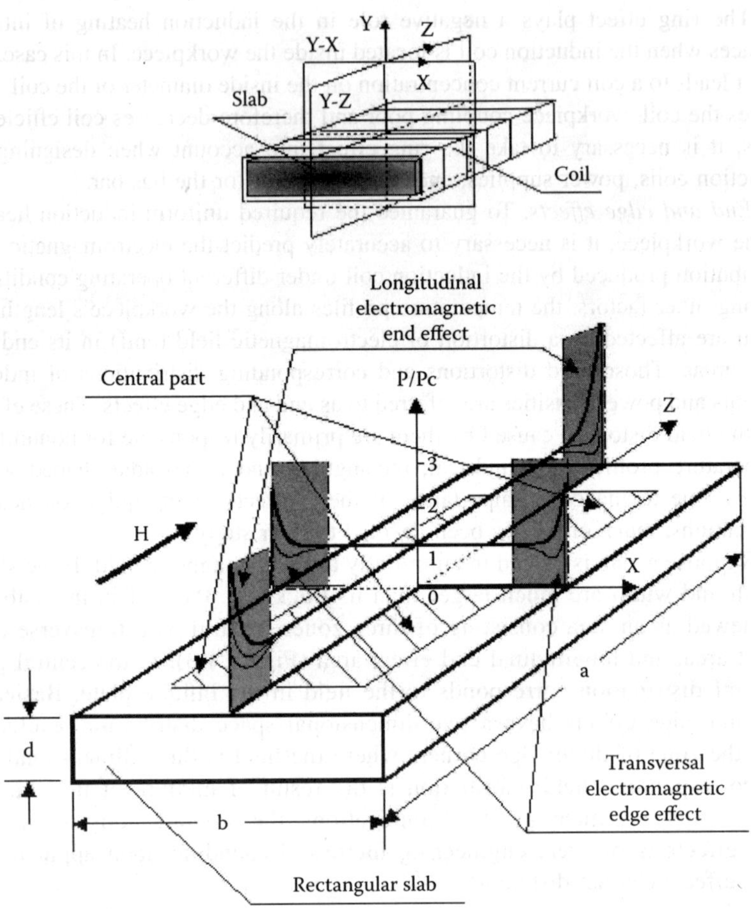

FIGURE 1.3 Electromagnetic end and edge effect of the slab.

Thermal conductivity, λ, designates the rate at which heat travels across a thermally conductive workpiece. A material with a high λ value will conduct heat faster than a material with a low λ value. The thermal conductivity is a nonlinear function of temperature.

According to Fourier's law, a large temperature difference between surface and core and a high value of thermal conductivity of the metal result in intensive heat transfer from the hot surface of the workpiece toward the cold core. Conversely, the rate of heat transfer by conduction is inversely proportional to the distance between regions with different temperatures.

In contrast to conduction, heat transfer by convection is carried out by fluid, gas, or air (i.e., from the surface of the heated workpiece to the ambient area). The well-known Newton's law can describe convection heat transfer. This law states that the heat transfer rate is directly proportional to the temperature difference between the workpiece surface and the ambient area:

$$q_{conv} = \alpha \cdot (t_{sur} - t_a) \,, \tag{1.3}$$

where
q_{conv} is heat flux density by convection, W/m²
α is the convection surface heat transfer coefficient, W/(m² · °C)
t_{sur} is the surface temperature, °C
t_a is ambient temperature, °C

The convection surface heat transfer coefficient is primarily a function of the thermal properties of the workpiece, the thermal properties of the surrounding fluid, gas, or air, and their viscosity or the velocity of the heat-treated workpiece if the workpiece is moving at high speed. It is particularly important to take this mode into account when designing low-temperature induction heating applications. In these applications, convection losses are equal to or exceed heat losses due to radiation.

With certain applications, the value of convection losses can vary dramatically, depending on the temperature of the workpiece and outside temperature, as well as workpiece geometry, surface conditions, and whether it is free or forced convection. In a number of induction heating applications, convection heat transfer cannot be considered as free convection.

In radiation mode of the heat transfer, the heat may be transferred from the hot workpiece into surrounding areas, including a nonmaterial region (vacuum). The effect of heat transfer by radiation can be introduced as a phenomenon of electromagnetic energy propagation due to a temperature difference. This phenomenon is governed by the Stefan–Boltzmann law of thermal radiation, which states that the heat transfer rate by radiation is proportional to a radiation loss coefficient, C_s, and the value of $t_{sur}^4 - t_a^4$.

The previously described determination of radiation heat loss is a valid assumption for mathematical modeling of a great majority of induction heating and heat treatment problems. However, in a few applications, the radiation heat transfer phenomenon can be complicated and such a simple approach would not be valid.

Complete details of all three modes of heat transfer can be found in several references.[2–9] In typical induction heating and heat treatment, heat transfer by convection and radiation reflects the value of heat loss. A high value of heat loss reduces the total efficiency of the induction heater.[1]

1.2 MATHEMATICAL MODELING OF INDUCTION HEATING PROCESSES

Mathematical modeling is one of the major factors in the successful design of induction heating systems. Theoretical models may vary from a simple hand-calculated formula to a very complicated numerical analysis that can require several hours of computational work using modern supercomputers. The choice of a particular theoretical model depends on several factors, including complexity of the engineering problem, required accuracy, time limitations, and cost.

Before an engineer starts to provide a mathematical simulation of any process, it is necessary to have a sound understanding of the nature and physics of that process. Engineers should also be aware of the limitations of applied mathematical models, assumptions, and possible errors and should consider correctness and sensitivity of the chosen model to poorly defined parameters such as boundary conditions, material properties, or initial temperature conditions. One model can work in certain applications, but give unrealistic results in another. Underestimation of features of the process or overly simple assumptions can lead to an incorrect mathematical model (including chosen governing equations) that will not provide the required accuracy.

It is important to remember that any computational analysis can, at best, produce only results that are derived from the governing equations. Therefore, the first and most important step in any mathematical simulation is to choose an appropriate theoretical model that will correctly describe the technological process or phenomenon.

As mentioned earlier, induction heating is a complex combination of electromagnetic, heat transfer, and metallurgical phenomena. These are tightly interrelated because the physical properties of heat-treated materials depend strongly on magnetic field intensity and temperature as well as chemical composition. This section concentrates on mathematical modeling of the electromagnetic field and thermal processes that occur during induction heating.[1]

1.2.1 Mathematical Modeling of Electromagnetic and Temperature Fields

In the general case, a space–time temperature distribution within a heated workpiece is described by a highly complicated system of interrelated Maxwell's and Fourier equations for electromagnetic and temperature fields:

$$curl\,\overline{H} = J + \frac{\partial \overline{D}}{\partial \tau}; \tag{1.4}$$

$$curl\,\overline{E} = -\frac{\partial \overline{B}}{\partial \tau}; \tag{1.5}$$

$$div\,\overline{B} = 0; \tag{1.6}$$

$$div\,\overline{E} = 0; \tag{1.7}$$

$$c(t)\gamma(t)\frac{\partial t}{\partial \tau} - div(\lambda(t) grad\, t) + c(t)\gamma(t)\overline{V} grad\, t = -div[\overline{E} \cdot \overline{H}], \qquad (1.8)$$

where

$\overline{E}$ is a vector of electric field intensity.

$\overline{D}$ is a vector of electric flux density.

$\overline{B}$ is a vector of magnetic flux density.

$\overline{H}$ is a vector of magnetic field intensity.
J is conduction current density.
t is temperature.
$\gamma(t)$ is specific density of the metal.
$c(t)$ is specific heat.
$\lambda(t)$ is thermal conductivity of the metal.

$\overline{V}$ is a vector of velocity.
τ is time.

Equation (1.4) to Equation (1.8) includes special notations *curl* and *div*. These notations are popular in vector algebra and are used to express particular differential operations. For example, in a rectangular coordinate system, *div* and *curl* represent the following operations:

$$divU = \frac{\partial U_X}{\partial X} + \frac{\partial U_Y}{\partial Y} + \frac{\partial U_Z}{\partial Z}; \qquad (1.9)$$

$$curlU = \begin{vmatrix} i & j & k \\ \frac{\partial}{\partial X} & \frac{\partial}{\partial Y} & \frac{\partial}{\partial Z} \\ U_X & U_Y & U_Z \end{vmatrix} = \qquad (1.10)$$

$$= i\left(\frac{\partial U_Z}{\partial Y} - \frac{\partial U_Y}{\partial Z}\right) + j\left(\frac{\partial U_X}{\partial Z} - \frac{\partial U_Z}{\partial X}\right) + k\left(\frac{\partial U_Y}{\partial X} - \frac{\partial U_X}{\partial Y}\right)$$

where
i, j, k are unit vectors in the 3 standard Cartesian directions.

The technique of calculating electromagnetic fields depends on the ability to solve Maxwell's equations (Equation 1.4 through Equation 1.7) for general time-varying electromagnetic fields.

Maxwell's equations not only have a purely mathematical meaning, but also have a concrete physical interpretation as well. From Equation (1.4), it follows that *curl*$\bar{H}$ always has two sources: conductive (J) and displacement $\left(\frac{\partial \bar{D}}{\partial \tau}\right)$ currents. A magnetic field is produced whenever electric currents are flowing in surrounding objects. From Equation (1.5), one can conclude that a time rate of change in magnetic flux density, $\bar{B}$, always produces the curling field, $\bar{E}$, and induces currents in the surrounding area. In other words, it produces an electric field in the area where such changes take place. The minus sign in Equation (1.5) determines the direction of that induced electric field. This fundamental result can be applied to any region of the space.

Let us consider how Equation (1.4) and Equation (1.5) can be used to support the basic explanation of some of the electromagnetic processes in induction heating provided in Section 1.1. The application of alternating voltage to the induction coil will result in the appearance of an alternating current in the coil circuit. According to Equation (1.4), an alternating coil current will produce in its surrounding area an alternating (changing) magnetic field that will have the same frequency as the source current (coil current). That magnetic field's strength depends on the current flowing in the induction coil, the coil geometry, and the distance from the coil. The changing magnetic field induces eddy currents in the workpiece and in other objects located near that coil.

According to Equation (1.5), induced currents have the same frequency as the source coil current; however, their direction is opposite to that of the coil current. This is determined by the minus sign in Equation (1.5). According to Equation (1.4), alternating eddy currents induced in the workpiece produce their own magnetic fields, which have opposite directions to the direction of the main magnetic field of the coil. The total magnetic field of the induction coil is a result of the source magnetic field and induced magnetic fields.

As one would expect from an analysis of Equation (1.4), there can be undesirable heating of tools, fasteners, or other electrically conductive structures located near the induction coil.

Equation (1.6) and Equation (1.7) have real significance in induction heating and heat treatment of an electrically conductive body. From these equations, it follows that the divergences of magnetic flux density and electric field intensity are equal to zero. It means that $\bar{B}$ and $\bar{E}$ lines have no source points at which they originate or end; in other words, $\bar{B}$ and $\bar{E}$ lines always form continuous loops.

The previously described Maxwell's equations (Equation 1.4 through Equation 1.7) are in indefinite form because the number of equations is less than the number of unknowns. These equations become definite when the relations between the field quantities are specified. The following constitutive relations are additional and hold true for a linear isotropic medium:

$$D = \varepsilon\varepsilon_0 E; \tag{1.11}$$

$$B = \mu_r \mu_0 E; \tag{1.12}$$

$$J = \sigma E; \tag{1.13}$$

where

the parameters ε, μ_r, and σ denote, respectively, the relative permittivity, relative magnetic permeability, and electrical conductivity of the material;

$\sigma = 1/\rho$, where ρ is electrical resistivity.

The constant μ_0 is the permeability of free space (the vacuum) and, similarly, the constant ε_0 is the permittivity of free space (see Section 1.1).

By taking Equation (1.11) and Equation (1.13) into account, Equation (1.14) can be rewritten as:

$$curl\overline{H} = \sigma\overline{E} + \frac{\partial(\varepsilon_0 \varepsilon E)}{\partial\tau}. \tag{1.14}$$

For most practical applications of the induction heating of metals when the frequency of current is less than 10 MHz, the induced conduction current density, J, is much greater than the displacement current density, $\frac{\partial\overline{D}}{\partial\tau}$; so the last term on the right-hand side of Equation (1.14) can be neglected. Therefore, Equation (1.14) can be rewritten as:

$$curl\overline{H} = \sigma\overline{E}. \tag{1.15}$$

The Fourier equation (1.8) describes in general the transient (time-dependent) heat transfer process in a metal workpiece. The heating process is caused by heat sources induced by eddy currents (so-called heat generation). The heat source density, F, induced by eddy currents per unit time in a unit volume can be obtained by solving the electromagnetic problem as:

$$F = -div[\overline{E} \cdot \overline{H}]. \tag{1.16}$$

The value of specific heat c (in Equation 1.8) indicates the amount of energy that must be absorbed by the workpiece to achieve the required temperature change. A high value of specific heat corresponds to a higher required power.

The specific heat, c, and thermal conductivity, λ, are nonlinear functions of temperature. Nevertheless, in the great majority of induction heating applications, a rough approximation of thermal conductivity in simulations of heating processes (i.e., assuming $\lambda = const$) will not lead to significant errors in temperature distributions. At the same time, regardless of application, a rough approximation of specific heat (i.e., an assumption of $c = const$) could create significant errors in obtaining the required coil power and temperature profile within the workpiece.

With suitable boundary and initial conditions, Equation (1.8) represents the three-dimensional temperature distribution at any time and at any point in the workpiece. Solution of the set of Equation (1.4) through Equation (1.8) can be obtained only by numerical methods. These methods are widely and successfully used in the computation of electromagnetic and heat transfer problems. For each problem or family of similar problems, certain software or numerical methods are preferred. No single universal computational method fits and is optimum for solving all induction heating problems.

Because of space limitations, we do not give an overview of the numerical methods available for electromagnetic field and heat transfer calculations. Many publications describe the features and applications of mathematical modeling methods. An interested reader can study the description of the most popular computational techniques used for simulation of heat transfer and electromagnetic processes in references 1 and 10 through 24.

1.2.2 Basic Model of the Induction Heating Process

Equation (1.5) through Equation (1.8) as well as Equation (1.15) are valid for general three-dimensional electromagnetic and thermal fields and allow one to find all of the required design parameters of the induction system. Although there is a considerable practical interest in three-dimensional problems, most engineering problems in induction heating tend to be reduced to consideration of two- or one-dimensional fields.

It can be shown that, for the great majority of induction heating applications, it is possible to simplify the mathematical model further by some typical assumptions. For example, it is possible to take the material properties as piecewise continuous and to neglect hysteresis and magnetic saturation. It should be mentioned here that for such induction heating applications as heating prior to forging, rolling, and extrusion, a heat effect due to hysteresis losses does not typically exceed 7% compared to the heat effect due to eddy current losses. Therefore, an assumption of neglecting the hysteresis is valid.

Assuming that the currents have a steady-state quality, we can conclude that the electromagnetic field quantities in Maxwell's equations are harmonically oscillating functions with a single frequency. Thus, a time-harmonic electromagnetic field can be introduced. In other words, an assumption of harmonically oscillating currents with a single frequency means that harmonics are absent in the impressed and the induced currents and fields.

For many induction heating applications, the quantities of the magnetic field (such as magnetic vector potential, electric field intensity, and magnetic field intensity) may be assumed to be entirely directed. This allows one to reduce the dimensionality of governing equations. Moreover, most mathematical models of induction heating tend to be handled with a combination of the following assumptions:

- Neglecting nonlinearity by averaging process parameters on the corresponding temperature intervals;
- Mathematical description of heated workpieces as regular bodies (plate, cylinder, rectangle, sphere);
- Simplified description of the geometric input data for induction heating system;
- Taking into account nonuniformity of temperature distribution only along one or two coordinate axes (reducing a three-dimensional temperature field to a one- or two-dimensional form).

Under these assumptions, the set of Equation (1.5) through Equation (1.8) and Equation (1.15) can be reduced to the following equations in differential form:

$$\nabla^2 \dot{\bar{H}}(x,\tau) = j\omega\mu_0\sigma\dot{\bar{H}}(x,\tau) \; ; \tag{1.17}$$

$$\frac{\partial t(x,\tau)}{\partial \tau} = a\nabla^2 t(x,\tau) + \frac{1}{c\gamma} F(x,\tau) \; . \tag{1.18}$$

Here,
∇^2 is the Laplacian;
x is a spatial coordinate;
$\omega = 2\pi f$, where f is a frequency of coil current;
a is average value of temperature conductivity of heated material.

Equation (1.17) is a one-dimensional linear Helmholtz's equation with respect to complex magnetic field intensity $\dot{\bar{H}}$. Equation (1.18) is a one-dimensional linear heterogeneous equation of heat transfer with respect to temperature $t(x,\tau)$ at velocity $V = 0$. Equation (1.17) and Equation (1.18) can be solved separately.

$F(x,\tau)$ is a function described distribution of internal heat source density induced by eddy currents per unit time in a unit volume. $F(x,\tau)$ can be obtained by solving Equation (1.17) as:

$$F(x,\tau) = \frac{1}{2}\sigma E_m^2 \, , \tag{1.19}$$

where

$$\dot{\bar{E}} = -\frac{1}{\sigma}\frac{\partial \dot{\bar{H}}}{\partial x}. \tag{1.20}$$

Here, E_m denotes an amplitude value of the electric field intensity.

By using well-known Expression (1.16) and solutions of Equation (1.17),[25] it is possible to obtain the following expressions with respect to $F(x,\tau)$ for the axially symmetric case of unlimited plate:

$$F_1(l,\ \xi,\ P_0) = \frac{P_0(\tau)}{X} W_1(\xi,\ l);\quad W_1(\xi,\ l) = \frac{ch(\sqrt{2}\xi l) - \cos(\sqrt{2}\xi l)}{sh(\sqrt{2}\xi) - \sin(\sqrt{2}\xi)}\sqrt{2}\xi; \tag{1.21}$$

and cylindrical workpiece of infinite length:

$$F_2(l,\ \xi,\ P_0) = \frac{P_0(\tau)}{X} W_2(\xi,\ l);\quad W_2(\xi,\ l) = \xi\frac{ber'^2(\xi l) + bei'^2(\xi l)}{ber(\xi)ber'(\xi) + bei(\xi)bei'(\xi)}. \tag{1.22}$$

Here,

$P_0(\tau)$ is active power absorbed by unit surface of heated body;
X is a cylinder radius or a half of slab thickness;
$ber(z)$, $bei(z)$, $ber'(z)$, $bei'(z)$ are Kelvin's functions and their first derivatives;
l is a relative value of spatial coordinate in a plate depth or cylinder radius ($l = x/X$);
ξ is a specific parameter defined as:

$$\xi = X\sqrt{2}/\delta, \tag{1.23}$$

where δ is a current penetration depth, which can be calculated as follows[10]:

$$\delta = \sqrt{2/(\mu_0\omega\sigma)}. \tag{1.24}$$

For the typical induction heating process frequency of coil current is constant ($\xi = const$). Then, total internal heat power absorbed by unit volume of heated body can be written as:

$$P(\tau) = P_0(\tau)/X. \tag{1.25}$$

There is a squared relationship between specific power $P(\tau)$ and inductor voltage.[10] The value of $P(\tau)$ is in the range of 0 to P_{max} corresponding to maximum inductor voltage:

$$0 \le P(\tau) \le P_{max}, \ \tau \ge 0 . \tag{1.26}$$

Substituting $F(x,\tau)$ in the forms of Equation (1.21) and Equation (1.22) into Equation (1.18), one can obtain the following one-dimensional linear heterogeneous equation of heat transfer considered further as a basic mathematical model of induction heating process:

$$\frac{\partial\theta(l,\varphi)}{\partial\varphi} = \frac{\partial^2\theta(l,\varphi)}{\partial l^2} + \frac{\Gamma}{l}\frac{\partial\theta(l,\varphi)}{\partial l} + W(\xi,l)u(\varphi), \ l \in [0,1]; \ \varphi \in [0; \varphi^0] . \tag{1.27}$$

Equation (1.27) is written with respect to relative units of temperature $\theta(l,\varphi)$ and power of internal heat sources $u(\varphi)$. These relative values can be calculated according to the formulas:

$$\theta(l,\varphi) = \frac{t(x, \tau) - t_b}{P_{max} X^2}\lambda ; \tag{1.28}$$

$$u(\varphi) = \frac{P(\tau)}{P_{max}} . \tag{1.29}$$

In Equation (1.27),

$\Gamma = 0$ or $\Gamma = 1$ for the plate or cylinder, respectively.

$W(\xi, l)$ is determined according to Equation (1.21) or Equation (1.22).

$\varphi = \dfrac{a\tau}{X^2}$ is the Fourier number.

φ^0 is the total time required for heating.

t_b is a basic temperature.

The formulation of a problem requiring the solution of a partial differential equation also requires the specification of appropriate boundary conditions and initial conditions. Specification of the dependent variable or its time derivative at time zero is referred to as an *initial condition*. The *initial temperature condition* refers to the temperature profile within the workpiece at time $\varphi = 0$:

$$\theta(l, 0) = \theta_0(l) . \tag{1.30}$$

The initial temperature distribution is usually uniform and corresponds to the ambient temperature. In some cases, the initial temperature distribution is nonuniform due to residual heat after the previous technological process. The condition in Equation (1.30) is required only when dealing with a transient heat transfer problem where the temperature is a function not only of the space coordinates but also of time.

For physical problems, the term "boundary" literally means on the physical boundary of the region in space in which the solution is sought. The three most common boundary conditions for induction heating problems are:

1. Value of temperature is specified on the boundary of the heated body. This boundary condition is also known as a Dirichlet condition or a boundary condition of the first kind.
2. Normal derivatives of temperature are specified on the boundary. This condition is known as a Neumann condition or a boundary condition of the second kind.
3. A linear combination of conditions 1 and 2 is specified on the boundary. Convective boundary conditions in heat transfer are of this type. This condition is known as a Robbins condition or a boundary condition of the third kind.

If the heated body is geometrically symmetrical along the axis of symmetry, the Neumann boundary conditions can be formulated as:

$$\frac{\partial\theta(0,\,\varphi)}{\partial l} = 0\ , \tag{1.31}$$

$$\frac{\partial\theta(1,\,\varphi)}{\partial l} = q(\varphi) < 0\ . \tag{1.32}$$

The condition in Equation (1.31) implies that the temperature gradient in a direction normal to the axis of symmetry is zero. In other words, no heat exchange takes place at the axis of symmetry. This boundary condition can also be applied in the case of a perfectly insulated body.

In Expression (1.32), $q(\varphi)$ represents a relative value of heat losses and can be determined as:

$$q(\varphi) = \frac{Q_s(\varphi)}{P_{\max} X}\ . \tag{1.33}$$

Here, $Q_s(\varphi)$ is a flow of heat loss from a surface of the heated body (i.e., during quenching or as a result of workpiece contact with cold rolls or water-cooled guides, etc.).

For most induction heating problems, boundary conditions include the heat losses due to convection. In this case, the boundary condition of the third kind can be expressed as:

$$\frac{\partial\theta(1,\varphi)}{\partial l} = Bi(\theta_a(\varphi) - \theta(1,\varphi)), \tag{1.34}$$

where

$\partial\theta/\partial l$ is the temperature gradient in a direction normal to the surface at the point under consideration.

$Bi = \alpha X/\lambda$ is the Biot number.

α is the convection surface heat transfer coefficient.

$\theta_a(\varphi)$ denotes a relative value of ambient temperature $t_a(\tau)$ calculated as:

$$\theta_a(\varphi) = \frac{t_a(\tau) - t_b}{P_{max}X^2}\lambda. \tag{1.35}$$

The heat losses at the workpiece surface are highly variable because of the nonlinear behavior of convection losses.

Equation (1.27) with boundary conditions in Equation (1.31), Equation (1.32), and Equation (1.34) are the most popular equations for mathematical modeling of the heat transfer processes in induction heating and heat treatment applications.

Let us assume that power of internal heat sources $u(\varphi)$ can be changed almost in an arbitrary way. In this case, the solution of Equation (1.27) under initial and boundary conditions (1.30) through (1.32) and (1.34) can be written in the form of Duhamel integral as[25]:

$$\theta(l,\varphi) = \Phi(l,\varphi) + \int_0^{\varphi} \frac{\partial\Lambda(l,\varphi-\tau)}{\partial\varphi}u(\tau)\,d\tau. \tag{1.36}$$

Taking into consideration appropriate boundary conditions of the second (Φ_2, Λ_2) and the third (Φ_3, Λ_3) kinds, functions $\Phi(l, \varphi)$ and $\Lambda(l, \varphi)$ can be calculated as[25]:

$$\Phi_2(l,\varphi) = (\Gamma+1)\left(\theta_{00} + \int_0^{\varphi} q(\tau)\,d\tau\right) + 2\sum_{n=1}^{\infty}\frac{K(\mu_n l)}{K^2(\mu_n)} \times \left[\theta_{0n}e^{-\mu_n^2\varphi} + K(\mu_n)\int_0^{\varphi} q(\tau)e^{-\mu_n^2(\varphi-\tau)}\,d\tau\right]; \tag{1.37}$$

$$\Phi_3(l,\varphi)=\sum_{n=1}^{\infty}\frac{2Bi^2K(\mu_n l)e_n^{-u_n^2\varphi}}{\left(Bi^2+\mu_n^2+(1-\Gamma)Bi\right)K_1^2(\mu_n)}\left[\theta_{0n}+\mu_n K_1(\mu_n)\int_0^{\varphi}\theta_\alpha(\tau)e^{\mu_n^2\tau}d\tau\right];\tag{1.38}$$

$$\Lambda_2(l,\varphi)=(\Gamma+1)W_0(\xi)\varphi+2\sum_{n=1}^{\infty}\frac{W_n(\xi)K(\mu_n l)}{\mu_n^2K^2(\mu_n)}\cdot\left[1-e^{-\mu_n^2\varphi}\right];\tag{1.39}$$

$$\Lambda_3(l,\varphi)=2\sum_{n=1}^{\infty}\frac{W_n(\xi)K(\mu_n l)}{\left(\mu_n^2+Bi^2+(1-\Gamma)Bi\right)K^2(\mu_n)}\cdot\left[1-e^{-\mu_n^2\varphi}\right].\tag{1.40}$$

In Expression (1.37) through Expression (1.40), the fundamental functions $K(\mu_n l)$, $K_1(\mu_n l)$ should be calculated using following expressions:

$$K(\mu_n l)\Big|_{\Gamma=0}=\cos\mu_n l;\ K_1(\mu_n)\Big|_{\Gamma=0}=\sin\mu_n\ ;\tag{1.41}$$

$$K(\mu_n l)\Big|_{\Gamma=1}=J_0(\mu_n l);\ K_1(\mu_n)\Big|_{\Gamma=1}=J_1(\mu_n),\tag{1.42}$$

where $J_0(\mu_n l)$ and $J_1(\mu_n)$ are Bessel functions of the zero and first orders, respectively.

Under boundary conditions of the third kind, fundamental numbers μ_n, n = 1, 2, … represent roots of the equation:

$$BiK(\mu)-\mu K_1(\mu)=0\ .\tag{1.43}$$

Under boundary conditions of the second kind, fundamental numbers μ_n, n = 0, 1, 2, … represent roots of the equation:

$$K_1(\mu)=0\ .\tag{1.44}$$

In Expression (1.37) through Expression (1.40):

$$W_n(\xi)=\int_0^1 W(\xi,l)l^{\Gamma}K(\mu_n l)dl;\quad \theta_{0n}=\int_0^1\theta_0(l)l^{\Gamma}K(\mu_n l)dl,\ n=0,1,2,\ldots.\tag{1.45}$$

The finite and sufficient number of terms of series should always be used in given expressions.

1.3 TYPICAL INDUSTRIAL APPLICATIONS AND FUNDAMENTAL PRINCIPLES OF INDUCTION MASS HEATING

This book is devoted to a discussion of the use of induction heating principles for a large group of applications referred to as mass heating. This term applies to a variety of applications in which metal is heated for forging, forming, extrusion, coating, etc. Typically, it is required to have a uniformly heated workpiece. However, it is sometimes necessary to heat certain areas of the workpiece selectively and care must be taken to provide a required temperature distribution.

Temperature greatly affects the formability of metals. Heating of a component to temperatures that correspond to the plastic deformation range creates a favorable condition for metal to be subsequently forced by various means into a desired shape. The most popular metal hot working processes for which induction heating is applied follow.[1]

- *Forging*. Billets or bars are heated fully or partially, in cut length or continuously, and are forged in presses, hammers (repeated blows), or upsetters (which gather and form the metal). Steel components by far represent the majority of forged parts. At the same time, aluminum, copper, brass, bronze, cobalt, nickel, and titanium, as well as some other metal alloys, are also inductively heated and forged for a number of commercial applications.
- *Forming*. Hot forming includes a variety of metal working operations generally encompassing bending, expanding, and spinning. The versatility of induction heating is that it can selectively heat through specific areas of the workpiece or can heat areas to different temperatures providing required temperature gradients, making it a popular choice for hot forming operations.
- *Extrusion*. Extruding is the process of forcing or squeezing metal through a die. Ferrous and nonferrous metals are heated by induction prior to extrusion.
- *Rolling*. Bars, billets, rods, slabs, blooms, strips, and sheets are processed in rolling mills. These components are made from ingots or continuous cast metals and their alloys.

The goal of using induction heating in all of these applications is to provide the metal workpiece at the hot working stage with the desired (typically uniform) temperature across its diameter/thickness as well as along its length and across its width. The required final temperature typically depends on the material chemical composition and particularities of the postheating metal processing operation. For example, for steel billets, the final temperature is in the range of 1000 to 1200°C and the required surface-to-core and noise-to-tail temperature uniformity in the billet is commonly specified as ±50°C for rolling and ±25°C for forming. For heating of aluminum alloy billets before extrusion, the final temperature is in the range of 450 to 480°C and required uniformity is in the range of ±10 to ±15°C. Certain applications may require a nonuniform end-to-end temperature profile (for example, heating providing required temperature gradients).

In some cases, the initial temperature of the workpiece prior to induction heating is the ambient temperature. In other cases, the initial temperature is not uniform — for example, due to uneven cooling of the slab, transfer bar, strip, or bloom as it progresses from the caster. Surface layers, and particularly the edge areas, become much cooler than the internal regions.

The power ratings of induction heating machines range from less than 100 kW up to dozens of megawatts. The success of an induction mass heating system is based on an in-depth understanding of the process features. This is imperative for developing the sophisticated design concepts and precise engineering that lead to the achievement of a reasonable, commercially acceptable compromise among often contradictory process requirements and design criteria.

Cylindrical and rectangular solenoid multiturn induction coils are most often used in induction mass heating applications. The four basic heating modes in induction mass heating (Figure 1.4) follow.[1]

- *Static heating*. In the static heating mode, a workpiece such as a billet or slab is placed into an induction heating coil for a given period of time while a set amount of power is applied until the component reaches the desired heating conditions. Upon reaching the required

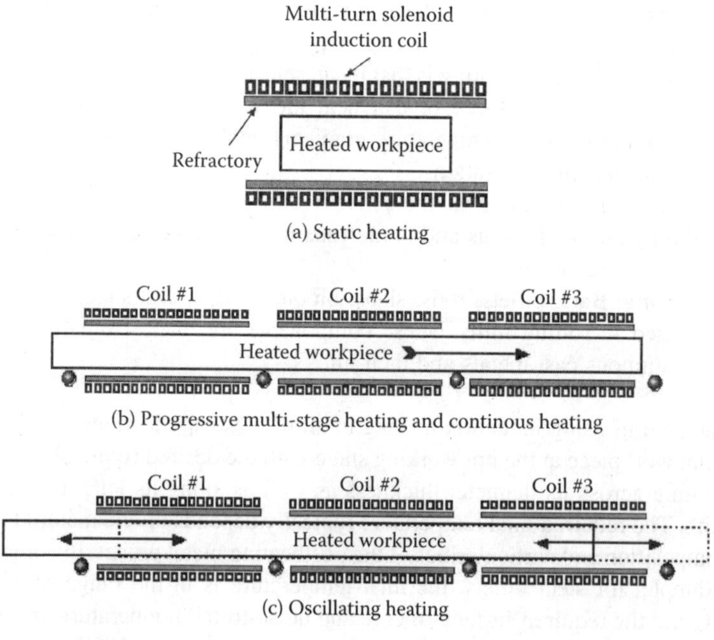

FIGURE 1.4 Four basic heating modes used in induction mass heating.

thermal conditions, the heated component is extracted from the induction heater and delivered to the metal forming station. The next cold workpiece is loaded into the coil and the process repeats.

- *Progressive multistage heating*. This heating mode occurs when two or more heated workpieces (e.g., billets) are moved (via pusher, indexing mechanism, walking beam, etc.) through a single-coil or multicoil induction heater. Therefore, components are sequentially heated (in a progressive manner) at certain predetermined heating stages inside the heater.
- *Continuous heating*. With the continuous heating mode, the workpiece is moved in a continuous motion through one or more induction heating coils. This heating mode is commonly used when it is required to heat long components such as bars, slabs, strips, tubes, wires, blooms, and rods.
- *Oscillating heating*. In this heating mode, a component moves back and forth (oscillates) during the process of heating inside a single-coil or multicoil induction heater with an oscillating stroke featuring a space-saving design approach.

When modern induction mass heating systems are designed, temperature uniformity of the heated component is only one of the goals. Additional design criteria include maximum production rate, minimum metal losses, and the ability to provide flexible and compact systems that have high electrical efficiency. Other important factors include quality assurance, process repeatability, automation capability, environmental friendless, reliability, and maintainability of the equipment. The last criterion, but not the least, is the competitive cost of an induction heating system.

One of the challenges in induction heating arises from the necessity to provide the required surface-to-core temperature uniformity. Due to the physics of the process, the workpiece core tends to be heated more slowly than its surface. The main reason for the heat deficit of the heated component is the skin effect discussed earlier.

It has been pointed out that, in induction heating, the heat transfer by convection and radiation reflects the value of surface heat losses. This heat loss varies with temperature. The analysis shows that the convection losses are the major part of the heat loss in low-temperature applications such as induction heating of tin, lead, and aluminum alloys. In hot working applications (including induction heating of steel, titanium, cobalt, and nickel), radiation losses are much greater than convection losses, representing the major portion of total heat loss from the workpiece surface.

It is typically much easier to provide surface-to-core temperature uniformity for metals with high thermal conductivity, such as aluminum, silver, or copper. Metals with poor thermal conductivity, including stainless steel, titanium, and carbon steel, require extra care to obtain the desired temperature uniformity. This

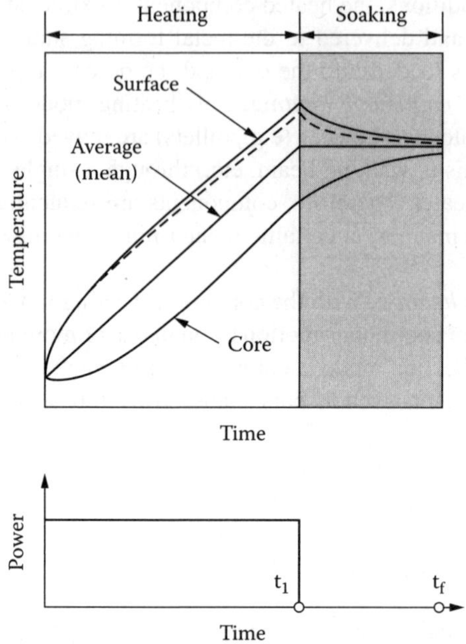

FIGURE 1.5 Time–temperature profile during static heating of a nonmagnetic solid cylinder.

"extra care" includes proper selection of heat mode, frequency choice, process time, and other parameters.

Figure 1.5 shows the typical time–temperature curve for static induction heating of a nonmagnetic solid cylinder. As one can see, immediately after heating begins, the surface temperature and average temperature begin to rise. In contrast, there is a delay before the core temperature starts to grow. If the materials' properties vary according to a linear function with temperature and surface heat losses are absent, there will be a linear region where all three temperatures are represented by three straight lines (Figure 1.5, solid lines). The surface-to-core temperature difference in this area is proportional to the power density during the heating cycle, the frequency, geometry, and material properties of the heated components.

As soon as power is cut off, the surface temperature decays rapidly, due to heat transfer toward a cooler core, and there is a corresponding rise in the core temperature. During the soaking stage, the surface-to-core temperature differential decreases and the heated component approaches the temperature uniformity required for hot working.

In reality, the surface-to-core temperature differential starts to decrease before the soaking stage begins. This temperature differential starts to decline during the heating stage (Figure 1.5, dotted curve representing surface temperature). This takes place due to the increasing surface heat loss with temperature and the depth of induction heating during the heating cycle. Because the electrical resistivity of most metals increases with temperature, current penetration depth increases.

It is important to note that the soaking stage can be performed when the heated workpiece is inside the induction coil and/or during the workpiece transfer stage to the hot forming machinery. The latter approach allows minimizing the total process time.

It should be mentioned that, in reality, the time–temperature profiles are more cumbersome than those shown in Figure 1.5. To improve the performance of an induction heating machine and reduce the heating time while providing surface-to-core temperature uniformity, power pulsing can be applied. Power pulsing refers to a technique that applies short bursts of power to maintain a desired surface temperature or a maximum allowable surface-to-core temperature difference. Pulse heating consists of a series of "Heat ON" and "Heat OFF" cycles until the desired uniformity is obtained (Figure 1.6). Depending upon the particular application, the process time reduction with pulse heating can exceed 40%.

The heating time can be reduced even more when applying accelerated heating (Figure 1.7). Obviously, to provide this type of heating, it is necessary to have a power supply that allows a gradual reduction of output power. The accelerated heating approach may have several modifications. Figure 1.8 shows

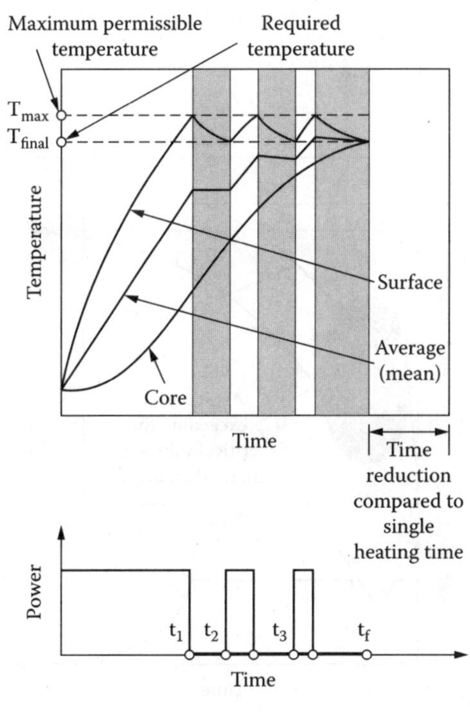

FIGURE 1.6 Power pulsing.

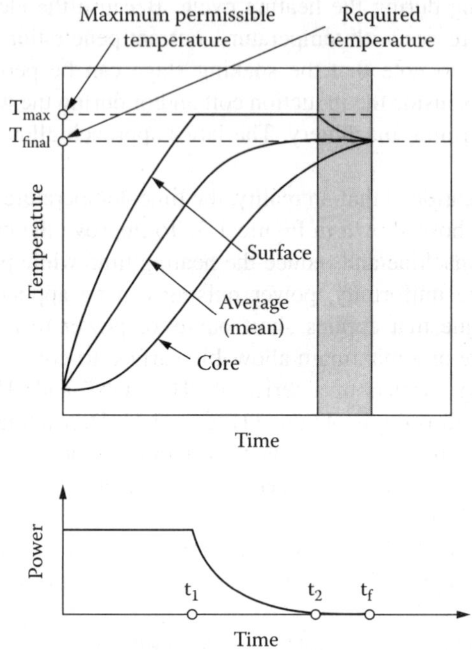

FIGURE 1.7 Accelerated heating.

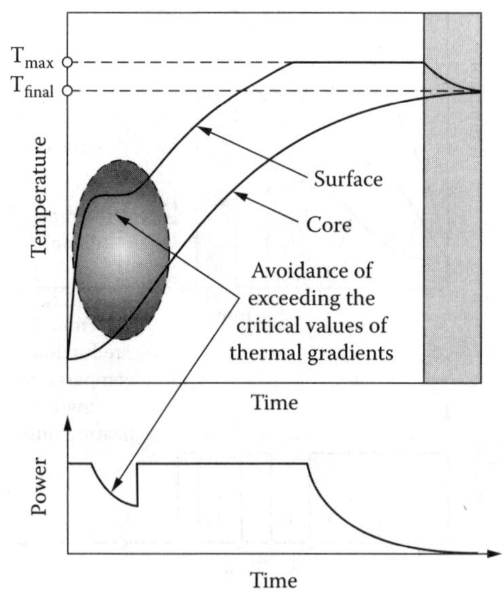

FIGURE 1.8 Modification of accelerated heating.

one of the modifications of accelerated heating that is particularly useful when heated bars, slabs, or billets have a tendency to crack. These cracks appear due to excessive thermal stresses (thermal shocks) that appear when thermal gradients exceed the maximum permissible level. These levels vary depending upon the metal's chemical composition, microstructure, and temperature.

Although it has been mentioned that the curves shown in Figure 1.5 through Figure 1.8 illustrate typical time–temperature profiles that take place during static heating, these curves are practically identical for progressive multistage heating and continuous heating modes as well. For multistage or continuous heating, the time axis represents the length of the induction heating line or coil length. Bursts of power can represent the power of inline coils that may have different length and windings, and/or can be individually fed from different power supplies with the ability to adjust the output power and frequency.[1]

1.4 DESIGN APPROACHES OF INDUCTION MASS HEATING

There are many different types of induction heating coils. The ability of any coil to establish a magnetic field depends upon the ampere-turns of the coil, geometry, and its electrical circuit. The selection of power, frequency, and coil length in induction heating is highly subjective, depending upon the type of heated metal, required temperature uniformity, time of heating, and so on. Frequency is one of the most critical parameters in these applications. If the frequency is too low, an eddy current cancellation within the heated body will take place, resulting in poor coil efficiency.

However, when the frequency is too high, the skin effect will be highly pronounced, resulting in a current concentration in a very thin surface layer compared to the diameter/thickness of the heated component. In this case, a lengthy heating time will be required to provide sufficient heating of the internal areas and the core. Prolonged heat time results in an increase of the radiation and convection heat losses that, in turn, reduce the thermal efficiency of the induction heater and diminish the main advantage of induction heating. Frequency is always a reasonable compromise.

The determination of the length of the coil line is another important step in specifying an induction heating system. In determining how long the coil line needs to be, the time required for heating the workpiece to an acceptable temperature condition is actually being determined. Heating time is a complex function of various factors, including the size of the workpiece, frequency, heating mode, power density, maximum permissible temperature, material properties, and the required surface-to-core temperature uniformity.

Coil design is, of course, an extremely important factor in developing an efficient induction heating system. Solenoid type (helical) multiturn coils are most often used in induction heating applications. The electrical efficiency and coil coupling have a marked effect on the coil's ability to deliver heating power to the workpiece. Smaller gaps between the surface of the heated workpiece and the coil result in better electromagnetic coupling and, consequently, higher coil efficiency. However, it is wise to remember that the total efficiency of the induction coil is a product of coil electrical efficiency and coil thermal efficiency.

Nearly all induction coils used for heating prior to hot forming of metals consist of thermal isolation (also called refractory or liners) between the coil and the heated workpiece protecting the coil windings from heat exposure and providing a thermal barrier. Thus, the chance of thermal shock is reduced and heat loss is minimized in water recirculating through the induction coil.

The thermal refractory can drastically decrease the heat losses from the surface of the heated workpiece. At the same time, the use of a refractory necessitates larger coil-to-workpiece gaps, which in turn deteriorates the electromagnetic coupling between the induction coil and the heated workpiece, leading to a reduction of coil electrical efficiency (Figure 1.9). Thus, the refractory allows one to improve coil thermal efficiency, but it also reduces coil electrical efficiency. In the great majority of mass heating applications and, in particular, in heating prior to hot working, it is advantageous to use refractory. Its minimum thickness is usually about 12 mm (0.5 in.). Different materials can be used for manufacturing refractory. These include castable (i.e., Visil), silicon carbide, Alumina, H-91, Zircar, and other fibrous ceramic materials.

The two principal approaches to building coils can be categorized as the open-wound or the refractory-encased approach. The open-wound method provides more convenient repair in the event of failure but the choice is never that simple. Replaceable refractory liners are commonly utilized in the construction of open-wound coils. Coils with replaceable liners have lower repair time and cost compared to cast coils.

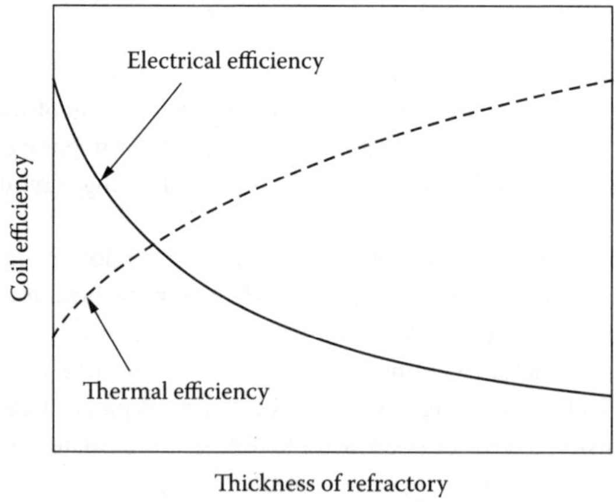

FIGURE 1.9 Coil electrical and thermal efficiencies vs. thickness of refractory.

Although the encased coil using a castable refractory offers durability and longer life, it still follows recommended maintenance routines that should not be neglected. To repair a cast coil, it must be shipped to a service center where it is broken down to the bare coil. Then the coil must be repaired, reinsulated, and recast. This process is time consuming considering that the operator can replace refractory liners on site in a matter of minutes.

Self-supporting internal water-cooled skid rails to protect linear surfaces must be robust enough to carry the full load of heated billets with minimal deflection; however, from another perspective, they should be transparent enough to the electromagnetic field that they will not be significantly heated. Water-cooled rails occupy space inside the coil and consume additional coolant, thus increasing the kilowatt losses. These rails terminate beyond the open-wound coil box envelope and extra expansion of longer rails must be accommodated.

To extend the life of water-cooled rails, they are wear coated. In the past, stellite was applied to the surface of the rails as a wear coating. This required an application process that works by fusing this material to the surface of the rail by heating with an oxygen acetylene torch. By replacing this with a less labor-intensive plasma coating process, it is possible to apply a wider variety of materials, including tungsten carbide and chrome carbide, providing longer rail life.

In addition to water-cooled alloy skid rails, various ceramic materials have been developed for use as wear plates. Ceramic liners are embedded into the refractory surface and have been successful in many applications. For high-wear/high-temperature applications, the liners are manufactured from special alloys.

In addition to improved alloys, new tubing may be provided with an extra heavy wall section that affords a much thicker wear area comparable to welding an additional layer of material to the top side of the water-cooled rail. These rails also utilize a plasma-applied wear coating on the surface to afford additional wear resistance and longer life.

In applications in which refractory liners are made from materials with extremely high wear-resistant properties, some end users actually skid or convey the billets across the surface of the ceramic liner, thus eliminating the use of skid rails and any cold spots resulting from their use.

Most manufactured liners are round. For cylindrical billets, a round shaped liner naturally suits the coil geometry. However, when heating square billets, this mismatch of shapes may result in a noticeable increase of the coil opening in addition to space allowances for the skid rails and thermal insulation. This reduces the electromagnetic coupling between the coil and billet. Longer coils call for a larger thickness of the refractory to ensure mechanical strength.

To optimize the design, the manufacturers of induction heating systems have developed different procedures. For example, Newelco windings are plastic coated for insulation purposes before they are set in a high-quality refractory cement. After the winding is positioned in a carefully prepared and assembled box, a former is centrally inserted through it to create the coil tunnel. The refractory is poured around the winding in such manner that air pockets are eliminated, after which the refractory is aged following controlled procedures. The profile of the former is such that recesses along its length allow for small-bore skid rails to be positioned prior to casting. Rails that are half sunken into the refractory directly take the load of the material being carried.

The end boards of the coil boxes are carefully recessed to enable rails to be terminated within the total envelope, so these coil units can be shorter and butted against each other. Gaps between individual coil units manufactured by Newelco are designed to be only 12 mm. This ensures sufficient latitude for changing from one unit to another while avoiding the larger energy-sapping gaps inevitable with open-wound coils. It also follows that minimal gaps and closeness of the tunnel walls to the billet decrease atmosphere attack and thus scale generation.

Sufficient radial clearance must be allowed for the largest size heated workpiece, taking into consideration the existence of the skid rails, the thermal expansion of the workpiece that is being heated, and its actual shape, because some workpieces can bow (i.e., long bars, rods, and slabs).

The coil space factor, K_{space}, is an important parameter of the coil design and should be as high as possible. The coil space factor represents how tightly the coil turns are wound. The space between turns should be as small as possible, yet large enough to leave room for electrical insulation. The coil turn space factor for a multiturn coil can be determined according to Figure 1.10 and is typically in the range of 0.7 to 0.9.

High-conductivity round and square copper tubing known as oxygen-free high-conductivity copper (OFHC) is commonly used for coil fabrication because it is naturally profiled for water cooling and because copper is a good electrical conductor with mechanical properties suited for coil fabrication. In some rare cases, copper tubing does not provide a large enough area for energy transfer. To compensate for tube constraints, a thick copper strip is sometimes brazed to the external water-cooling copper tube.

Coil windings are designed to accommodate the tube size, coil geometry, number of turns, and overall length with the workpiece size, shape, production rate, and load matching with the power supply.

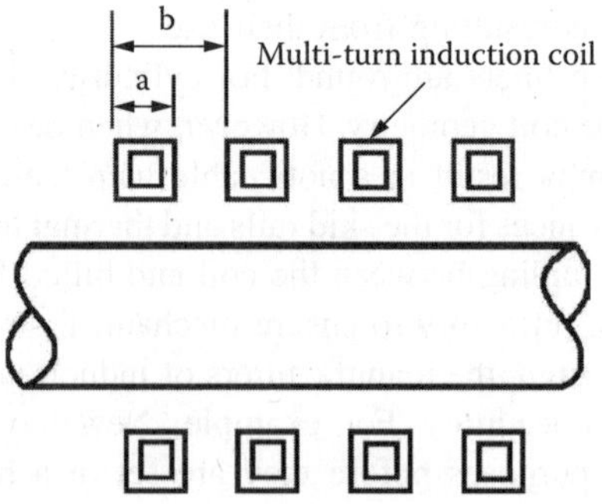

FIGURE 1.10 Coil turn space factor, $K_{space} = a/b$.

Coil cooling is a very important aspect of a heater design. It is obvious that coil cooling should be as close as possible to the heating face. Tube wall thickness is chosen based on the system operating frequency. The heating face wall thickness should increase as the frequency decreases. For example, a system with low operating frequency requires a thicker wall tube than a high-frequency system. This fact is directly related to the current penetration depth in the copper (δ_1) and holds true for tubing coils and coils made from copper sheet. The coil electrical losses will be minimum if the copper tubing wall thickness is greater than $1.6\delta_1$. A coil tubing wall smaller than $1.6\delta_1$ results in a reduction in coil efficiency and an increase of coil tubing power losses.

In some cases, the tubing wall may be thicker than that calculated according to the previously mentioned recommendation. This is because it may not be mechanically reliable to use wall tubing that is too thin due to the mechanical flexing caused by electromagnetic forces. As the frequency is lowered, more attention must be paid to coil support because there is more vibration at lower frequencies, especially at the turns near both ends of multiturn solenoid coils.

High-dielectric epoxies and some nylons are often used as dielectric insulating materials that eliminate arcing between coil windings. A fluidized bed process or an electrostatic coating process can provide dielectric insulation. In certain cases, some ceramic coatings afford protections from high-temperature exposure and suitable dielectric stress.[1]

1.5 TECHNOLOGICAL COMPLEX "HEATER–EQUIPMENT FOR METAL HOT WORKING"

The principal distinctive feature of the system approach is a necessity to take into consideration the whole sequence of the interconnected technological processes (preheating, transfer, hot working) as a technological complex being controlled. Overall performance indexes can be considered when optimizing technological complex heater–hot working equipment as a whole.

For a particular technological complex, it is important to reveal, by means of valid decomposition, local cost functions for each stage of the technological process. Along with these criteria, it is possible to find proper optimal control inputs, to determine optimal parameters for considered technologies, and to set optimal initial data values for design solutions. Such an optimization approach offers some advantages in comparison with typical approaches that simply rely on independent control of the individual processes.

Obtained results lead control algorithms' abilities out beyond the framework of traditional tasks of "serving to" technology, thus ensuring their active participation in its formation. This objective is fulfilled first by means of incidental solution of principal development problems for optimal technological modes and, second, by optimal design of production equipment from the point of view of its further optimal operating.

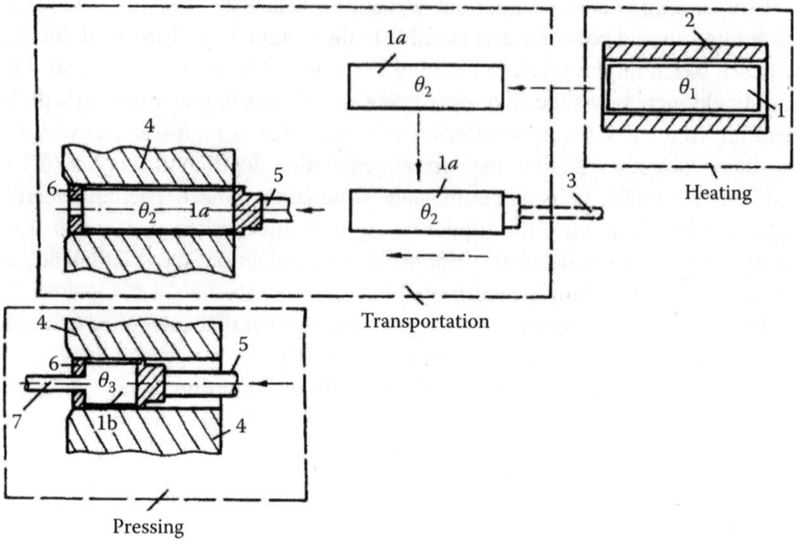

FIGURE 1.11 Technological complex "heater–hot working equipment." 1: Heated billet; 1a: billet in process of transportation; 1b: billet in process of hot working; 2: induction heater; 3: feeder mechanism; 4: press; 5: compression ram; 6: die hole; 7: press-product.

Let us consider the problem of optimization of a technological complex for induction heating of cylindrical billets and their subsequent hot forming (for example, extrusion) on hydraulic presses (Figure 1.11). Thermal conditions of billet and temperature uniformity are imperative for this application. A technological complex "heater–hot working equipment" can be considered as a set of thermal treatment processes on all stages of technological cycle, including heating in inductor, workpiece delivery to hot working equipment (transportation stage), and pressure processing (i.e., extrusion).

Billet transfer time, Δ_T, often results in appreciable distortion of a workpiece temperature distribution after exiting the inductor. The value Δ_T is determined by how far induction heating installation (IHI) is located from the hot working equipment and by design of the feeding mechanism. Using the control inputs, it is possible to modify time required for heating, but time Δ_T is beyond the control. This means that the controlled stage of process in IHI is supplemented by the uncontrollable transfer stage. Thus, it is necessary to complete a mathematical model of the heating process (Section 1.2) by equations describing the workpiece temperature field evolving over transfer time. The temperature distribution after the transfer stage can be regarded as an output-controlled function because an ultimate control goal is to provide required temperature state of the workpiece at hot working operations.

The final product of technological complex heater–hot working equipment is a press product with fixed length Z end. In this case, the pressing–extrusion stage should be considered as a final stage of a total process affecting the required billet temperature state. An appropriate mathematical model is needed to describe how the temperature field is changed over the stage of processing by pressure.

Hot working operation imposes the two most general requirements on the temperature field. First, maximum value, $\theta_{\max}(\varphi)$, of the temperature within the heated workpiece volume is restricted by the certain limiting value, θ_{adm}. Second, the temperature differences within the heated workpiece should be restricted during the heating process so that the maximum value, $\sigma_{\max}$, of expanding thermal stresses $\sigma(l, \varphi)$ (that appeared due to temperature differences) would not exceed prescribed admissible value σ_{adm} that corresponds to ultimate stress limit of the heated material. Temperature within the heated workpiece during pressing is restricted by maximum admissible temperature in the die hole, θ_{cr}.

Heating power, $F^*(t)$, and pressing velocity, V_p, can be considered as control inputs for heating and pressing processes. The values of these control inputs are also restricted by certain limiting admissible values, $F^*_{\max}$ and $V_{p\max}$.

Under previously described conditions, the statement of an optimal control problem can be formulated for technological complex heater–hot working equipment as follows. It is required to get a final press product of desired length and hold cost criteria at extremum value by time-varying controls that is restricted by a preassigned set and that satisfies the constraints on a temperature distribution.

When maximum productivity is required, a minimal total heating time, τ_{end}, can be considered as a cost criterion. When a criterion of minimum cost is required, then, depending upon the situation, different performance indexes could be considered. Particular OCP formulations for the technological complex and methods to determine optimal control inputs will be discussed in Chapter 6 of the book.

1.6 TECHNOLOGICAL AND ECONOMIC ADVANTAGES OF INDUCTION HEATING

Induction heating installations are of considerable current use in industry. In the past three decades, heating by induction has become more popular. A major reason is the ability to create high heat intensity very quickly at well-defined locations on the part. This leads to low process cycle time (high production rate) with repeatable quality. Induction heaters offer several advantages over analogue equipment.

Induction heating is particularly useful when highly repetitive operations are performed. Once an induction heating machine is properly adjusted, part after part is heated with identical results. Usually, no further attention is required, except the loading and unloading of the workpieces. The ability of induction heating to heat successive parts identically means that the process is adaptable to completely automatic operation, in which the workpieces are loaded and unloaded mechanically with no operator present.

Induction heating can be utilized in line with other technological operations. This saves the time of transporting heated workpieces from one part of the factory to another. Another attractive characteristic of induction heating is its ability to heat only a certain part of a workpiece if required.

Induction heating is more energy efficient and inherently more environmentally friendly than most other heat sources, including gas-fired furnaces, salt and lead baths, carburizing, or nitriding systems.[26–28] Any smoke and fumes that may occur due to residual lubricants or other surface contaminants can be easily removed. A considerable reduction of heat exposure is another factor that contributes to the environmental friendliness of induction heaters.

Induction heating provides much better surface quality of heated metal in comparison with competitive technologies. A significant reduction of scale results in substantial metal savings.

Induction systems usually require far less startup and shutdown time and lower labor cost for machine operators. Other important factors of induction heating machinery include quality assurance, automation capability, high reliability, and easy maintainability of the equipment. In many cases, induction heating will require minimum shop floor space and produce less distortion in the workpiece.

In the past, fuel-fired furnaces that utilized natural gas, fuel oil, or liquid petroleum gases were often used because of the low cost of fuel. However, in recent decades, producers have been shifting their preference toward induction heating systems and this tendency continues to grow at an increasing pace. There are several reasons for this shift.

First, fuel-fired furnaces demand a very long heating tunnel to achieve the desired temperature uniformity. The large required space often presents a problem in plants due to the limited space available on the shop floor, particularly when the heating systems must be incorporated into an already existing production line. For example, it is frequently necessary to locate a slab or transfer a bar edge reheater into the limited space between an existing caster and rolling mill.

On the other hand, heating in fuel-fired furnaces results in a significant metal loss and poor surface quality (due to scale formation, decarburization, oxidation, coarse grains, etc.). Scale reduction and improved surface quality of heated parts lead to longer die life and minimum postprocessing operations. Unlike fuel heating in which heat is transferred by radiation, the induction-type energy transfer generates the heat directly in the workpiece.

Finally, fuel-fired heating increasingly faces environmental restrictions (air pollution) and restrictions from an ergonomic perspective because operators are exposed to hot air blasts and unstable work flow.

These are only some factors that have resulted in induction heating becoming a more popular approach for thorough heating of slabs, blooms, billets, bars, tubes, strips, wires, rods, and other components made of ferrous and nonferrous metals.[1]

REFERENCES

1. Rudnev, V.I. et al., *Handbook of Induction Heating,* Marcel Dekker, New York, 2003.
2. Gebhart, B., *Heat Transfer,* 2nd ed., McGraw–Hill, New York, 1971.
3. Patankar, S., *Numerical Heat Transfer and Fluid Flow,* Hemisphere, New York, 1980.
4. Incropera, E.P. and Dewitt, D.P., *Fundamentals of Heat Transfer,* Wiley, New York, 1981.
5. Myers, R.F., *Conduction Heat Transfer,* McGraw–Hill, New York, 1972.
6. Rohsenow, W.M. and Hartnett, J.P., *Handbook of Heat Transfer,* McGraw–Hill, New York, 1973.
7. Siegel, R.I. and Howell, J.R., *Thermal Radiation Heat Transfer,* 2nd ed., McGraw–Hill, New York, 1980.
8. Wiebelt, J.A. and Howell, J.R., *Engineering Radiation Heat Transfer,* Holt, Rinehart & Winston, New York, 1980.
9. Leslie, W.C., *The Physical Metallurgy of Steels,* McGraw–Hill, New York, 1981.
10. Sluhotskii, A.E., Nemkov, V.S., and Pavlov, N.A., *Induction Heating Installations,* Energoatomizdat, St. Petersburg, 1981.
11. Nemkov, V.S., Polevodov, B.S., and Gurevich, S.G., *Mathematical Modeling of High Frequency Heating Equipment,* Energoatomizdat, St. Petersburg, 1991.
12. Sluhotskii, A.E. et al., *Induction Heating Equipment,* Energoatomizdat, St. Petersburg, 1981.
13. Nemkov, V.S. and Demidovich, V.B., *Theory and Computation of Induction Heating Installations,* Energoatomizdat, St. Petersburg, 1988.
14. Rudnev, V.I., Mathematical simulation and optimal control of induction heating of large-dimensional cylinders and slabs, Ph.D. thesis, Electrical Engineering University, St. Petersburg, 1986.
15. Samarskii, A.A., *Theory of Finite Difference Schemes,* Nauka, Moscow, 1977.
16. Demirchian, K.S. and Chechurin, V.L., *Computational Methods of Electromagnetic Field Simulations,* MEI, Moscow, 1986.
17. Muehlbauer, A., Software for modeling induction heating equipment by using finite element method, *Appl. Therm. Eng.,* 23, 1647, 1991.
18. Boergerding, R. and Muehlbauer, A., Numerical calculation of temperature distribution and scale formation in induction heaters for forging, *Proc. UIE Sem. "Simulation and Identification of Electroheat Processes,"* 134, 1997.
19. Muehlbauer, A., *Aktuelle Forschungsbeiträge zur EMV — Magnetische Streufelder in der Umgebung von Induktionsanlagen,* EMC Kompendium, Muenchen, 1999.
20. Jin, J.M., *The Finite Element Method in Electromagnetics,* Willey, New York, 1993.
21. Lowther, D.A. and Silvester, P.P., *Computer-Aided Design in Magnetics,* Springer, Berlin, 1986.
22. Livesley, R.K., *Finite Elements: an Introduction for Engineers,* Cambridge University Press, New York, 1983.
23. Tamm, I.E., *Fundamentals of the Theory of Electricity,* Moscow, Metallurgy, 1981.
24. Hayt, W.H., *Engineering Electromagnetics,* McGraw–Hill, New York, 1981.
25. Rapoport, E.Ya., *Optimization of Induction Heating of Metals,* Metallurgy, Moscow, 1993.

26. Muehlbauer, A., *Industrielle Elektrowärmetechnik,* Vulkan Verlag, Essen, 1992.
27. Baake, E., Joern, K.-U., and Muehlbauer, A., *Energiebedarf und CO2-Emission Industrieller Prozessverfahren,* Vulkan–Verlag, Essen, 1996.
28. Tudbury, C.A., *Basics of Induction Heating*, Rider, New York, 1960.

2 Optimization Problems for Induction Heating Processes

This chapter provides a detailed conceptual description of the optimization problems for induction heating processes (IHPs) prior to metal hot working operations. The optimization problems are discussed from the standpoint of the modern theory of optimal control with respect to systems with distributed parameters. An induction heating process can be considered an example of a system under control. The typical cost functions and mathematical models are provided in this chapter. Some explanations will be given to facilitate understanding such terms as controlled outputs and control inputs, disturbances, constraints, etc.

An overview of the most typical and important process constraints taking place in induction heating will be provided in this chapter as well. Considered constraints include but are not limited to power supply limitations, technological requirements, and specifics of the heater design. Requirements to final temperature distribution within a billet will also be discussed in this chapter. The optimization problem with respect to IHP simplified models will be formulated in terms typically used in the theory of applied optimal control.

The accurate and approximated approaches are presented here to transform the complicated optimal control problem into a much simpler task of determining unknown parameters of sought control functions. These parameters make clear physical sense and define controls uniquely. This procedure is called parameterization. A highly effective special engineering parameterization technique has been developed and will be demonstrated. As an example, this unique technique will be illustrated by solving a typical optimization problem for an IHP that will allow maximizing production rate of the heating system.

2.1 OVERVIEW OF INDUCTION HEATING PRIOR TO METAL HOT WORKING OPERATIONS AS A PROCESS UNDER CONTROL

The main purpose of induction heating installations (IHIs) is to provide a required workpiece temperature state prior to further stages of the technological process (i.e., hot forming, forging, rolling, extrusion, etc.). An IHI should provide a metal workpiece (i.e., billet, slab, or bar) at the hot working stage with the desired

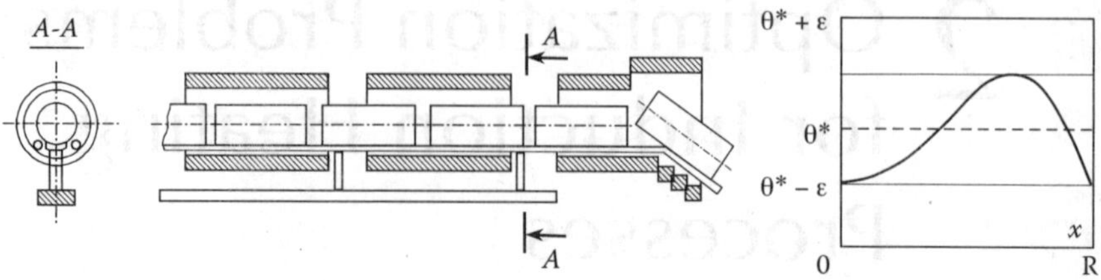

FIGURE 2.1 Induction through heater and qualitative outlet temperature distribution in the billet cross-section.

(typically uniform) temperature distribution across its diameter/thickness as well as along its length and across its width.

There are several ways to estimate a tolerable deviation of final temperature distribution from a desired one. It is typically necessary to provide the fixed maximum absolute value of temperature variation. This means that at the end of the heating the temperature in any point of the billet should deviate not more than by prescribed value ε from desired temperature θ^* (Figure 2.1). Another approach might deal with an estimation of the root mean square error between the required temperature distribution and the actual final temperature distribution.

Certain applications may require a nonuniform end-to-end temperature profile (for example, heating providing required temperature gradients). However, the variation of final temperature distribution from desired one has the same meaning as in the case of uniform heating and could be estimated in the same manner. Thus, the ultimate goal of the heating process prior to consequent hot working operation is to obtain a desired temperature profile within a workpiece that would correspond to particular heating requirements.

In such a case, the induction heating process can be considered as a dynamic controlled system, evolving over time. An ability to control the heating process means a capability to influence a temperature field in order to achieve a desired goal. Therefore, temperature distribution can be defined as *an output controlled function* of the process. It is important to recognize that a temperature field varies with time and spatial coordinates (within workpiece volume). Temperature distribution can be affected through a set of *control inputs* (or *controls*). By choosing in every instant the value of each control input, one can modify temperature distribution and dynamic behavior of an induction heating system.

As the ultimate goal, required final temperature distribution could be achieved by different means, including different operating modes of heating systems, coil design specifics, etc. To choose certain control properly, one needs to determine a cost function (performance index, cost criterion), i.e., a function reflecting technical and economical efficiency of an induction heating system. Therefore, we are interested in choosing a space-time-varying control input that provides a required final temperature profile of a workpiece and minimizes (or maximizes if appropriate) the cost function.

Control function resulting from a solution of the described problem is said to be *optimal control*, and the considered problem is known as an *optimal control problem* (OCP) for a heating process. Such a formulation should be considered as a very general description of OCP. To obtain more suitable practical solutions from engineering perspectives, it is necessary to apply an exact formulation of an optimal control problem. An explanation of optimal control theory terms must also be emphasized and defined more precisely.

First, it is necessary to choose a specific performance index of IHI depending on the major features of a particular industrial system being optimized. Proper control inputs (one or more) need to be defined with regard to different IHI designs. When choosing a particular set of control inputs, it is necessary to take into account the possibilities of their practical implementation. Obviously, to describe control input/output relationships, it is necessary to have an appropriate mathematical model of the heating process.

Often, an engineer is limited in his choice of controls by the set of constraints that defines the set of admissible control inputs. These constraints are referred to as *constraints on control inputs* and are of fundamental importance in the OCP formulation. Moreover, the technology of heating and hot working imposes a number of the requirements to a temperature field evolving over a heating process. These requirements make a sense of technological constraints and add a complexity to the solution of an optimal control problem. Physical constraints imposed by the environment in which the induction heating is to be done can also play an important role.

Factors that cause an unavoidable adverse effect on a temperature field can be considered as *disturbances*. Fluctuations of the controlled system characteristics from their calculated deterministic models can be related to disturbances as well. Such fluctuations are mainly caused by, first, inadequately determined characteristics of a process to be controlled and, second, by uncertainty in a definition of initial data (i.e., material properties, etc.). In any case, existence of disturbances results in deviations of controlled output response on control input from the expected one. These deviations can be significant, and sometimes unacceptable. Due to the stochastic character of disturbances, it is often difficult to evaluate such deviations with required accuracy. It is assumed that disturbances can be taken into account with respect to solving optimal control problems.

Therefore, to have a proper formulation of the optimal control problem for an induction heating process, it is necessary to define the following factors properly (Figure 2.2):

- Cost function for optimization of induction heating installation
- Mathematical model of the process
- Set of admissible control inputs
- Constraints imposed on control inputs
- Additional technological constraints imposed on behavior of a temperature field of heated workpieces

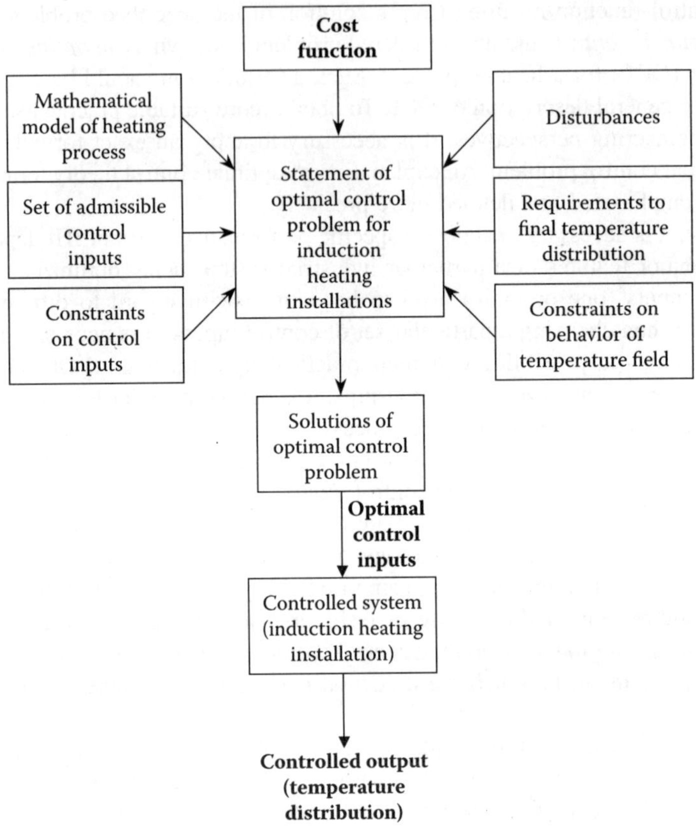

FIGURE 2.2 Basic data for statement of optimal control problem for induction heating process.

- Models that describe major disturbances
- Requirements to final temperature distribution within the heated workpieces

Descriptions and classifications of all of these factors are presented next, as well as their definition with respect to solving optimal control problems.

2.2 COST CRITERIA

Optimal control represents a control strategy that is the best or more favorable in terms of a particular sense or criterion — i.e., in terms of a defined performance requirement or specification. Therefore, the statement that a certain control is optimal is fairly meaningless unless the sense in which the control is optimal is also stated. Often, the performance specification is formulated in terms of mathematical function. In this case, the optimal control minimizes (or maximizes if appropriate) a particular function or parameter. A cost function that defines performance could be described by integral cost criteria considered over the time interval from beginning to final time. An integral cost criterion is used whenever the average performance over the chosen time duration is of primary importance.

In many induction heating applications, the selection of cost function, i.e., the specified integral criterion, cannot be easily determined. A system approach should be used when choosing the most appropriate cost function for a particular IHI optimal control problem. An overall performance index can be considered when optimizing technological complex "heater–hot working equipment" as a whole. For particular technological processes, a justified choice of optimization criterion is carried out as a result of common OCP decomposition into an optimal control problem for a heating process and a problem of hot working process optimization. Simultaneous solution of these problems provides optimal operation mode for the complex heater–hot working equipment.

When maximum productivity is required, a minimal total heating time, τ_{end}, can be considered as a cost function. Using the integral form of τ_{end} representation, the integral cost criterion, I_1, may be written as:

$$I_1 = \int_0^{\tau_{end}} d\tau = \tau_{end} \rightarrow \min . \tag{2.1}$$

The cost criterion (Equation 2.1) is often called a *time-optimal criterion*, and the corresponding optimal control problem is called a *minimal time-optimal control problem.*

When a criterion *of minimum cost is* required, different performance indexes can be considered, depending upon the situation. An alternative cost function could be the one in which heating accuracy is seen as the most important specific parameter that represents the error (deviation) between actual final temperature distribution, $t(x,\tau_{end})$ (within volume Ω of heated body), and desired distribution, $t^*(x)$ (where x is the spatial coordinates of points within volume Ω).[1] Decreasing this deviation (i.e., increasing heating accuracy) leads to a preferable condition during metal hot working operation and cost reduction at this stage.

The most common optimal control approach might deal with an estimation of heating accuracy by the norm of deviation of $t(x,\tau_{end})$ from $t^*(x)$ — for example, by root mean square error or by absolute maximum deviation within volume, Ω, of the heated body. Then, given a cost criterion, I_2 can be defined as:

$$I_2 = \left\| t(x,\tau_{end}) - t^*(x) \right\| \rightarrow \min . \tag{2.2}$$

This means a determination of the control inputs that would lead to a minimum deviation of actual temperature distribution from the desired one. In Equation (2.2), $\left\| t(x,\tau_{end}) - t^*(x) \right\|$ denotes the norm of error.

In some cases, the most essential part of the process effectiveness deals with cost of energy/power used for heating. In this case, it is proper to consider a cost function, I_3, which can be represented in the integral form as:

$$I_3 = \int_0^{\tau_{end}} P(\tau)d\tau \rightarrow \min , \tag{2.3}$$

where $P(\tau)$ is consumed time-dependent electrical power. Minimization of the cost function (Equation 2.3) leads to minimization of the overall required energy.

In many cases, an essential part of the cost is material expenses. For high-temperature heating, the appreciable portion of material expenses represents a metal loss due to scale formation. As shown in Butkovskij et al.,[2,3] the value of metal losses can also be considered as optimization criterion I_4 and could be written in the integral form as follows:

$$I_4 = \int_0^{\tau_{end}} f(t_{sur}(\tau))d\tau \rightarrow \min , \tag{2.4}$$

where $f(t_{sur}(\tau))$ is the known nonlinear time-dependent function of surface temperature $t_{sur}(\tau)$.

An interesting and commonly encountered special case arises when it is necessary to take into account all the previously mentioned components (or more than one) of the cost. It appears appropriate to combine cost functions (Equation 2.1 through Equation 2.4). The following overall function can be used as a combined criterion of the optimal control problem:

$$I_{\Sigma} = \sum_{i=2}^{4} C_i I_i + C_{\tau}\tau_{end} \rightarrow \min . \tag{2.5}$$

Criterion (2.5) includes all cost components (Equation 2.2 through Equation 2.4) with weighting coefficients, C_i, making sense of a relative costing and representing the importance of each of the previously considered items of the overall cost. Increase of the time required for heating results in the corresponding increase of expenses according to Criterion (2.5) with correspondent weight coefficient C_{τ}.

It should be mentioned here that other expenses can be included in I_{Σ} when needed — for example, cost of transmission power losses in an electric circuit "a power supply inductor."[1] When minimized by appropriate choice of control input, the cost function (Equation 2.5) results in the necessity to make a reasonable compromise between certain conditions. In specific cases (in particular, if

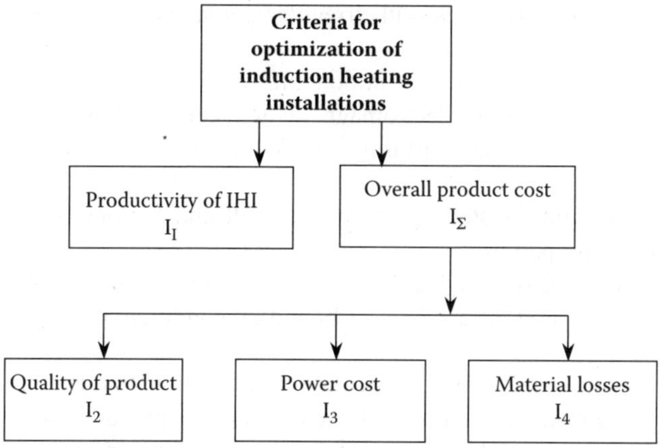

FIGURE 2.3 Basic cost criteria for optimization of induction heating process.

weighting coefficients C_i cannot be easily determined a priori), the problem of multicriteria optimization is posed with a number of separate cost functions described previously.

The considered cost criteria take into account all major IHI performance characteristics, including *productivity* (time-optimal criterion I_1), *quality* of product (heating accuracy criterion I_2), *power cost* (energy consumption criterion I_3), *material losses* (cost function I_4 minimizing metal losses due to scale formation), and *overall product cost* (combined cost criterion I_Σ) (Figure 2.3). All described cost functions are related to steady-state operation modes of induction heating installations in the technological complex "heater–hot working equipment."

The most typical transient operation modes include but are not limited to changes of production rate, variation of workpiece sizes and heated metals properties, start-up and shut-down modes, technological breaks, etc. The ultimate goal of transient mode optimization is minimizing losses due to variations from steady-state modes.[1] Appropriate optimization problems differ by essential specificity and are not reviewed further in this book.

2.3 MATHEMATICAL MODELS OF A HEATING PROCESS

To solve an induction heating optimization problem, specifications or descriptions of a heating process must be put into a form amenable to analysis and evaluation. The basic descriptions or mathematical models of heating processes usually represent a system of differential equations and other mathematical relations.

Generally speaking, a space–time temperature distribution within an inductively heated body — controlled output function — is described by a highly complicated system of interrelated Maxwell and Fourier equations for electromagnetic and temperature fields (see Chapter 1). These equations allow consideration of all essential specifics of IHI operation, such as:

- Heating of workpieces with irregular geometry as well as complex inductor geometry
- Essentially three-dimensional nonuniform distribution of internal heat sources caused by an electromagnetic field generated by induction coil, which results in nonuniform temperature distribution within workpieces
- Nonlinear interrelated nature of Maxwell and Fourier equations and material properties
- Thermal radiation and heat convection from the surface of a heated workpiece (nonlinear nature of boundary conditions)

These equations are used for calculation of the induction heating processes. Quite often, cumbersome numerical methods are required to solve nonlinear interrelated electrothermal equations.[4,5] At the same time, in many cases certain assumptions could be made to obtain a reasonable simplification of the mathematical model. For a large number of induction heating systems, these approximations and simplifications lead to mathematical models in the form of linear ordinary differential equations (see Chapter 1). Even though models studied in this chapter are linear, they are important for analysis of nonlinear systems and choosing control strategies.

Such models usually provide maximum opportunities to analyze general specifics of optimal processes. Besides that, experience shows that these models provide the engineering means of quantitative estimation of temperature fields. In addition, the qualitative overall conclusions with respect to simplified models can be extended to more complicated models. Finally, results obtained for simplified models can be used as initial approximations for calculations based on complex numerical models.

Most mathematical models of induction heating tend to be handled with a combination of the following assumptions:

- Neglecting nonlinearity by averaging process parameters on the appropriate temperature intervals
- Concentration on mathematical description of regular-shaped workpieces (i.e., plate, cylinder, rectangle, parallelepiped, sphere)
- Simplified description of the geometric input data for an induction heating system
- Taking into account nonuniformity of temperature distribution only along one or two coordinates (reducing a three-dimensional temperature field to one- or two-dimensional forms)

Thus, it is possible to obtain analytical solutions for electromagnetic and thermal problems separately. Thermal problems can be reduced to the linear nonhomogeneous partial differential Fourier equation with corresponding linear boundary conditions (see Chapter 1). This equation describes a time-dependent heat transfer process in a metal workpiece and can be considered as a mathematical model of heating process.

The right side of the Fourier equation (heat conduction equation) consists of the function that represents a spatial time-dependent distribution of intensity of internal electromagnetic heat sources. This function can be obtained as a result of solving a linear Helmholtz equation for complex magnetic field intensity as a simplified case of the electromagnetic problem. Depending upon the particular induction heating process, the typical simplified models can be introduced as:

- One-, two-, and three-dimensional temperature fields describing static or continuous modes of heating (see Chapter 1)
- Multistage processes of heat treatment, including a heating stage as well as a soaking/cooling stage that takes place during workpiece transportation to the metal forming operation

In this chapter, a linear one-dimensional model of static induction heating of the regular-shape bodies (infinitely long wide plate or cylinder of infinite length (see Chapter 1) will be considered as a basic model. Furthermore, the opportunity of obtained results' extension to other IHP models is investigated, taking into account all marked essential properties of actual processes.

The basic model describes one-dimensional processes of heat distribution along the plate thickness or cylinder radius due to heating by internal sources. It is assumed that temperature nonuniformity along other coordinates will be neglected. The last assumption does not result in essential errors in most practical applications because of unessential nonuniformity of heat source distribution along the workpiece length and relatively small heat losses from the workpiece butt-ends. An appropriate heat conduction equation and typical boundary conditions of the second kind can be written for generality, using dimensionless units (*relative units*), as Equation (1.27) through Equation (1.32).

Heat transfer by convection is carried out by gas or air (i.e., from the surface of the heated workpiece to the ambient area) and heat loss flow density, $q(\varphi)$, should be considered in Equation (1.32) according to Expression (1.34). It is important to mention at this point that the right side of Equation (1.27) includes the function $W(\xi,l)u(\varphi)$ that represents an internal nature of the heat sources inherent to induction heating.

Internal heat sources are caused by an electromagnetic field generated by induction coil. Total specific power of these internal heat sources will be designated as $u(\varphi)$. According to Expression (1.21) through Expression (1.22), the function $W(\xi,l)$ should be considered as a distribution of internal heat sources. Equation (1.27) with boundary conditions (Equation 1.31 and Equation 1.32) describes temperature distribution $\theta(l,\varphi)$ along coordinate l (i.e., along the plate thickness or radius of cylinder) at any moment $\varphi > 0$. Temperature distribution, $\theta(l,\varphi)$, is obtained under conditions of arbitrary variables in time and within the workpiece volume internal heating power $W(\xi,l)u(\varphi)$. It is assumed that there are appreciable heat losses, $q < 0$, from the workpiece surface $l = 1$. It is possible to

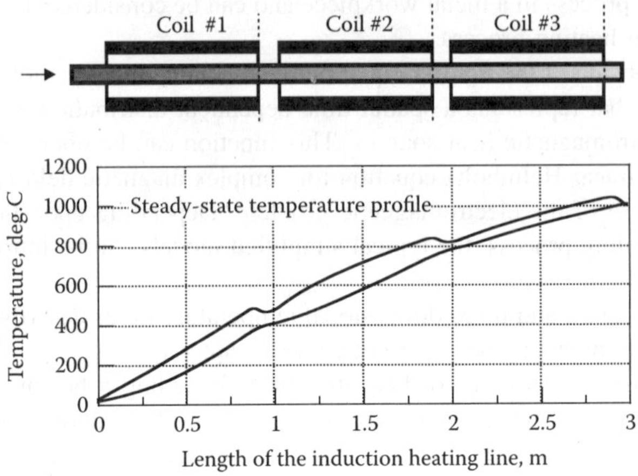

FIGURE 2.4 Inline induction heating.

find an explicit form of the expression $\theta(l,\varphi)$ thanks to obtaining an exact analytical solution of a governing equation taking into account known functions $W(\xi,l)$ and $u(\varphi)$.

Despite many simplified assumptions, the model (Equation 1.27 through Equation 1.35) provides a reasonable fit for the real process of induction heating on the appropriate temperature intervals. Such an approach is often used for description of temperature fields during induction heating of nonferrous cylindrical billets. It is also proper to use this model for computation of an induction heating of magnetic billets during a "hot" stage (above the Curie temperature).[1,6,7]

Equation (1.27) through Equation (1.35) describe a temperature field in steady-state continuous heating under assumption that an effect of heat transfer in the axial direction (along coordinate y) by heat conduction is negligibly small (Figure 2.4). The distance covered by the bar inside the heater — i.e., the value of longitudinal spatial coordinate y — represents a position of the particular area of heated bar inside induction coil. The value y varies from zero to the length of the induction heater and is determined thus for each time step, φ, by the simple expression:

$$y = V\varphi , \tag{2.6}$$

where V is a workpiece velocity.

Therefore, for a continuously operating heater, substitution of variable $\varphi = y/V$ in Equation (1.27) results in an equation describing steady-state distribution of temperature in an axial direction (along coordinate y) as well as in a radial direction (along coordinate l). In this case, function $u(\varphi)$ in Equation (1.27) describes steady-state distribution of heating power along the inductor length. Thus, the basic model (Equation 1.27 through Equation 1.35) can be applied for static and continuous induction heating modes. For continuous induction heating, Expression (2.6) should be taken into account as well.

2.4 CONTROL INPUTS

One of the most important features of a modern induction heating system is the ability to control the critical process parameters effectively. The control system should allow presetting a number of system input parameters with the expectation that, via a specified control algorithm, the result will be the desired system output. In a mass heating application, the final temperature distribution represents the main controlled output function.

An internally generated heating power is the most significant process parameter that affects temperature distribution. Proper control of a heating process involves appropriate choice of the internal power density simulated by function $F(\xi,l,\varphi) = W(\xi,l)u(\varphi)$. As shown earlier, the right side of Equation (1.27) consists of this function. Therefore, in a general case, the given function should be considered as a space–time control function applied to a mathematical model of a heating system in the form of differential equations (Equation 1.27 through Equation 1.35). However, such control functions in many cases are too complex to be practically applied. Attempts to implement required function $F(\xi,l,\varphi)$ were not successful as well. Therefore, it is important to develop a set of particular control inputs that could be relatively easily implemented in engineering practice.

At the first look, this may seem like a relatively simple process that calls for the total consumed heating power, $u(\varphi)$, as control input. In this respect, voltage applied to an induction coil can be considered as a required control function. A complication to the control process occurs due to the difficulties in providing desired spatial distribution of internal heat sources. It could be impractical to realize certain power distribution along the length of an induction coil. Many of the existing means to control power distribution face strict limitations or merely cannot produce the required spatial distribution of internal heat sources when heating certain metals. For example, when power supplies have appreciably limited frequency adjustability (so-called inverters with fixed frequency $\xi = const$), the dependence $W(\xi,l)$ along l (i.e., heat source distribution along billet radius) is uniquely defined by the law of electromagnetic wave propagation in metals (Equation 1.22). In this case, a spatial distribution of internal heat sources cannot be controlled over a heating interval.

It is possible to consider a frequency as a control input parameter that affects spatial distribution of internal heat sources when using adjustable frequency supplies. However, it is only possible to do this as long as heat source distribution depends upon frequency (in other words, $W(\xi,l)$ depends upon parameter ξ). Generally speaking, it is possible to control a depth of the subsurface heat-generating layer over time, i.e., a variable current penetration depth (see Chapter 1).

In static heating, the spatial distribution of heat sources along heated workpiece length can be controlled using electromagnetic longitudinal end effects (Figure 1.3). Proper coil overhang is only one of the possible design approaches that can provide required (i.e., uniform) temperature distribution along the workpiece length.

The majority of induction billet heaters utilize a heating mode in which billets are moved end to end through a heater. As applied to steady-state continuous heating, a variety of design concepts of an induction heater provide required heat generation power distribution, $u(y/V)$, along a heater length.

A multicoil inductor design is one of the most popular approaches (Figure 1.4b). In this case, an appropriate number of inline coils is used. Each coil may have different diameter, length, and number of turns. In addition, each coil can have different frequency and power. Therefore, piecewise constant dependence, $u(y/V)$, can be implemented for each coil. This dependence represents a piecewise power approximation along the heating line. The power of each coil varies within certain limits and can be different from coil to coil. Air gaps between coils and end effects can be taken into consideration in a similar manner. Different circuit connections, variable coil windings can also be considered among other means of the heating control along the length on an induction heater (see Chapter 1).

It should be mentioned at this point that all of these mentioned means of the longitudinal power control have certain physical and technological limitations in providing a particular distribution of heat sources along the length of an induction heater. Generally speaking, all control inputs suitable for induction heating processes can be divided into three groups (Figure 2.5).

The first group consists of control inputs that do not depend on space coordinates. These control inputs are time-dependent parameters (lumped control inputs), including:

- Voltage, current, or frequency of operating current
- Power of each coil
- Production rate of progressive multistage heaters or speed of continuous heating installations

All these control inputs make common sense, according to which the term "control input" is associated with certain "steering wheels"; operating position of these "wheels" can be changed in time in a desirable way.

The lumped control inputs are used to control static heaters. They also can be utilized to control continuously operating IHIs during transient modes of operation (see Chapter 1). A time-dependent control of the coil voltage can serve as a good example of the control that can be relatively simply realized using modern semiconductor switching devices. Inductor voltage can be easily changed almost in an arbitrary way but within certain prescribed bounds.

Coil power represents another effective way to control heat sources. This control input may be conducted relatively easily by appropriate variation of coil voltage. Remember that coil power is proportional to the second power of voltage. Therefore, power density of electromagnetic heat sources $u(\varphi)$ can be considered as a convenient time-dependent control function typically used for an induction heating process.

The selection of operating frequency can also be considered as an alternative control input. However, a necessity to be able to adjust frequency of the power

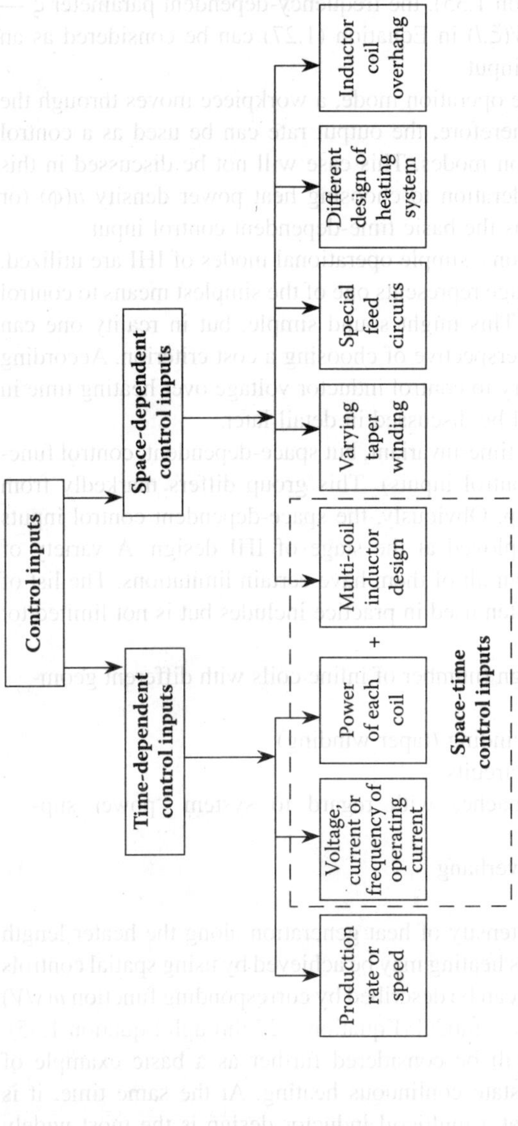

FIGURE 2.5 Control inputs used for induction heating processes.

supply could lead to undesirable system complexity and additional cost.[1,4,7] All these factors noticeably limit the use of frequency as an effective control input for a majority of induction mass heating applications. At the same time, in some unique applications (i.e., dual frequency gear hardening) adjustable frequency allows obtaining heat source distribution and correspondent temperature profile that could not be obtained by any other means. Therefore, in the process model (Equation 1.27 through Equation 1.35), the frequency-dependent parameter ξ — an argument of the function $W(\xi,l)$ in Equation (1.27) can be considered as an example of frequency control input.

According to a steady-state operation mode, a workpiece moves through the heater with constant speed. Therefore, the output rate can be used as a control input only in transient operation modes. This case will not be discussed in this text; we shall limit our consideration to choosing heat power density $u(\varphi)$ (or appropriate inductor voltage) as the basic time-dependent control input.

In many practical applications, simple operational modes of IHI are utilized. Stabilizing an optimal coil voltage represents one of the simplest means to control an induction heating process. This might sound simple, but in reality one can face certain problems from a perspective of choosing a cost criterion. According to several criteria, it is necessary to control inductor voltage over heating time in quite a complex way. This will be discussed in detail later.

The second group includes time-invariant, but space-dependent, control functions (spatial or distributed control inputs). This group differs markedly from control inputs of the first group. Obviously, the space-dependent control inputs represent design solutions employed at the stage of IHI design. A variety of inductor designs is available, but all of them have certain limitations. The list of appropriate design solutions often used in practice includes but is not limited to:

1. Multicoil inductor design (number of inline coils with different geometry and power)
2. Proper choice of coil winding (taper winding)
3. Applying special feed circuits
4. Different design approaches with regard to system "power supply–inductor–metal"
5. Proper choice of coil overhang

Required distribution of intensity of heat generation along the heater length during a steady-state continuous heating may be achieved by using spatial controls 1–4. All of these control means can be described by corresponding function $u(y/V)$ with regard to the heating process model (Equation 1.27 through Equation 1.35). The control function $u(y/V)$ will be considered further as a basic example of distributed control for steady-state continuous heating. At the same time, it is important to underline here that a multicoil inductor design is the most widely used because of its comparatively simple implementation and flexibility.

Several spatial control inputs can be applied to static heating modes (for example, a choice of coil overhang or taper coil winding). Obviously, there is no single best approach to control an induction heating process and its recipe can be modified depending upon the applications and particular requirements. In some applications, it is advantageous to use a combination of time-dependent controls of the first group and space-dependent controls of the second group.

Control means that combine both features — time varying and varying along the space coordinates — refer to the third group of space–time control functions, or distributed control functions. One of the most typical implementations of space–time control is a multicoil design of a continuous heater with independently controlled voltage of separate coils. Such an approach provides the desired change of heating power in time and along the heater length. It is also one of the most widely used design solutions to control transient heating modes of an induction heater.

In this text, only the preceding explanation is made about spatial–time controls; the appropriate control problems are not considered further.

2.5 CONSTRAINTS

As mentioned in Section 2.1, care should be taken in defining the sets of admissible values of controlled outputs and control inputs. These sets depend on physical parameters of the heating system, technological requirements, power supply limitations, IHI design parameters, and some other specifics. If control input is restricted or other process parameters are specified, the control technique becomes more complicated from a formal viewpoint and from a technical viewpoint. The list of the most typical process constraints taking place in induction heating includes but is not limited to (Figure 2.6) constraints imposed on:

- Control inputs
- Billets' temperature distribution evolving over time of heating
- Heating process parameters related to specifics of subsequent metal working operations

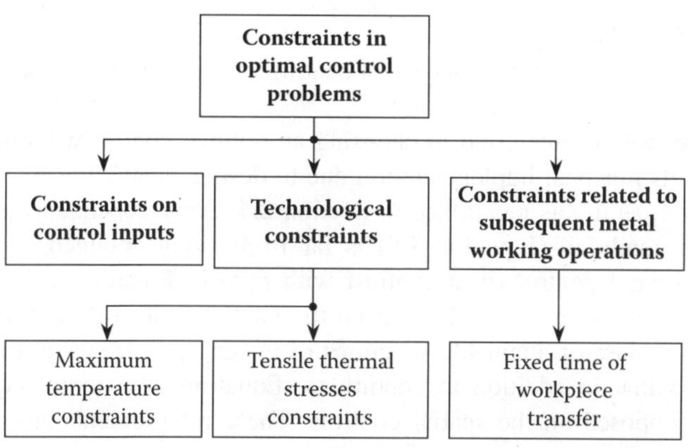

FIGURE 2.6 Typical constraints in optimal control problems for induction heating process.

2.5.1 Constraints on Control Inputs

A reasonable consideration for the optimal control is the requirement that the control input remain within certain prescribed limits — i.e., satisfy the prescribed restrictions. For example, during static heating, the maximum allowable value of heat generation power $u(\varphi)$ (considered as control input) is always restricted by maximum value U_{max} (known a priori) defined by power supply limitations. On the other hand, the minimum value $u(\varphi)$ is always equal to zero (the power supply is switched off). Thus, it will be reasonable to select control input $u(\varphi)$ with values on the restraint interval:

$$0 \le u(\varphi) \le U_{max},\ 0 < \varphi \le \varphi^0 . \tag{2.7}$$

As a result, the constraint on heating power $u(\varphi)$ is defined in a rather simple form of admissible limits $[0;U_{max}]$. Within these limits, the control input can vary arbitrarily during the whole heating cycle.

It is important to underline that constraint (Equation 2.7) is of primary importance, and it is necessary to take it into account in OCP formulation. As will be seen later, only extreme values $u(\varphi)$ are applied as control inputs in the majority of optimal control algorithms.

In a similar manner, the constraint on spatial control $u(y/V)$ can be written with regard to a continuous heating process:

$$0 \le u\left(\frac{y}{V}\right) \le U_{max},\ 0 < y \le y^0 , \tag{2.8}$$

Here,

U_{max} is maximum heating power predetermined a priori according to preliminary technical and economical analysis

$y^0 = V\varphi^0$ is a heater length (in relative units) corresponding to the time φ^0 required for a successful heating of the workpiece

In this case, we are interested in choosing an optimal control without any constraints on its practical implementation due to design capabilities.[1]

However, as it was mentioned in Section 2.4, every dependence $u(y/V)$ satisfying the condition (Equation 2.8) is far from being obtained by means of realizable spatial control (in a contrast with $u(\varphi)$ in Equation 2.7). Therefore optimal control function (resulting from the OCP solution) should be approximated in the class of attainable functions of heating power spatial distribution. This means that in addition to conditions (Equation 2.8), more requirements should be imposed on the spatial controls. These requirements are difficult to formalize.

The transition to the class of attainable control functions (Section 2.4) results in another treatment of spatial controls $u(y/V)$ and another constraint on their behavior. The voltages U_1, U_2, …, U_N for N coils of induction heater can be considered as spatial control inputs for multicoil inductor design (control input in Section 2.4). The number of sections, N, and their sizes are defined by the given IHI design. Extended statement of the problem (with a priori unknown number and lengths of coils) will not be considered further.

In this case, the spatial control $u(y/V)$ is prescribed as a set (vector) of sought parameters:

$$u\left(\frac{y}{V}\right) = \left(U_1, U_2, ..., U_N\right). \tag{2.9}$$

When sought parameters (Equation 2.9) are concerned in the design of a system "the power supply–IHI," the optimal control problem is solved with relaxing constraints on spatial control of the type (Equation 2.8). This means that, at the first stage of problem solving, the values of sought parameters are not limited at all. In some cases, the optimal values of powers and voltages for one or more coils do not satisfy any requirements. In this case, it is necessary to re-solve the optimal control problem, taking into account restraint, which was not satisfied. For example, if the required value of voltage for any coil exceeds maximum output voltages U_{max} of power supplies, then recalculation of the control input (Equation 2.9) should be done to satisfy the constraint:

$$U_k \leq U_{max}, \; k = 1, 2, ..., N. \tag{2.10}$$

2.5.2 Technological Constraints on Temperature Distribution during the Heating Process

Technology of metal heating processes (before hot working operations) imposes two of the most general requirements on the temperature distribution during the heating process. The first requirement demands that maximum value $\theta_{max}(\varphi)$ of the temperature within a heated workpiece should be below a certain admissible value θ_{adm}. If this value will be exceeded, then undesirable, irreversible adverse changes in material structural properties and even metal melting could take place.

Taking into consideration the model (Equation 1.27 through Equation 1.35) for static heating, this constraint can be written in the following form:

$$\theta_{max}(\varphi) = \max_{l \in [0,1]} \theta(l, \varphi) \leq \theta_{adm}, \; 0 < \varphi \leq \varphi^0. \tag{2.11}$$

For steady-state processes of continuous heating, the similar condition should be complied with all cross-sections of a workpiece (i.e., in all points $l \in [0,1]$) along the inductor length (i.e., in all points $y \in [0,y^0]$):

$$\theta_{\max}(y) = \max_{l\in[0,1]} \theta\left(l, \frac{y}{V}\right) \le \theta_{adm}, \quad 0 \le y \le y^0 . \tag{2.12}$$

Second, the temperature differences within the whole volume of a heated workpiece should be restricted during heating in such a way that the maximum value, $\sigma_{\max}$, of tensile thermal stresses $\sigma(l,\varphi)$ (due to thermal gradients) would not exceed prescribed admissible value, σ_{adm}, that corresponds to ultimate stress limit of the heated material. Therefore, somewhat similar to Equation (2.11) and Equation (2.12), the following constraints on the thermal stress field should be satisfied:

$$\sigma_{\max}(\varphi) = \max_{l\in[0,1]} \sigma(l,\varphi) \le \sigma_{adm},\ 0 < \varphi \le \varphi^0 , \tag{2.13}$$

$$\sigma_{\max}(y) = \max_{l\in[0,1]} \sigma\left(l, \frac{y}{V}\right) \le \sigma_{adm}, \quad 0 \le y \le y^0 . \tag{2.14}$$

Violation of these conditions could result in irreversible spoilage of a product, i.e., crack development.

Let us note that thermal stresses depend in rather complicated ways on distribution of temperatures within a workpiece. The simplest case arises for a temperature field model (Equation 1.27 through Equation 1.35) when the value $\sigma_{\max}$ appears to be proportional to the difference of average over cross-section temperature and temperature in a workpiece center, $l = 0$.[1–3]

$$\sigma_{\max}(\varphi) \cong \gamma\left[(\Gamma+1)\int_0^1 \theta(l,\varphi) l^{\Gamma} dl - \theta(0,\varphi)\right] \tag{2.15}$$

Here, γ is a coefficient determined by elastic properties of a heated material.

A substitution of Expression (2.15) into Equation (2.13) results in a complicated constraint on a temperature field of the following type:

$$\gamma\left[(\Gamma+1)\int_0^1 \theta(l,\varphi) l^{\Gamma} dl - \theta(0,\varphi)\right] \le \sigma_{adm}, 0 \le \varphi \le \varphi^0. \tag{2.16}$$

Referring to Equation (2.14), Expression (2.16) can be rewritten as:

$$\gamma\left[(\Gamma+1)\int_0^1 \theta\left(l,\frac{y}{V}\right)l^{\Gamma}dl-\theta\left(0,\frac{y}{V}\right)\right]\leq\sigma_{adm},0\leq y\leq y^0. \tag{2.17}$$

Additional technological constraints (Equation 2.11, Equation 2.16 or Equation 2.12, Equation 2.17) add complexity to an OCP solution and appropriate control functions. For simplification purposes, these constraints can be neglected during the first stage of the process of finding the OCP solution.

If the subsequent analysis proves that a temperature field during optimal heating process does not violate technological requirements, then those technological restraints are complied with "automatically."

If obtained values θ_{max} and/or σ_{max} exceed permissible limits, then constraints on θ_{max} and σ_{max} (or only one of them) should be taken into consideration. In most practical cases, such preliminary analysis allows simplifying the problem by excluding at least one of those complex technological constraints, in particular the constraint on thermal stresses.

2.5.3 Constraints Related to Specifics of Subsequent Metal Working Operations

One of the most typical technological constraints is imposed by a metal forming operation (Figure 1.11). It often takes an appreciable time to deliver a heated workpiece from induction heater to forming station. During workpiece transportation on air, the temperature distribution of the heated workpiece can be significantly distorted. This transportation stage represents the last stage of the operation of an IHI. Under certain circumstances, a transportation time and respected surface cooling phenomenon can also be considered as a specific technological constraint.

The total heating time, φ_1^0, of a workpiece is a sum of time, φ^0, required for heating, and fixed time, Δ_T, needed for workpiece delivery to the metal forming station:

$$\varphi_1^0=\varphi^0+\Delta_T\,. \tag{2.18}$$

The value Δ_T is determined by how far the IHI is located from the hot working equipment and also by design features of the part-feeder mechanism.

Equation (2.18) can be regarded as a constraint under the following conditions:

- If the transfer time appears sufficient for considerable distortion of a workpiece temperature field after it exits induction heater.
- If an ultimate control goal is to provide a required temperature state of a workpiece just before hot working operations.

In this case, the constraint in Equation (2.18) can be imposed on total time φ_1^0.

A proper choice of control inputs would typically allow one to modify the time φ^0 required for heating. In contrast, transportation time, Δ_T, is usually beyond control. This means that a controlled process stage (heating stage in this case) is supplemented by the uncontrollable stage (workpiece transfer stage). Equation (2.18) cannot be written in the form of inequalities (Equation 2.8 through Equation 2.17).

Regarding the constraint in Equation (2.18), it is necessary to supplement the model (Equation 1.27 through Equation 1.35) with equations describing temperature distortion during workpiece transportation. In this case, according to Equation (2.18), the transfer stage should be considered as a final stage of a total process affecting billets' temperature state. Particular OCP formulations and methods to determine control inputs satisfying Equation (2.18) will be discussed in subsequent chapters.

2.6 DISTURBANCES

It is assumed that admissible control inputs can properly influence workpiece temperature distribution only if available information regarding an induction heating process is accurate and full. With regard to a basic model of an object (Equation 1.27 through Equation 1.35), this means that the following parameters should be precisely known:

- All electromagnetic and thermal properties of the heated material: penetration depth of current, specific heat, density, thermal conductivity, and temperature conductivity, etc.
- Density of heat loss flow $q(\varphi)$ from the workpiece surface that affects boundary conditions (Equation 1.32 and Equation 1.34).
- Initial temperature distribution $\theta^0(l)$ of the workpiece represented in Equation (1.30).

In other words, it is necessary that initial data regarding the heating system be precisely predetermined and no fluctuations of the controlled characteristics from their calculated deterministic values occur. However, practically speaking, presence of various uncertainties is unavoidable, and the system description is only known to a certain degree of accuracy. The incompleteness of the information is caused at first by an imperfection of our knowledge about a heating system (for example, uncertainties regarding physical properties). On the other hand, there are uncertainties with respect to particular conditions of the technological process.

Often, not well determined information includes only a limited range of possible variations, i.e., only minimum and/or maximum values, and within the said limits these values could vary in arbitrary way. In many cases, the conventional statistical estimations could not be applied. Therefore, conditions of limited uncertainty are typical for an induction heating process. These conditions cannot be "corrected" in advance.

It is possible to consider all the previously discussed characteristics as unknown deviations from expected or computational values. We shall call all these factors *disturbing actions* or, simply, *disturbances.* Disturbances affect controlled output regardless of control inputs. That is why they distort a temperature field of a workpiece — i.e., they change the expected response of a heating process on control inputs. Under conditions of considerable uncertainty, these distortions could be quite essential. This fact results in giving proper weight to disturbances in OCP formulation.

Within the context of selected mathematical models of an induction heating process, typical disturbances may be classified into two main groups: *functional* disturbances and *parametric* disturbances (Figure 2.7). The most typical functional disturbances are:

1. Heat loss variation with time that affects boundary conditions (Equation 1.32 and Equation 1.34) (these variations are not completely known a priori).
2. Distribution of initial temperature $\theta_0(l)$ along spatial coordinate in Equation (1.30).
3. Instability of power supply voltage or frequency of operating current, etc.

The third disturbances (3) are the second most important ones, that is why they will not be taken into consideration here.

All the previously mentioned uncertainties in electromagnetic and thermal characteristics of the heated material could belong to a group of parametric disturbances. Disturbances that are included in boundary conditions (Equation 1.32 and Equation 1.34) of the selected mathematical model of an induction heating process can be prescribed in the parametric form as well. In particular, for convection heat losses from a surface of heated workpiece to a surrounding area of ambient temperature $\theta_a(\varphi)$, according to Equation (1.34), approximate value of similarity criterion *Bi* can be considered as a parametric disturbance. If initial temperature distribution is uniform:

$$\theta_0(l) = \theta_0 = const, 0 \le l \le 1, \tag{2.19}$$

then only θ_0 would represent a parametric disturbance.

Hereafter we limit our consideration to a simpler problem of considering only disturbances in the parametric form.

Uncertain factors *Bi* and θ_0 that are involved in boundary conditions of a model (Equation 1.27 through Equation 1.35) can be expressed by the following inequalities:

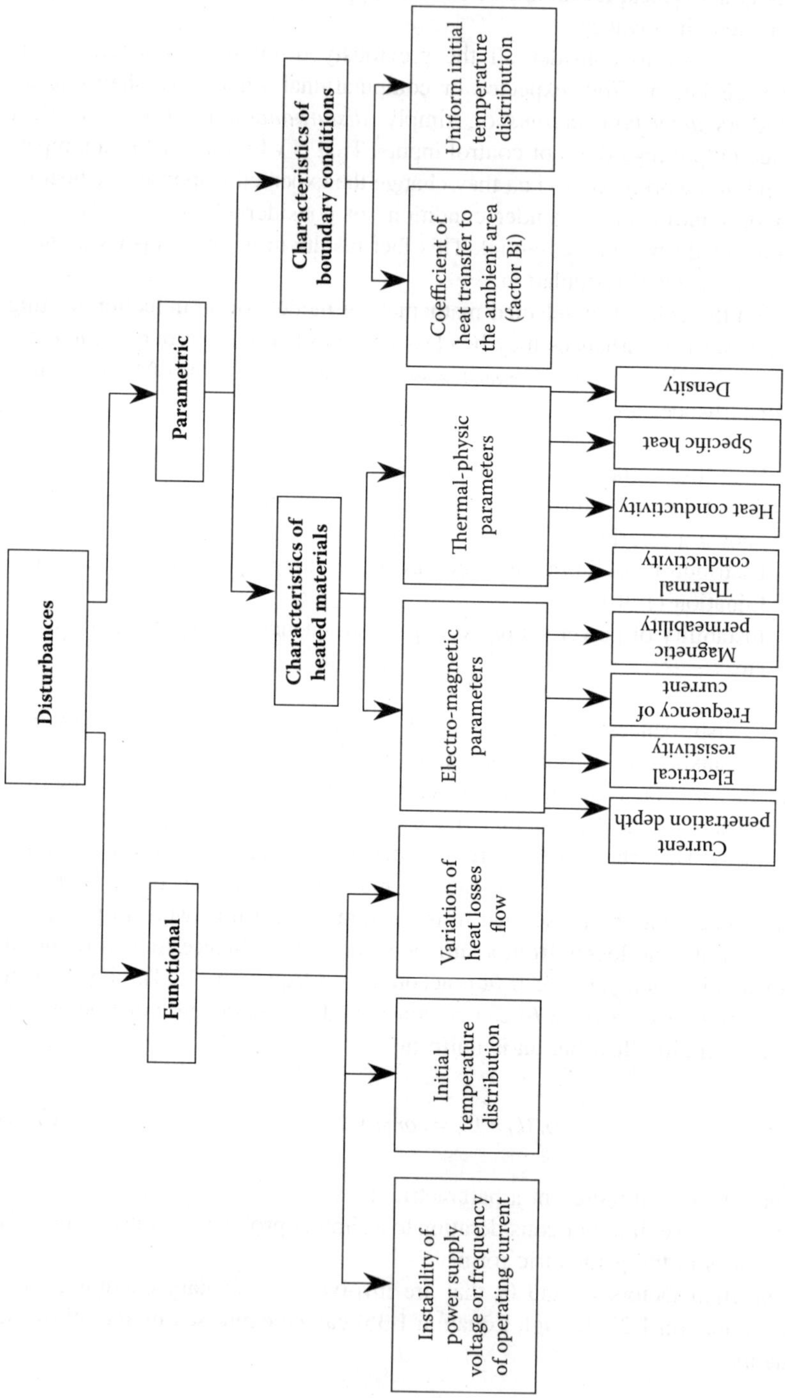

FIGURE 2.7 Typical disturbances in induction heating systems.

$$Bi_{\min} \leq Bi \leq Bi_{\max} \ ; \theta_{0\min} \leq \theta_0 \leq \theta_{0\max} \ . \tag{2.20}$$

Values of these disturbances are limited only by maximum and minimum values $Bi_{\min}$, $Bi_{\max}$ and $\theta_{0\min}$, $\theta_{0\max}$, which usually can be calculated simply enough from upper and lower limits of these parameters. Similar treatment should be given to other parametric disturbances involved in OCP formulation.

2.7 REQUIREMENTS OF FINAL TEMPERATURE DISTRIBUTION WITHIN HEATED WORKPIECES

The ultimate goal of an induction heating process is obtaining desired temperature distribution within workpieces before subsequent technological operations. That is why proper formulation of the requirements to these temperature states is of primary importance. In most cases of induction heating of metals prior to hot working, it is necessary to have uniform temperature distribution within the heated workpiece. Desired temperature depends on metal properties and specifics of metal hot working operations.

Mathematically speaking, this requirement of providing a uniform heating can be formulated as following: it is necessary to provide temperature uniformity at the level of prescribed value $\theta^* = const$ at the end of heating process $\varphi = \varphi^0$:

$$\theta(l, \varphi^0) \equiv \theta^* ; 0 \leq l \leq 1 . \tag{2.21}$$

However, due to the nature of induction heating, it is normally not possible to obtain a perfectly uniform temperature distribution. One reason for that is an existence of surface heat losses $q(\varphi) < 0$. Therefore, requirement (Equation 2.21) with regard to a perfectly uniform final temperature distribution is unrealizable. Indeed, if inequality $q(\varphi^0) < 0$ in Equation (1.32) practically always remains true, then:

$$\frac{\partial \theta(1, \varphi^0)}{\partial l} < 0 \ , \tag{2.22}$$

That is why the temperature gradient on a workpiece surface always is negative at the end of a heating process.

On the other hand, according to Equation (2.21), this gradient should be equal to zero for all values l, including $l = 1$ as well, which leads to contradiction. This property of an unattainability of final uniform temperature distribution (property of an uncontrollability of induction heating process) is a specific feature of an induction heating process due to existence of internal heat sources.[1]

The condition of the uniform temperature distribution (Equation 2.21) appears theoretically attainable for workpiece heating in installations with external heat exchange (gas-fired furnaces, infrared and electric resistance furnaces, etc.). Therefore, the OCP for induction heating with the requirement (Equation 2.21) is unsolvable and should be considered as an ill-defined problem.

Evidently, it is necessary to "weaken" the condition in Equation (2.21), taking into account practically always existing tolerances on the error (deviation) between actual final temperature distribution and desired distribution. In this case, a set of allowable temperature states (corresponding to given tolerance) should be preset instead of fixed desired temperature distribution (Equation 2.21).

To provide a proper mathematical description of such a set, it is necessary to define how the deviation from temperature distribution (Equation 2.21) should be estimated. The most common approach might deal with estimation by root mean square errors β_{ms} (within a volume of a heated body):

$$\beta_{ms} = \int_0^1 \left[\theta(l,\varphi^0) - \theta^*\right]^2 l^\Gamma dl \le \beta^*_{ms}, \tag{2.23}$$

where β^*_{ms} is an admissible value of β_{ms}, defined by given tolerances.

Moreover, estimation of accuracy by value β_{ms} is not in good compliance with industrial technology requirements. In the most typical technological processes, maximum admissible value ε of absolute deviation of temperature distribution $\theta(l,\varphi^0)$ from the required outlet temperature θ^* is prescribed. This means that, at the end of a heating cycle, the temperature in any point of the billet volume should deviate not more than by value ε from required temperature θ^*:

$$\left|\theta(l,\varphi^0) - \theta^*\right| \le \varepsilon, \;\; 0 \le l \le 1. \tag{2.24}$$

Thus, minimizing β_{ms} does not exclude inadmissible local deviations of final temperature arising from required one. This means that if Equation (2.23) is met, there is no assurance that necessary Equation (2.24) is complied with in some points $l \in [0;1]$ (as a rule, on a surface of the heated body). At last, there are no effective means to estimate acceptable values β_{ms}, starting from a practically required condition (Equation 2.24).

These statements prove that it is necessary to formulate requirements in the form of Equation (2.24) as close to actual technology as possible. Let us underline that, from the formal point of view, Equation (2.24) appears to be more complicated than the inequality in Equation (2.23). That is why only Expression (2.23) is used sometimes in engineering practice to simplify this problem. Instead of Equation (2.24), the following inequality for maximum deviation on the interval [0,1] can be employed:

$$\max_{l\in[0,1]} \left|\theta(l,\varphi^0) - \theta^*\right| \le \varepsilon. \tag{2.25}$$

Regarding the basic process model (Equation 1.27 through Equation 1.35), we shall consider requirements to temperature distribution at the end of a heating cycle in the form Equation (2.25) everywhere below. Such distribution, $\theta(l,\varphi^0)$, we shall call sometimes for short "final temperature distribution." Substituting the variable in Equation (2.6) into Equation (2.25) results in the following form of the condition:

$$\max_{l\in[0,1]}\left|\theta\left(l,\frac{y^0}{V}\right)-\theta^*\right|\le\varepsilon. \tag{2.26}$$

Here, the radial temperature distribution (along coordinate l) at the heater exit $y = y^0$ is considered as final temperature distribution in steady-state mode of continuous heater.

If requirements to final temperature distribution are included into cost function, this means that an optimal control problem is reduced to minimizing deviation ε. In this case, it is not necessary to demand Equation (2.25) or any additional requirements of similar type.

Let us remark also that, for a number of technological processes, requirements applied to heating accuracy are weakened compared to Equation (2.25). In particular, they can be limited to only a certain part of a heated workpiece (for example, to surface layer) and/or required heating accuracy depends on spatial coordinate l. In such cases, Equation (2.25) is stated as:

$$\max_{l\in[l^*,l^{**}]}\left|\theta\left(l,\varphi^0\right)-\theta^*\right|\le\varepsilon(l). \tag{2.26a}$$

Here, the interval $[l^*; l^{**}]$, $0 \le l^* < l^{**} \le 1$ specifies depth and location of a layer within the boundaries of a heated workpiece for which requirements to final temperature state are set. Deviation $\varepsilon(l)$ is a given (for example, linear) function of coordinate l on interval $[l^*; l^{**}]$.

2.8 GENERAL PROBLEM OF TIME-OPTIMAL CONTROL

In this section, we shall be concerned with the time-optimal control problem for the heating process described previously. Application of obtained results to different and more complex OCP statements is left to the following sections.

At first, we shall consider the most general optimal control problem, assuming that complete information is available about the heating process and imposed requirements. Let us also assume that there are no disturbances affecting the heating system and the time of workpiece transfer to hot working equipment is negligible. We will also consider that an induction heating process is properly described by Equation (1.27) through Equation (1.35). The workpiece temperature distribution will be treated as an output controlled function of the process. Control input function is chosen from the set of admissible controls to influence temperature distribution and dynamic behavior of an induction heating system.

Let the control problem be formulated based upon the data presented in Section 2.2 through Section 2.7 as follows:

- The initial temperature state is defined as Equation (1.30) and required temperature distribution is described by Equation (2.25).
- The control constraint set is given in a form of Equation (2.7), Equation (2.8), or Equation (2.10).
- The constraint on a temperature field can be written in one of the forms (Equation 2.11 through Equation 2.17).
- The cost function for each control input can be estimated by chosen criteria from the set (Equation 2.1 through Equation 2.5).

Then the statement of an optimal induction heating control problem can be formulated such as following. We wish to steer an initial workpiece temperature state to a desired target state and hold cost criteria at extremum value by space-time-varying control restricted by a preassigned set and satisfied by the constraint on a temperature field. The control input functions resulting from the preceding OCP solution are called *optimal controls.* To simplify the control problem, it will be assumed that initial temperature distribution is uniform: $\theta_0(l) \equiv 0 \equiv const$. A similar statement can be made for large numbers of optimal control problems of technology and science.

Let us consider the simplest, most typical, and representative problem of optimal control for a static heating process. A mathematical model is described by Equation (1.27) through Equation (1.35). Process time duration (Equation 2.1) can be considered as a cost criterion (problem of time-optimal control). Time-dependent heating power $u(\varphi)$ is chosen as lumped control input restrained by (Equation 2.7). Technological constraints (Equation 2.11 through Equation 2.17) are not taken into account because it is assumed that they are complied with all admissible controls. This control problem represents optimization of IHIs when the ultimate goal is to provide a maximum productivity under a prescribed temperature state of a heated workpiece.

Therefore, it is necessary to select a control function $u(\varphi) = u^*(\varphi)$ that provides steering workpiece initial temperature distribution to desired temperature θ^* with prescribed accuracy ε (according to Equation 2.25) in minimal optimal process time. Control input is bounded by minimum and maximum allowable values U_{max} and $U_{min} = 0$ (see Equation 2.7), and temperature field is described by Equation (1.27) through Equation (1.35).

Conventional induction heating technology represents heating when inductor voltage is constant during the whole process. For the model (Equation 1.27 through Equation 1.35), this heating mode corresponds to constant value of heating power $u(\varphi) = u_0 = const$. Thus, the value u_0 should be chosen in such a

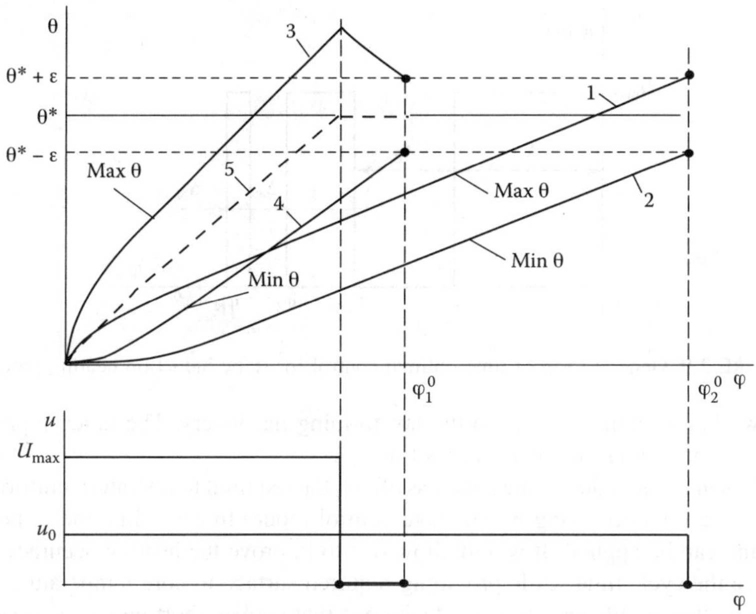

FIGURE 2.8 Time-temperature curves for one- and two-stage heating modes.

way that, at the end of a heating cycle, the temperature difference across billet thickness does not exceed 2ε as Equation (2.25) requires (Figure 2.8, curves 1, 2).

Because the surface-to-core temperature difference, $\theta(1,\varphi^0) - \theta(0,\varphi^0)$, is proportional during heating cycle to the power u_0, we are forced to choose sufficiently low heating power (smaller than U_{max}) to obtain allowable deviations ε from θ^*. This results in increasing the required heating time φ^0 (approximately in inverse proportion to u_0), leading to a process that would not be optimal with respect to the criterion in Equation (2.1). Heating time can be reduced when a two-stage process is applied. Figure 2.8 shows the typical time–temperature curves for two-stage static induction heating. As one can see, immediately after the beginning of heating with $u(\varphi) \equiv U_{max} > u_0$, the surface temperature (curve 3) and average temperature (curve 5) begin to rise. In contrast, there is a time delay before the core temperature (curve 4) starts to grow.

As soon as power is switched off during the second process stage, ($u(\varphi) \equiv 0$), the surface temperature rapidly decreases due to heat soaking towards a cooler core. A corresponding rise in the core temperature takes place during this stage. During the soaking stage, the surface-to-core temperature differential decreases and the heated component approaches the temperature uniformity required for hot working.

The advantage of a two-stage process in comparison to a single-stage process deals with lower value of ε (Figure 2.8), and optimal heating time decreases with increasing value U_{max}. It is important to note that the soaking stage can be performed when the heated workpiece is inside the induction coil and/or during

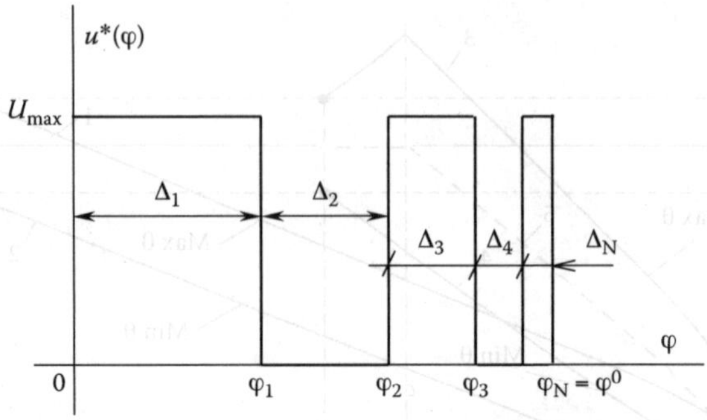

FIGURE 2.9 General form of time-optimal control for static induction heating process.

the workpiece transfer stage to the hot forming machinery. The latter approach allows minimizing the total process time φ^0.

In some cases, due to the heat loss effect, the required temperature uniformity cannot be obtained using a two-stage control mode. In cases like these, power pulsing can be applied. It would allow one to improve the heating accuracy and reduce the cycle time while providing required surface-to-core temperature uniformity. Power pulsing refers to a technique that applies short bursts of power to maintain a desired surface temperature or a maximum allowable surface-to-core temperature difference. Pulse heating consists of a series of "heat ON" and "heat OFF" cycles until the desired uniformity is obtained (Figure 1.6).

It is possible to prove mathematically[1] that the time-optimal control consists of alternating stages of heating with maximum power $u \equiv U_{max}$ (heat ON) and subsequent soaking under $u \equiv 0$ (heat OFF) cycles. The number $N \geq 1$ of stages is defined uniquely by given heating accuracy ε and it increases with decreasing ε. Therefore, the shape of the optimal control algorithm is known, but the number of stages N and durations $\Delta_1, \Delta_2, \ldots, \Delta_N$ of those stages (Figure 2.9) remains unknown.

For any particular process, the number N and the values $\Delta_1, \Delta_2, \ldots, \Delta_N$ should be determined during subsequent calculation. Thus, the problem of time-optimal control is parameterized. *Parameterization of the control problem* means the specification of control input function by means of one or more variables that are allowed to take on values in a given specified range. Therefore, the initial problem is reduced to searching for parameters Δ_i, $i = 1, 2, \ldots, N$, uniquely specifying optimal control input $u^*(\varphi)$. Now search for control function can be written as:

$$u^*(\varphi) = \frac{U_{max}}{2}\left[1 + (-1)^{j+1}\right], \; \varphi_{j-1} < \varphi < \varphi_j, \; j = 1, 2, \ldots, N. \tag{2.27}$$

Here, φ_j is the end point of the jth stage (jth point of control switch). Obviously, $\varphi_j = \sum_{i=1}^{j} \Delta_i$; we assume that $\varphi_0 = 0$. According to Equation (2.27), during any stage with an odd number (i.e., with odd number j), $u^*(\varphi) = U_{\max}$ and, during an even stage, $u^*(\varphi) = 0$ (Figure 2.9). In particular, for the appropriate values ε, Expression (2.27) describes a two-stage process of heating if $N = 2$, $\Delta_1 = \varphi_1$, $\Delta_2 = \varphi^0 - \varphi_1$. It will be shown further that, for conventional heat losses from the workpiece surface and typical requirements to heating accuracy, the two-stage control algorithm (Equation 2.27) usually represents the time-optimal heating mode.

If initial temperature and heat losses are known and assuming applying a control input (Equation 2.27), at the end of the heating cycle (at $\varphi = \varphi^0 = \varphi_N$) the temperature in any point $l \in [0,1]$ depends only on values Δ_i, $i = 1, 2, \ldots, N$. This means that $\theta(l,\varphi^0)$ in Equation (2.25) is described by relation $\theta(l,\Delta)$, where $\Delta = (\Delta_i)$, $i = \overline{1,N}$ is a set of Δ_i (Δ — vector of time intervals of all control stages [Figure 2.9]). By substituting $u(\varphi)$ in Equation (2.27) and taking into consideration initial and boundary conditions (Equation 1.30 through Equation 1.34), this expression can be obtained in an explicit form as a solution of the heat conduction equation (Equation 1.27) for $\varphi = \varphi_N$.

In a case (Equation 1.34 and Equation 2.19) when initial temperature distribution is assumed to be uniform, $\theta_0(l) \equiv \theta_0 = \theta_a = const$ (this corresponds to a constant ambient temperature θ_a), the solution can be written as[1]:

$$\theta(l, \varphi^0) \equiv \theta(l, \Delta) = \theta_0 + U_{\max} \sum_{m=1}^{N} (-1)^{m+1} \Lambda\left(l, \sum_{i=m}^{N} \Delta_i \right). \tag{2.28}$$

Here, $U_{\max} \Lambda\left(l, \sum_{i=m}^{N} \Delta_i \right)$ is a temperature in the point l at the time $\varphi = \sum_{i=m}^{N} \Delta_i$ when the control function takes on maximum allowable value (i.e., heating power is maximum: $u(\varphi) \equiv U_{\max}$) and $\theta_0 = 0$. Function $\Lambda(l,\varphi)$ is called the *response function* — i.e., the output from a heating process (Equation 1.27 through Equation 1.35), which is obtained in response to applied input $u(\varphi) \equiv 1$ under $\theta_0 = 0$.

For the model (Equation 1.27 through Equation 1.35), response function $\Lambda(1,\varphi)$ can be calculated using the relation in Equation (1.40) as:

$$\Lambda(l, \varphi) = \sum_{n=1}^{\infty} \frac{2W_n(\xi)K(\mu_n l)}{(\mu_n^2 + Bi^2 + (1-\Gamma)Bi)K^2(\mu_n)} (1 - e^{-\mu_n^2 \varphi}). \tag{2.29}$$

Therefore, the response function represents the infinite sum of exponents with different factors for each value l. All notations used in the expression for these factors were explained in Section 1.2.2.

It is clear that in engineering practice, only a finite number of members in convergent series (Equation 2.29) is always regarded. Thus, the condition (Equation 2.25) required for a given final temperature state can be rewritten as:

$$\Phi(\Delta) = \max_{l \in [0,1]} \left| \theta(l,\, \Delta) - \theta^* \right| \leq \varepsilon \,, \tag{2.30}$$

where $\theta(l,\Delta)$ is defined by Expression (2.28) and Expression (2.29).

The problem now is reduced to determination of such time intervals, Δ_i, $i = \overline{1, N}$, of alternating heating and soaking stages that provide satisfying requirements (Equation 2.30) in minimal possible time. Total time is equal to the sum of all Δ_i. Then, a cost criterion can be determined as a following sum:

$$I(\Delta) = \sum_{i=1}^{N} \Delta_i \rightarrow \min_{\Delta} \,, \tag{2.31}$$

where the minimum is taken over all admissible vectors Δ.

From the formal point of view, the optimal control problem is reduced to a mathematical programming problem minimizing an object function (Equation 2.31) of N variables, Δ_i, where restraint on a set of admissible values, Δ_l, is prescribed in the form of Equation (2.30). As it was shown earlier, the inequality (Equation 2.30) represents a set of infinite number of constraints in a form of Equation (2.24) for each of values $l \in [0,1]$. Therein lies a principal difference of a problem (Equation 2.30 and Equation 2.31) compared to classical mathematical programming problems, in which only the finite constraints number is considered.

Let us recall also that a priori unknown number N of sought parameters should be found in the course of the problem (Equation 2.30 and Equation 2.31) solution; this number N depends on prescribed value ε in Equation (2.30). It adds additional complexity to this problem compared to a conventional one.

For a steady-state process of continuous heating described by Equation (1.27) through Equation (1.35) and Equation (2.6), substitution of a variable (Equation 2.6) results in optimal spatial distribution of power $u^*(y/V)$ along heater length. The curve shown in Figure 2.9 is practically identical for continuous heating modes. In case of continuous heating, the time axis represents the length of the induction heating line or coil length. Bursts of power can represent the power of inline coils, which may have different length, windings, and/or can be individually fed from different power supplies. Under constraint (Equation 2.8), optimal heat power distribution can be represented as a stepwise function in the form (Equation 2.27) (Figure 2.10):

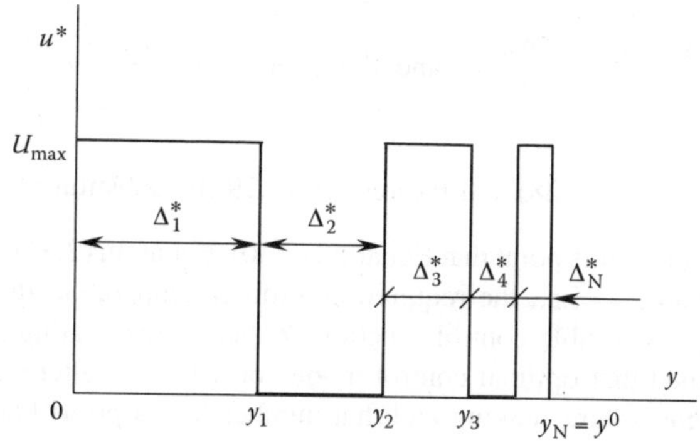

FIGURE 2.10 Optimal heat power distribution along the length of continuous induction heater.

$$u^*\left(\frac{y}{V}\right) = \frac{U_{max}}{2}\left[1+(-1)^{j+1}\right], \ \frac{y_{j-1}}{V} < \frac{y}{V} < \frac{y_j}{V}; \ j = 1, 2, ..., N. \quad (2.32)$$

Here, unknown parameters represent the lengths $\Delta_i^* = \Delta_i \cdot V$ of alternating inline coils with heat ON and heat OFF (the lengths of the active and passive intervals).

The spatial control function (Equation 2.32) can be considered as a time-dependent optimal control (Equation 2.27) scanning along the heater length. Then cost function (Equation 2.31) becomes equal to y^0/V — i.e., it represents total heater length y^0 with constant factor $1/V$.

The optimal control problem is reduced to minimizing of the heater length when velocity, V, of the workpiece movement is prescribed or to obtaining maximum velocity V (maximum output rate of the heater) for given y^0. In this case, instead of Equation (2.30) and Equation (2.31) and taking into account transition from the condition in Equation (2.25) to the constraint in Equation (2.26), one can obtain the following mathematical programming problem:

$$\begin{cases} I\left(\dfrac{\Delta^*}{V}\right) = \displaystyle\sum_{i=1}^{N} \frac{\Delta_i^*}{V} \to \min_{\left(\frac{\Delta^*}{V}\right)}; \\ \Phi\left(\dfrac{\Delta^*}{V}\right) = \displaystyle\max_{l\in[0,1]} \left|\theta\left(l, \frac{\Delta^*}{V}\right) - \theta^*\right| \le \varepsilon, \end{cases} \quad (2.33)$$

where $\frac{\Delta^*}{V} = \left(\frac{\Delta_1^*}{V}, \frac{\Delta_2^*}{V}, \ldots, \frac{\Delta_N^*}{V}\right)$, and the expressions for $\theta\left(l, \frac{\Delta^*}{V}\right)$ are determined by Expression (2.28) and Expression (2.29) by substituting $\Delta_i = \frac{\Delta_i^*}{V}$.

Let us assume at this point that Equation (2.30), Equation (2.31), and Equation (2.33) are solvable — i.e., the required controlled temperature distribution can be obtained by admissible control function applied to the heating system. Here, it is still assumed that optimal control mode consists of N alternating stages of heating and temperature soaking and that number N is a priori known. Furthermore, we shall discuss how N can be defined depending on required permissible value ε and how the conditions of attainability are set. We shall also extend our results to the solution of similar problems. Advantages of an optimal control method compared to known numerical methods will be shown as well.

2.9 MODEL PROBLEMS OF OPTIMAL CONTROL RESPECTIVE TO TYPICAL COST FUNCTIONS

In the previous section, it has been shown that the general problem of time-optimal control of an induction heating process described by Equation (1.27) through Equation (1.35) is reduced to a special problem of mathematical programming in the form of Equation (2.30) and Equation (2.31):

$$\begin{cases} I(\Delta) = \sum_{i=1}^{N} \Delta_i \to \min_{\Delta}; \\ \Phi(\Delta) = \max_{l \in [0,1]} \left|\theta(l, \Delta) - \theta^*\right| \le \varepsilon. \end{cases} \tag{2.34}$$

As we shall see, a whole class of optimization problems can be solved in the manner indicated: problems that involve the typical cost criteria while the process is described by the model (Equation 1.27 through Equation 1.35).

2.9.1 Problem of Achieving Maximum Heating Accuracy

Among typical cost criteria (accepted for static and continuous induction heating processes), an important cost function is defined as error between the required temperature distribution and the actual one obtained at the end of the heating cycle (Section 2.2). Everywhere else such error is called *heating accuracy*, which, according to Equation (2.25) (see Section 2.7), is estimated by absolute deviation, ε, from the desired temperature state.

If Equation (2.25) is true for certain $\varepsilon = \varepsilon_1$, then it remains true for all $\varepsilon > \varepsilon_1$ at the same value of φ^0. Therefore, minimal heating time $\varphi^0_{\min}$ decreases as ε

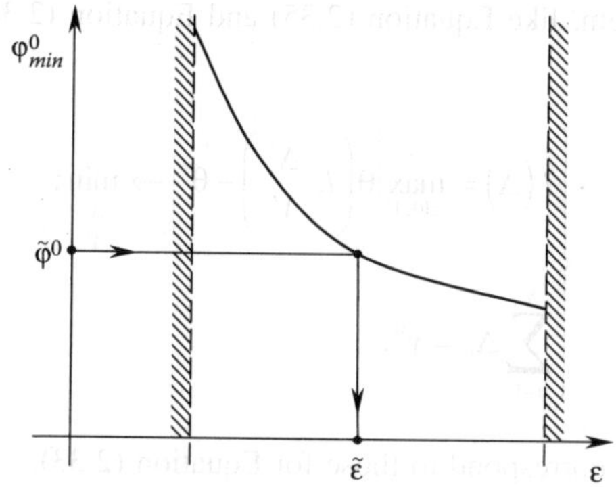

FIGURE 2.11 Dependence of minimal process time on required heating accuracy.

grows (Figure 2.11). Then, for prescribed value $\varphi^0 = \tilde{\varphi}^0$, it is impossible to obtain absolute deviation ε from a desired temperature state less than $\varepsilon = \tilde{\varepsilon}$ attainable at the time $\varphi^0_{\min}(\tilde{\varepsilon}) = \tilde{\varphi}^0$ (Figure 2.11) because it needs additional time.

This implies that, for a static heating process, the optimal control (that provides maximum heating accuracy at prescribed time $\varphi^0 = \tilde{\varphi}$) coincides with time-optimal control resulting in such accuracy $\varepsilon = \tilde{\varepsilon}$, which satisfies equality $\varphi^0_{\min}(\tilde{\varepsilon}) = \tilde{\varphi}^0$ (Figure 2.11). It means that minimal value of absolute error, attainable in $\tilde{\varphi}^0$, appears equal to $\tilde{\varepsilon}$.

Thus, in the problem of maximum absolute heating accuracy for a given time, the optimal control has the same form (Equation 2.27) (Figure 2.9) as in the time-optimal control problem. Instead of Equation (2.34), this problem is reduced to the following problem of mathematical programming minimizing cost function (Equation 2.35) with constraint (Equation 2.36):

$$\begin{cases} I(\Delta) = \max\limits_{l \in [0,1]} \left| \theta(l, \Delta) - \theta^* \right| \to \min\limits_{\Delta}; & (2.35) \\ \Phi(\Delta) = \sum\limits_{i=1}^{N} \Delta_i = \tilde{\varphi}^0. & (2.36) \end{cases}$$

Solution of this problem coincides with the solution of Equation (2.34) for value ε being a root of the equation $\varphi^0_{\min}(\varepsilon) = \tilde{\varphi}^0$. Solution of a number of time-optimal problems (Equation 2.34) for different values ε allows one to define dependence $\varphi^0_{\min}(\varepsilon)$. Therefore, solution of Equation (2.35) and Equation (2.36) on a minimum ε for a given time is determined uniquely by solving the number of time-optimal problems for different values ε.

It can be shown that, for steady-state continuous heating (in the inductor of given length y^0 with given throughput), a problem of maximum accuracy is reduced to solving a number of problems (Equation 2.33) for different values ε.

In total, this problem, like Equation (2.35) and Equation (2.36), can be written as follows:

$$\begin{cases} I(\Delta) = \max_{l \in [0,1]} \left| \theta\left(l, \dfrac{\Delta^*}{V}\right) - \theta^* \right| \to \min_{\frac{\Delta^*}{V}}; \\ \sum_{i=1}^{N} \Delta_i^* = y^0. \end{cases} \tag{2.37}$$

Here, all notations correspond to those for Equation (2.33).

Chosen voltages $U_1, U_2, \ldots, U_N$ (or corresponding heat powers $u_1, u_2, \ldots, u_N$) for all N coils of the multicoil continuous heater can be considered as parametric spatial controls in Equation (2.9) (Section 2.4 and Section 2.5). Then, the following modification of the problem of maximum accuracy can be obtained for steady-state continuous heating similarly to Equation (2.35) and Equation (2.37):

$$I(U) = \max_{l \in [0,1]} \left| \theta(l, U) - \theta^* \right| \to \min_{U}. \tag{2.38}$$

Here, $U = (U_1, U_2, \ldots, U_N)$; voltages $U_1, U_2, \ldots, U_N$ (or powers $u_1, u_2, \ldots, u_N$ for separate coils) are sought parameters instead of $\Delta_1, \Delta_2, \ldots, \Delta_N$ in Equation (2.35) and Equation (2.36) or Δ_i^*/V in Equation (2.37). At least at the first stage, additional constraints are not taken into account, in contrast to Equation (2.35) through Equation (2.37).

The function $\theta(l,U)$ in Equation (2.38) is similar to $\theta(l,\Delta^*/V)$ in Equation (2.37). This expression represents a temperature distribution along radial coordinate l at the heater exit $y = y^0$. It is determined uniquely by vector of parameters U. Similarly to Equation (2.28), after substitution (2.6), this relation can be obtained in explicit form as solution of the heat conduction equation (Equation 1.27) with boundary and initial conditions (Equation 1.30 through Equation 1.34).

For the linear process model (Equation 1.27 through Equation 1.35) and neglecting any air gaps between induction coils, a distribution, $u(y)$, of internal heat power along the inductor length can be represented as a step function shown in Figure 2.12.

For constant voltage on each coil, the intensity of heat generation caused by an electromagnetic field is assumed constant along the whole coil length. In this case, the value of heating power for each coil is considered as unknown parameter. At the same time, the lengths of each coil are predetermined by coordinates $y_1, y_2, \ldots, y_N$. This case differs from the one presented in Figure 2.10, in which for prescribed value U_{max}, it is necessary to define lengths $\Delta_1^*, \Delta_2^*, \ldots, \Delta_N^*$ of intervals with given heat powers.

It should be mentioned that the control algorithm (Equation 2.32) shown in Figure 2.10 allows using only on–off power control because there are only two

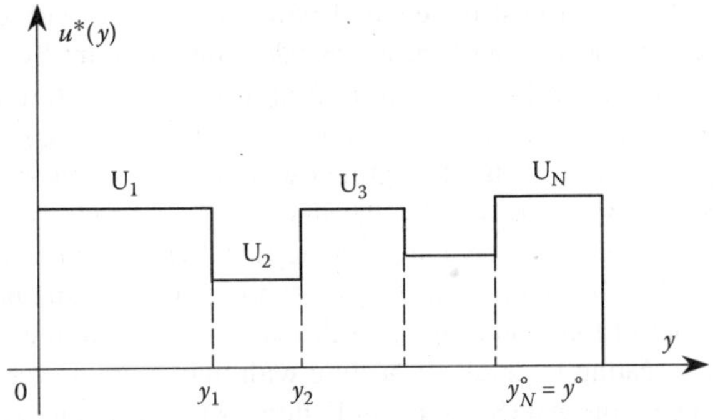

FIGURE 2.12 Heat power distribution along the length of multicoil continuous heater.

allowable values of power: U_{max} or 0. Control input shown in Figure 2.12 represents a multistage control with different values of heating power for different coils. A series of separate coils can be connected to a power supply in parallel. In this case, all separate coils have the same voltage and can be considered as one controlled coil. Therefore, the number, N, of sought parameters can be equal to or less than a number of inductor coils.

2.9.2 Problem of Minimum Power Consumption

Another important cost criterion is power consumption (Equation 2.3) for a workpiece heated up to a given temperature with required accuracy (Section 2.2). Generally speaking, power cost is uniquely defined by required value of increment of average temperature, θ_{av}, process time, φ^0, and level of heat losses, $q(\varphi)$.

Let us consider time-optimal control applied to a model (Equation 1.27 through Equation 1.35) of a static heating process under assumptions accepted in Section 2.8. For time-optimal control algorithms, the simple equation of heat balance can be written in relative units as the following expression:

$$\theta_{av} = \int_0^{\varphi^0} \left[u(\varphi) + q(\varphi) \right] d\varphi . \tag{2.39}$$

Therefore, power consumption I_P is determined using prescribed value $\theta_{av} = \theta^*_{av}$ and $q(\varphi)$ by the simple relation:

$$I_P = \int_0^{\varphi^0} u(\varphi) d\varphi = \theta^*_{av} - \int_0^{\varphi^0} q(\varphi) d\varphi , \tag{2.40}$$

where $q(\varphi) < 0$ in (Equation 1.32) for all $\varphi \in [0, \varphi^0]$.

Thus, it follows that cost of required power is reduced with decreasing of process time due to decreasing heat losses taken into account by an integral in the right side of Equation (2.40). Starting from Equation (2.40), this means that the heating mode with constant maximum permissible power $u(\varphi) = u^*(\varphi) \equiv U_{max}$ during the whole process would be optimal with respect to criterion I_P.

As has been shown in Section 2.8, the algorithm $u^*(\varphi) \equiv U_{max}$ does not always provide the final required temperature accuracy ε. Obviously, in this case, control input, $u^*(\varphi)$, would be optimal with respect to energy consumption at given accuracy ε if control function $u^*(\varphi)$ is of the same form as well as time-optimal control with alternating intervals of heating with maximum power $u^*(\varphi) \equiv U_{max}$ and soaking stage $u^*(\varphi) \equiv 0$ (Section 2.8, Figure 2.8). During intervals with $u^*(\varphi) \equiv 0$, power consumption is equal to 0 and, during intervals with $u^*(\varphi) \equiv U_{max}$, appropriate increase of average temperature is reached in the fastest possible way — i.e., with minimum power consumption.

Generally speaking, the similar mode is optimal with respect to criterion I_P under required heating accuracy ε. Presented substantiation is proved by mathematical analysis,[1] which determines single-type (ON–OFF) optimal control by power of internal heat sources for static induction heating processes. Control input of this type is optimal with respect to both criteria: minimum time and minimum power consumption. However, generally speaking, it does not mean that these optimal controls completely coincide for the same prescribed values ε.

Note that, in many cases, time optimal and power-optimal control inputs in the form of Equation 2.27 correspond to different time periods Δ_i, $i = 1, 2, \ldots, N$ and different number N of constancy intervals (Figure 2.9). For example, if required accuracy $\varepsilon = \varepsilon_0$ is reached in a "one-stage" heating process with constant power, $u(\varphi) \equiv U_{max}$, over total process time, then this control would be time-optimal for this value ε. The same accuracy, but with less power consumption, could be achieved in "two-stage" mode by decreasing duration of the first interval with $u(\varphi) \equiv U_{max}$ and adding the second interval for temperature leveling with $u(\varphi) \equiv 0$, when power consumption is zero. In this case, the temperature drop at the end of heating will decrease until the required level owing to increase of the total process time.

As one can see, considered optimal control algorithms that belong to one class of control inputs can differ from one another by particular values of appropriate parameters. As will be shown further, this does not exclude situations when there is a full coincidence of time-optimal control with power-optimal control for certain requirements to heating accuracy.

Under heating power control in Equation (2.27), power costs (Equation 2.40) are proportional to the sum of odd control intervals instead of the sum (Equation 2.31) of all control intervals for the time-optimal problem. Then, an optimization criterion could be written as:

$$I_P = \sum_{i=1,3,5,\ldots,N_1}^{N_1} \Delta_i \to \min_{\Delta}, \tag{2.41}$$

$$N_1 = N \text{ for odd } N, N_1 = N-1 \text{ for even } N.$$

Therefore, the problem of the workpiece heating up to required temperature with required (according to Equation 2.30) accuracy is reduced (similarly to Equation 2.34) to the following mathematical programming problem of minimum cost function (Equation 2.41), with constraint (Equation 2.30):

$$\begin{cases} I_P = \sum\limits_{i=1,3,5,\ldots,N_1}^{N_1} \Delta_i \to \min\limits_{\Delta}, \\ \Phi(\Delta) = \max\limits_{l\in[0;1]} \left|\theta(l,\Delta) - \theta^*\right| \le \varepsilon; \ \Delta = (\Delta_1, \Delta_2, \ldots, \Delta_N). \end{cases} \tag{2.42}$$

Similarly (as was done earlier in Section 2.8 and Section 2.9.1), it is possible to extend Equation (2.42) to problems of optimization of energy consumption for steady-state continuous heating processes.

In conclusion, it should be mentioned that a number of optimization problems (that can be parameterized) are reduced to mathematical programming problems (similar to those discussed in Section 2.8 and Section 2.9).[1] Some of those problems we shall review next.

Parameters Δ_i, uniquely describing required control inputs, can have the most different physical sense corresponding to each specific problem. Intervals Δ_i can be considered as periods of time or coordinates along inductor length (intervals of control input constancy) as well as voltages on different coils of the heater. Generally speaking, parameters Δ_i can have quite different physical meaning. Other examples will be discussed in the subsequent chapters of this text. Depending upon meaningful characteristics of vector Δ, there will be different relations $\theta(l,\Delta)$. These relations are defined by solution of equations of mathematical model in the appropriate class of control inputs.

Expressions for chosen optimization criterion could be modified and complicated in comparison with the simplest linear functions in Equation (2.34) and Equation (2.42). As was already pointed out in Section 2.8, in all cases, principal specificity of any similar problem results in necessity of taking into account constraint in the form of Equation (2.30). This constraint is governed by the requirements to final temperature state. This specificity lies also in necessity of minimization of similar type cost functions (see, for example, Equation 2.35, Equation 2.37, and Equation 2.38). These features result in appreciable difficulties while calculating optimal induction heating processes; the calculations are reduced to solution of mathematical programming problems.

Known formal techniques for such solution are reduced to relatively complicated numerical methods; their general orientation is not connected to specific content of the problem. An alternative effective approach to this problem is suggested in the following chapter. It is based on physical properties of investigated heating systems and offers many opportunities for development of simple and effective engineering techniques of optimized process computation.[1,8]

REFERENCES

1. Rapoport, E.Ya., *Optimization of Induction Heating of Metals,* Metallurgy, Moscow, 1993.
2. Butkovskij, A.G., Malyj, S.A., and Andreev, Yu.N., *Optimal Control of Metals Heating,* Metallurgy, Moscow, 1972.
3. Butkovskij, A.G., Malyj, S.A., and Andreev, Yu.N., *Control of Metals Heating*, Metallurgy, Moscow, 1981.
4. Rudnev, V.I. et al., *Handbook of Induction Heating,* Marcel Dekker, New York, 2003.
5. Nemkov, V.S. and Demidovich, V.B., *Theory and Computation of Induction Heating Installations,* Energoatomizdat, St. Petersburg, 1988.
6. Pavlov, N.A., *Engineering Thermal Computation of Induction Heaters,* Energia, Moscow, 1978.
7. Sluhotskii, A.E. et al., *Induction Heating Installations,* Energoatomizdat, St. Petersburg, 1981.
8. Rapoport, E.Ya., *Alternate Method for Solving Applied Optimal Control Problems,* Nauka, Moscow, 2000.

3 Method for Computation of Optimal Processes for Induction Heating of Metals

This chapter delves into basics of a new optimal control method that allows one to solve a wide range of optimization problems for induction heating of metals prior to hot forming. The new optimization technique is called the "alternance method." This method is based on qualitative features of temperature distribution within the heated workpiece at the end of the optimal heating process. These features have clear physical meaning and are similar to properties of the best uniform approximation of given functions to zero.[1] To explain new optimal control techniques in clear and simple form it is reasonable to consider only the general problem of time-optimal control for the static heating process[2–5] (see Section 2.8). Subsequent chapters will show how to extend the suggested method to the solution of other optimization problems.

3.1 UNIVERSAL PROPERTIES OF TEMPERATURE DISTRIBUTION WITHIN WORKPIECES AT END OF TIME-OPTIMAL INDUCTION HEATING PROCESSES

As has been discussed in Section 2.8, the time-optimal control algorithm $u^*(\varphi)$ of the static induction heating process is defined as a step function (Equation 2.27) with $N \geq 1$ intervals of constancy (Figure 2.9), where $u^*(\varphi)$ is an internal heat power. Therefore, the time-optimal control consists of alternating stages of heating with maximum power $u \equiv U_{max}$ ("heat ON") and subsequent soaking ("heat OFF") under $u \equiv 0$ cycles. The number of stages ($N \geq 1$) is defined uniquely by given heating accuracy ε. Smaller ε requires a larger number of stages, N. Therefore, the shape of the optimal control algorithm is known, but the number N and time durations $\Delta_1, \Delta_2, \ldots, \Delta_N$ of the described stages (Figure 2.9) remain unknown.

For any particular process, the number N and the values $\Delta_1, \Delta_2, \ldots, \Delta_N$ could be determined during subsequent computations. Particular values of these variables depend on required accuracy ε of approximation to prescribed final temperature distribution (according to Equation 2.30).

Let notation $\varepsilon_{\min}^{(N)}$ denotes minimum value of ε attainable by the application of control inputs that are described in the form of Equation (2.27). It means that $\varepsilon_{\min}^{(N)}$ represents maximum achievable heating accuracy or minimum temperature deviation for the considered class of control inputs. That is why $\varepsilon_{\min}^{(N)}$ is called *minimax*. If the last time interval is equal to zero, then a control function (Equation 2.27) that consists of N power pulsing intervals can be examined as a "particular case" of the similar control input with $N+1$ intervals. It follows here that in this case, the expression $\varepsilon_{\min}^{(N)} \geq \varepsilon_{\min}^{(N+1)}$ is true for any $N \geq 1$.

Detailed analysis shows[2] that the following sequence of inequalities takes place:

$$\varepsilon_{\min}^{(1)} > \varepsilon_{\min}^{(2)} > \ldots > \varepsilon_{\min}^{(N)} > \varepsilon_{\min}^{(N+1)} > \ldots > \varepsilon_{\min}^{(N^*)} = \varepsilon_{\inf} \geq 0\ . \tag{3.1}$$

Here, $\varepsilon_{\min}^{(k)}, k = \overline{1, N^*}$ are minimax values for control functions with k intervals of constancy; $\varepsilon_{\inf}$ is the best attainable heating accuracy in the class of step-function control inputs (Equation 2.27) with any number of constancy intervals. (Therefore, $\varepsilon_{\inf}$ represents the least attainable value ε in Equation 2.30.) It is obvious that $\varepsilon = 0$ when final temperature distribution coincides with the required one. If it is not possible to obtain desired temperature distribution exactly, then $\varepsilon_{\inf} > 0$. Speaking in terms of control theory, this means that it is impossible to steer an initial temperature state to target temperature state without error.

From a practical perspective, this situation is typical for induction heating applications where target temperature distribution is stipulated by process charts (see Section 2.7). Here, $\varepsilon_{\inf}$ becomes equal to minimax $\varepsilon_{\min}^{(N^*)}$, attainable in the class of control functions with N^* constancy intervals, where number N^* is unknown a priori.

As shown earlier, it is proper to assume that optimal control input (Equation 2.27) can consist of N intervals of constancy only in the case when required values of ε in Equation (2.30) and Equation (2.34) are not less than the minimax $\varepsilon_{\min}^{(N)}$:

$$\varepsilon \geq \varepsilon_{\min}^{(N)}\ . \tag{3.2}$$

One can prove[2,5] that, in the case of one-dimensional models of heating process (Equation 1.27 through Equation 1.35), the following constitutive relation holds true:

$$N = S \ \ for\ all\ \ \varepsilon:\ \varepsilon_{\min}^{(S)} \leq \varepsilon < \varepsilon_{\min}^{(S-1)}\ . \tag{3.3}$$

Depending upon the place of ε in the sequence (Equation 3.1), the value of parameter N can be defined according to Equation (3.3). For example, optimal control function will consist of $N = 2$ constancy intervals if the prescribed heating accuracy is equal to $\varepsilon_{min}^{(2)}$ or it is within the limits between $\varepsilon_{min}^{(2)}$ and $\varepsilon_{min}^{(1)}$ (i.e., $\varepsilon_{min}^{(2)} \le \varepsilon < \varepsilon_{min}^{(1)}$). If the required value ε equals $\varepsilon_{min}^{(1)}$, then, according to the same rule (Equation 3.3), the value of N will be as $N = 1$.

The values of minimaxes in the sequence of inequalities (Equation 3.1) are unknown a priori. Therefore, it is impossible to determine in advance that the conditions in Equation (3.2) and Equation (3.3) will be satisfied for the prescribed value of ε. For that reason, the values $\varepsilon_{min}^{(N)}$ for different N should be included in the set of unknown parameters of optimal process.

Optimal control, being the solution of the considered time-optimal problem, has definite set of intervals $\Delta_1^0, \Delta_2^0, \ldots, \Delta_N^0$. Therefore, the vector $\Delta^0 = (\Delta_1^0, \Delta_2^0, \ldots, \Delta_N^0)$ represents the solution of the mathematical programming problem in Equation (2.34). All control inputs consisting of N constancy intervals with time periods $\Delta \ne \Delta^0$ should not be considered as optimal. Spatial temperature distribution $\theta(l,\Delta^0)$ at the end of an optimal process is a response to vector Δ^0 and should satisfy the constraint (Equation (2.30)) in Equation (2.34). The temperature distribution $\theta(l,\Delta^0)$ differs from all other final temperature distributions $\theta(l,\Delta)$ (for $\Delta \ne \Delta^0$) by several important properties. The suggested method takes these properties for solving any time-optimal problems of the type of Equation (2.34) into consideration.

Restraint (Equation 2.30), imposed on temperature distribution $\theta(l,\Delta^0)$, means that absolute deviations $|\theta(l,\Delta^0) - \theta^*|$ of final temperature from the required one should not exceed the prescribed value of ε for all spatial coordinates, $l \in [0; 1]$. These deviations prove to be strictly less than ε on interval $[0; 1] \ni l$ (Figure 3.1, curve 1), or they become strictly equal to ε in separate points (one or more) of maximum deviations within the same interval (Figure 3.1, curves 2 through 4).

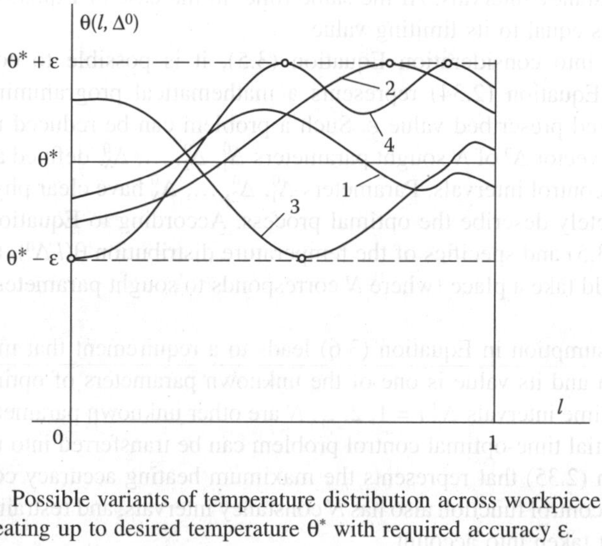

FIGURE 3.1 Possible variants of temperature distribution across workpiece thickness at the end of heating up to desired temperature θ^* with required accuracy ε.

The following basic feature of the final temperature distribution $\theta(l,\Delta^0)$ represents the main idea of the suggested optimal control method. The number of unknown parameters of the optimal control algorithm does not exceed the number of points within the heated workpiece, where the maximum admissible absolute deviations ε of final temperature from required one are reached.[2,5] In other words, if the considered optimal control problem is reduced to search for R unknown parameters, then such R points exist at the end of the optimal process that, for coordinates l_j^0 of these points, the following expressions will take place:

$$\left|\theta(l_j^0,\Delta^0)-\theta^*\right|=\varepsilon,\ j=1,2,\ldots,R;\ \Delta^0=\left(\Delta_1^0,\Delta_2^0,\ldots,\Delta_N^0\right);$$
$$0\le l_1^0<l_2^0<\ldots<l_R^0\le 1. \tag{3.4}$$

The following relations can be written according to Equation (3.3):

$$\varepsilon_{\min}^{(N)}<\varepsilon<\varepsilon_{\min}^{(N-1)} \tag{3.5}$$

or

$$\varepsilon=\varepsilon_{\min}^{(N)}. \tag{3.6}$$

This means that in the case of Equation (3.5), the desired heating accuracy ε does not reach minimum limiting value for the class of admissible control functions with N constancy intervals. At the same time, in the case of Equation (3.6), the value of ε is equal to its limiting value.

Taking into consideration Equation (3.5), it is possible to conclude the following. Equation (2.34) represents a mathematical programming problem with the fixed prescribed value ε. Such a problem can be reduced to the computation of vector Δ^0 of N sought parameters Δ_1^0, Δ_2^0, …, Δ_N^0, defined as durations of optimal control intervals. Parameters Δ_1^0, Δ_2^0, …, Δ_N^0 have clear physical sense and completely describe the optimal process. According to Equation (3.4) and Equation (3.5) and specifics of the temperature distribution $\theta(l,\Delta^0)$, the equality $R = N$ should take a place (where N corresponds to sought parameters of Δ_1^0, Δ_2^0, …, Δ_N^0).

The assumption in Equation (3.6) leads to a requirement that minimax $\varepsilon_{\min}^{(N)}$ is unknown and its value is one of the unknown parameters of optimal process (unknown time intervals Δ_i^0, $i = 1, 2, \ldots N$ are other unknown parameters). In this case, an initial time-optimal control problem can be transferred into the problem in Equation (2.35) that represents the maximum heating accuracy control problem. Then, control function also has N constancy intervals, and restraint (Equation 2.36) is not taken into account.

Almost without exception, this problem has a single solution that represents the solution of the initial time-optimal control problem. Therefore, taking into consideration Equation (3.6), Equation (2.34) can be transformed into the form of Equation (2.35) requiring one to search for vector Δ^0 and value of $\varepsilon_{\min}^{(N)}$. This means that it is necessary to determine $N+1$ parameters: $\Delta_1^0, \Delta_2^0, \ldots, \Delta_N^0$ and $\varepsilon_{\min}^{(N)}$. As one can see, the considered property of temperature distribution $\theta(l,\Delta^0)$ satisfies an expression $R = N + 1$ that takes place in Equation (3.4).

Taking into consideration expressions (in Equation (3.5) and Equation (3.6)), after supplementing Equation (3.4) by relations between number R and number N, the formulation of the basic property of final temperature distribution $\theta(l,\Delta^0)$ could be written as:

$$\begin{cases} \left|\theta(l_j^0,\, \Delta^0) - \theta^*\right| = \varepsilon,\ j = 1,2,\ldots,R;\ 0 \le l_1^0 < l_2^0 < \ldots < l_R^0 \le 1;\ \Delta^0 = \left(\Delta_1^0, \Delta_2^0, \ldots, \Delta_N^0\right); \\ R = \begin{cases} N,\ if\ \varepsilon_{\min}^{(N)} < \varepsilon < \varepsilon_{\min}^{(N-1)}; \\ N+1,\ if\ \ \varepsilon = \varepsilon_{\min}^{(N)}. \end{cases} \end{cases} \tag{3.7, 3.8}$$

Mathematically rigorous proof of Equation (3.7) and Equation (3.8) is provided in Rapoport.[2]

The fundamental importance of these expressions deals with the fact that they represent a system closed in the mathematical sense with respect to all parameters of the optimal process. In other words, according to Equation (3.8), the number of equalities (Equation 3.7) proves to be equal to the number of all sought parameters that completely define this process. This provides potential capability to transform a set of equalities into set of equations that ought to be solved with respect to unknown parameters that leads to the final solution of the initial OCP.

However, there might be some difficulties in finding an effective engineering technique to compute optimal induction heating processes. These difficulties arise due to transformation of a set of equalities (Equation 3.7 and Equation 3.8) into a set of equations. Different variants of the final temperature distribution correspond to the single set in Equation (3.7) and Equation (3.8). These variants differ from each other by the shape/form of temperature profile along axis l and by coordinates of points l_j^0.

For example, there might be a variety of temperature profiles representing $R = 3$ condition. Some of these profiles are shown in Figure 3.2. Though all temperature profiles shown in Figure 3.2 satisfy Equation (3.7) formally, at the same time they have different combinations of points l_j^0, $j = 1, 2, 3$. This results in the sets of equations distinguishing by signs of deviations $\theta(l_j^0,\Delta^0) - \theta^*$, coordinates l_j^0, and, of course, optimal vectors Δ^0. Therefore, it is necessary to introduce the single set of equations that would reveal a particular temperature profile. For temperature profiles shown in Figure 3.2a through Figure 3.2c (solid curves), the corresponding sets of equations will be as follows.

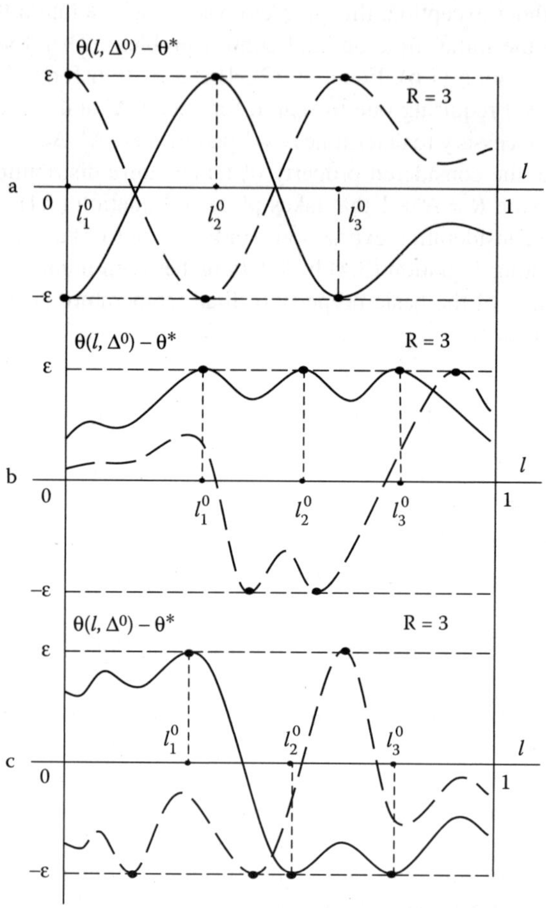

FIGURE 3.2 Variants of temperature profiles at the end of time-optimal heating process for $R = 3$, $\varepsilon_{min}^{(3)} < \varepsilon < \varepsilon_{min}^{(2)}$.

For the profile shown in Figure 3.2a:

$$\begin{cases} \theta(l_1^0, \Delta^0) - \theta^* = -\varepsilon; \\ \theta(l_2^0, \Delta^0) - \theta^* = +\varepsilon; \\ \theta(l_3^0, \Delta^0) - \theta^* = -\varepsilon. \end{cases}$$

For the case in Figure 3.2b:

$$\begin{cases} \theta(l_1^0,\ \Delta^0) - \theta^* = +\varepsilon; \\ \theta(l_2^0,\ \Delta^0) - \theta^* = +\varepsilon; \\ \theta(l_3^0,\ \Delta^0) - \theta^* = +\varepsilon. \end{cases}$$

For the profile shown in Figure 3.2c:

$$\begin{cases} \theta(l_1^0,\ \Delta^0) - \theta^* = +\varepsilon; \\ \theta(l_2^0,\ \Delta^0) - \theta^* = -\varepsilon; \\ \theta(l_3^0,\ \Delta^0) - \theta^* = -\varepsilon. \end{cases}$$

Even for the same temperature profile $\theta(l,\Delta^0) - \theta^*$ (having greater than R number of extremum points, which are "candidates" on the role of points l_j^0), different sets of equations could be written (see dotted curves in Figure 3.2b through Figure 3.2c).

The proper choice of a particular temperature distribution $\theta(l,\Delta^0)$ allows transforming Equation (3.7) into a corresponding set of equations. This choice can only be performed using additional information about a specific of temperature distribution (based on physical properties of induction heating processes). The following sections will discuss the ways to accomplish this task.

3.2 EXTENDED DISCUSSION ON PROPERTIES OF FINAL TEMPERATURE DISTRIBUTION FOR TIME-OPTIMAL INDUCTION HEATING PROCESSES

To choose a particular variant of temperature distribution $\theta(l,\Delta^0)$, it is necessary to apply additional information regarding qualitative properties of a temperature field that will be obtained at the end of the optimal heating process. The proper choice should be performed according to the required value of ε.

Detailed analysis shows that maximum admissible temperature deviations $\theta(l_j^0,\Delta^0) - \theta^*$ of the final temperature distribution from the required one (see Equation 3.7) arise with different signs for each pair of successively located points l_j^0.[2] It means that these deviation signs are alternating in points l_j^0, where $0 \le l_1^0 < l_2^0 < \dots l_R^0 \le 1$.

Taking this feature into consideration, Equation (3.7) and Equation (3.8) can be rewritten as:

$$\begin{cases} \theta(l_j^0,\ \Delta^0) - \theta^* = (-1)^j\ \psi\varepsilon,\ j = 1,2,\ldots,R;\ \ \Delta^0 = \left(\Delta_1^0, \Delta_2^0, \ldots, \Delta_R^0\right), \psi = \pm 1 \\ 0 \le l_1^0 < l_2^0 < \ldots < l_R^0 \le 1;\ \ R = \begin{cases} N,\ if\ \ \varepsilon_{\min}^{(N)} < \varepsilon < \varepsilon_{\min}^{(N-1)}, \\ N+1,\ if\ \ \varepsilon = \varepsilon_{\min}^{(N)}. \end{cases} \end{cases} \tag{3.9}$$

Here, a multiplier coefficient $(-1)^j$ provides a sign alternation of temperature deviations at points l_j^0 that are consecutively located within the interval [0, 1]. Coefficient ψ equals to +1 or –1 and provides signs "+" or "–" for temperature deviations at each point l_j^0.

Equation (3.9) significantly restricts the set of admissible temperature curves $\theta(l,\Delta^0)$. For example, Equation (3.7) and Equation (3.8) are true for all temperature profiles in Figure 3.2; however, only curves shown in Figure 3.2a meet requirement (Equation 3.9). The solid curve (Figure 3.2a) corresponds to $\psi = -1$ and, respectively, the dashed one corresponds to $\psi = +1$ in Equation (3.9).

The set of points l_j^0 where equalities (Equation 3.9) hold true are called *Chebyshev alternance*[1] in approximation theory. In this case, the rule (Equation 3.9) represents the fundamental alternance property of the final temperature distribution in time-optimal processes of induction heating. This is the reason why the newly proposed optimal control method is called *alternance method.*

It is important to note that the rule (Equation 3.9) does not yet allow determining the shape/form of curve $\theta(l,\Delta^0)$. Additional information with regard to a number of extremum points on the interval [0,1] is required. Figure 3.3 shows variants of different temperature profiles $\theta(l,\Delta^0)$ for the case of $R = 5$. All curves satisfactorily fit condition (3.9). If the curve has a minimal number of extremum points, then its form is unambiguously fixed as shown in Figure 3.3a. If the number of extremum points exceeds the required minimum ($R = 5$), then a variety of curves $\theta(l,\Delta^0)$ exists. The difference would also be in locations of points l_j^0. Some of these temperature profiles and corresponding points are presented in Figure 3.3b.

Based on physical properties of induction heating temperature fields, it can be shown that maximum possible number, $M_{\max}$, of extremum points for curve $\theta(l,\Delta^0)$ is determined according to the following expression[2,5]:

$$M_{\max} = \begin{cases} N,\ if\ N\ is\ even\ \ number; \\ N+1,\ if\ N\ is\ odd\ number. \end{cases} \tag{3.10}$$

for model Equation (1.27) through Equation (1.35) with uniform initial temperature distribution (Section 2.8), that will be considered further.

Simple physical explanation of Equation (3.10) can be offered. Taking into consideration constant maximum power $u(\varphi) \equiv U_{\max}$, the temperature distribution $\theta(l,\Delta_1)$ at the end of induction heating is well known (Figure 3.4). It has two

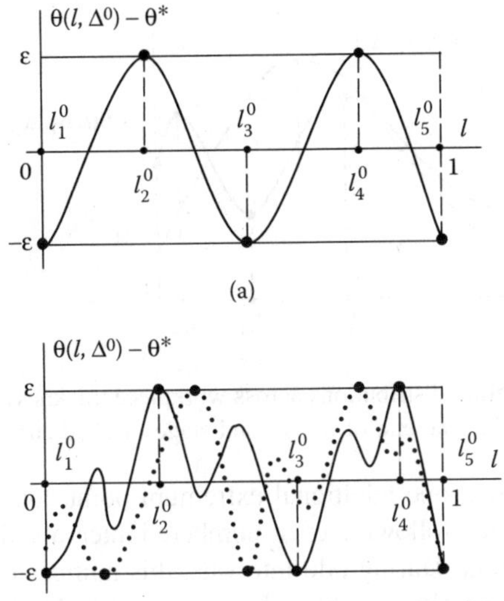

FIGURE 3.3 Variants of temperature profiles at the end of time-optimal heating process for $R = 5$.

extremum points with minimum and maximum temperatures accordingly: $l = l_{e1} = 0$ and $l = l_{e2}$, $0 < l_{e2} < 1$. The first point corresponds to a boundary condition (Equation 1.31) for $l = 0$. The second point is formed by negative temperature gradient on the surface $l = 1$ due to heat losses, according to the boundary condition (Equation 1.32) at $l = 1$. The shape of curve $\theta(l,\Delta_1)$ is invariable over the second bounded control interval under $u(\varphi) \equiv 0$ (Figure 3.4). During the third short interval, a "peak" of temperature is formed within the billet surface layer.

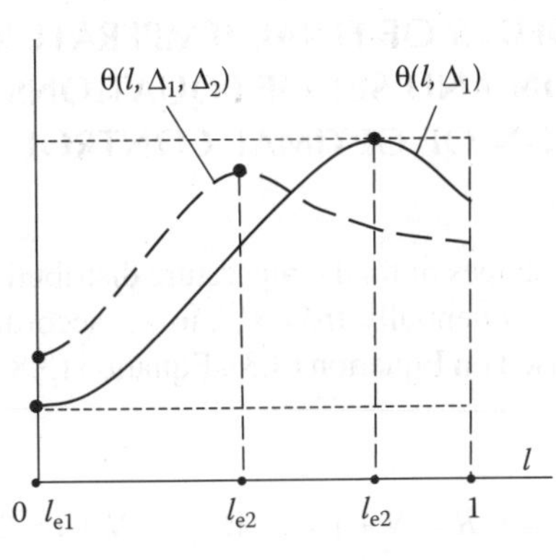

FIGURE 3.4 Temperature distribution across workpiece thickness at the end of heating with constant maximum power (solid curve) and after soaking stage (dashed curve).

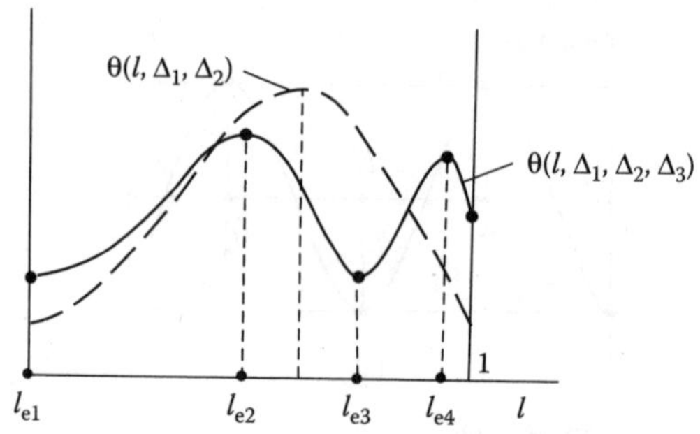

FIGURE 3.5 Temperature distributions across workpiece thickness during the third short stage (solid curve) and at the end of the second stage (dashed curve).

Therefore, at most, one additional extremum point could be added (Figure 3.5). Similarly, over the following even-numbered intervals, the number of extremum points is constant. During odd intervals, this number increases at most by one. As a result, it is possible to obtain Equation (3.10). Mathematically rigorous proof of this statement is provided in Rapoport.[2,5]

Rule (3.10) in combination with the alternance property (Equation 3.9) allows determining an unambiguous form of curve $\theta(l,\Delta^0)$ or a quite certain set of possible variations of temperature profiles. As a result, the basic system of correlations (Equation 3.7 and Equation 3.8) can be transformed into a corresponding set of equations.

The following sections describe the ways of determining the temperature profiles and methods of building the system of equations according to prescribed heating accuracy ε using stated properties of temperature fields at the end of optimal heating processes.

3.3 TYPICAL PROFILES OF FINAL TEMPERATURE DISTRIBUTION AND SET OF EQUATIONS FOR COMPUTATION OF OPTIMAL CONTROL PARAMETERS

Let us consider typical shapes of final temperature distribution $\theta(l,\Delta^0)$ for different values of ε decreasing sequentially from $\varepsilon_{\min}^{(1)}$ to $\varepsilon_{\inf}$ according to Equation (3.1).

If $\varepsilon = \varepsilon_{\min}^{(1)}$, then, based on Equation (3.3), Equation (3.8), and Equation (3.10), it is possible to write:

$$N = 1;\ R = N + 1 = 2;\ M_{\max} = N + 1 = 2.$$

This means that, for $\varepsilon = \varepsilon_{\min}^{(1)}$, there is only one possible temperature distribution $\theta(l, \Delta_1^0)$ (see Figure. 3.4) that is characterized by two extremum points $l = l_{e1} = 0$ $l = l_{e2} \in (0,1)$. Both points represent maximum admissible deviation

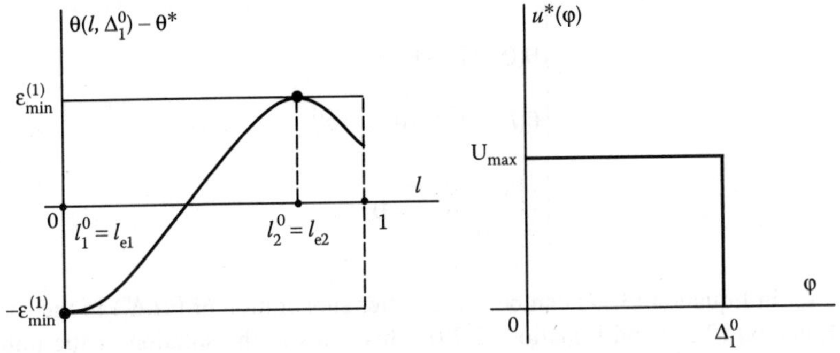

FIGURE 3.6 Optimal control $u^*(\varphi)$ and temperature distribution $\theta(l,\Delta_1^0)$ at the end of time-optimal, one-stage control process with limiting possible heating accuracy.

of final temperature from the required one. In this case, minimum and maximum differences $\theta(l,\Delta_1^0) - \theta^*$ are reached in the points l_{e1} and l_{e2}, respectively, that match the rule of sign alternating (Equation 3.9) for $\psi = +1$. Such distribution takes place due to small enough heat losses from the surface of a heated workpiece when a surface temperature remains within admissible limits. As a result, one can obtain the single variant of temperature curve $\theta(l,\Delta_1^0) - \theta^*$ shown in Figure 3.6.

It is possible to conclude at this point that at the end of a time-optimal, one-stage control process with limiting possible heating accuracy, the temperature profile has a minimum admissible temperature in the center of the billet. At the same time, the region with a maximum temperature is located within the billet's volume at a certain internal point. In this case, the inequality $\varepsilon_{\min}^{(1)} > \varepsilon_{\inf}$ always holds true.

Instead of the set in Equation (3.7), for the temperature profile shown in Figure. 3.6, the following set of two equations can be written:

$$\begin{cases} \theta\left(0,\Delta_1^0\right) - \theta^* = -\varepsilon_{\min}^{(1)}; \\ \theta\left(l_{e2},\Delta_1^0\right) - \theta^* = +\varepsilon_{\min}^{(1)}, \end{cases} \tag{3.11}$$

which includes two unknown parameters of optimal process: duration Δ_1^0 of heating stage and value of minimax $\varepsilon_{\min}^{(1)}$. Coordinate l_{e2} of a single internal point where temperature deviation $\theta(l,\Delta_1^0) - \theta^*$ reaches its maximum should be considered as a third unknown.

However, the observation that the temperature gradient is equal to zero at the point of extremum can be used to define unknown coordinate l_{e2}. Therefore, instead of Equation (3.11), the following set of three equations with respect to three unknowns — Δ_1^0, $\varepsilon_{\min}^{(1)}$, and l_{e2} — can be written:

$$\begin{cases} \theta(0, \Delta_1^0) - \theta^* = -\varepsilon_{\min}^{(1)}; \\ \theta(l_{e2}, \Delta_1^0) - \theta^* = +\varepsilon_{\min}^{(1)}; \\ \dfrac{\partial\theta(l_{e2}, \Delta_1^0)}{\partial l} = 0. \end{cases} \tag{3.12}$$

The set in Equation (3.12) can be solved after substitution of $\theta(l,\Delta_1^0)$ in the form of Equation (2.28) and Equation (2.29). This leads to the solution of the initial time-optimal problem for $N = 1$ and $\varepsilon = \varepsilon_{\min}^{(1)}$.

Let us assume that prescribed value ε in the subsequence of inequalities in Equation (3.1) belongs to the following range:

$$\varepsilon_{\min}^{(2)} < \varepsilon < \varepsilon_{\min}^{(1)} . \tag{3.13}$$

Then, according to Equation (3.3), Equation (3.8), and Equation (3.10), it is possible to write:

$$N = 2;\ R = N = 2;\ M_{\max} = 2. \tag{3.14}$$

This means that, for these values of ε, an optimal control consists of two stages. For this type of control inputs the shape of temperature distribution $\theta(l,\Delta_1^0,\Delta_2^0)$ is similar to single-stage control (Figure 3.4).

In this case, only the single set of points l_j^0 in Equation (3.7) that meets equalities in Equation (3.14) exists. The combination of these points is similar to the case of $\varepsilon = \varepsilon_{\min}^{(1)}$ (Figure 3.7) and the following system of three equations could be written instead of the system in Equation (3.12):

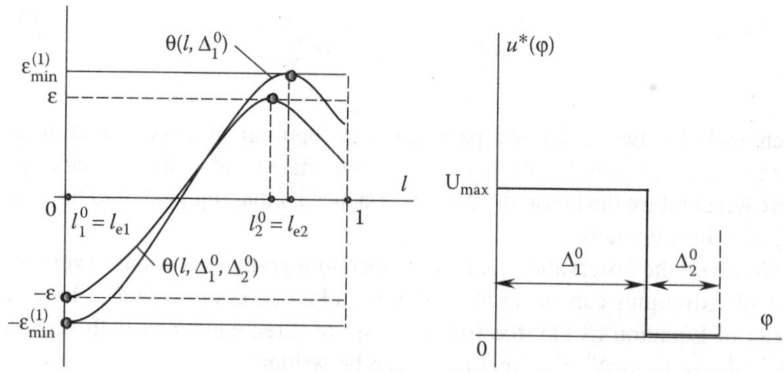

FIGURE 3.7 Optimal control $u^*(\varphi)$ and final temperature profiles for $\varepsilon = \varepsilon_{\min}^{(1)}$ ($\theta(l,\Delta_1^0)$) and for $\varepsilon_{\min}^{(2)} < \varepsilon < \varepsilon_{\min}^{(1)}$ ($\theta(l,\Delta_1^0,\Delta_2^0)$).

$$\begin{cases} \theta(0, \Delta_1^0, \Delta_2^0) - \theta^* = -\varepsilon; \\ \theta(l_{e2}, \Delta_1^0, \Delta_2^0) - \theta^* = +\varepsilon; \\ \dfrac{\partial\theta(l_{e2}, \Delta_1^0, \Delta_2^0)}{\partial l} = 0. \end{cases} \tag{3.15}$$

If the value of ε is preset, then this system should be solved with respect to three other unknown variables: optimal durations Δ_1^0 and Δ_2^0 of two control stages and coordinate $l_2^0 = l_{e2}$ of the single internal point where maximum temperature takes place. In contrast to $\theta(l,\Delta_1)$ in Equation (3.12), here $\theta(l,\Delta_1,\Delta_2)$ is a complicated function of two parameters, Δ_1 and Δ_2, of optimal control and spatial coordinate l.

Taking into account Equation (2.28) and Equation (2.29), the solution of the system in Equation (3.15) leads to the solution of the initial time-optimal problem for $N = 2$ and given values ε in Equation (3.13).

If $\varepsilon = \varepsilon_{\min}^{(2)}$, then according to Equation (3.3), Equation (3.8), and Equation (3.10), and in contrast to Equation (3.14) the following equalities take place:

$$N = 2,\ R = 3,\ M_{\max} = 2.$$

These equalities mean that there are $R = 3$ points, l_j^0, $j = 1, 2, 3$: two extremum points, $l_1^0 = 0$, $l_2^0 = l_{e2}$, and one additional point on the billet surface, $l_3^0 = 1$. The point l_3^0 appears due to the fact that the value of ε decreases from $\varepsilon > \varepsilon_{\min}^{(2)}$ to $\varepsilon = \varepsilon_{\min}^{(2)}$. This results in an existence of only one temperature profile, shown in Figure 3.8. As one can see, the rule in Equation (3.9) of sign alternating holds true for $\psi = +1$.

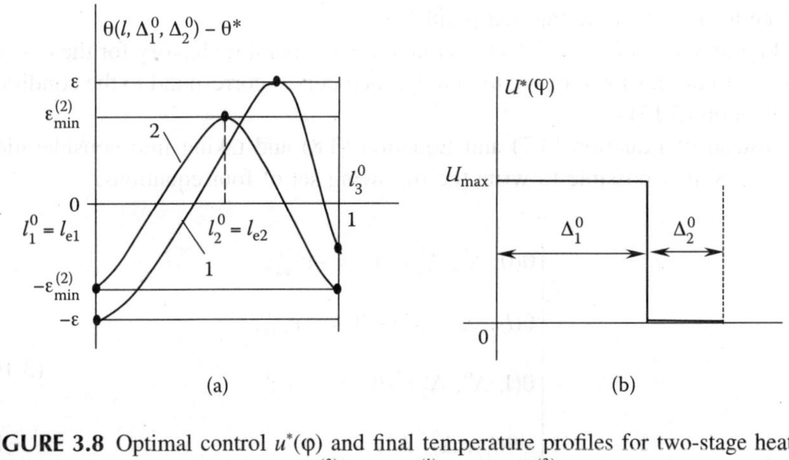

FIGURE 3.8 Optimal control $u^*(\varphi)$ and final temperature profiles for two-stage heating cycle with maximum accuracy. 1: $\varepsilon_{\min}^{(2)} < \varepsilon < \varepsilon_{\min}^{(1)}$; 2: $\varepsilon = \varepsilon_{\min}^{(2)}$.

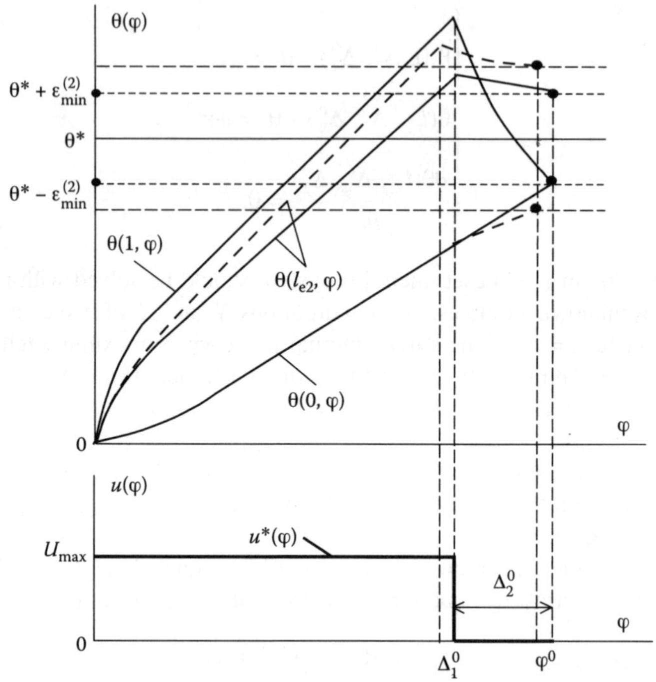

FIGURE 3.9 Optimal control $u^*(\varphi)$ and time–temperature history for two-stage heating cycle: $\varepsilon = \varepsilon_{\min}^{(2)}$ (solid curves) and $\varepsilon \in (\varepsilon_{\min}^{(2)}, \varepsilon_{\min}^{(1)})$ (dashed curves).

The achieved result is not obvious and means that for $\varepsilon = \varepsilon_{\min}^{(2)}$ at the end of time-optimal, two-stage process with limiting possible heating accuracy, the temperature profile has minimum admissible temperatures in the center of the heated billet and on its surface. At the same time, the area with the maximum temperature will be located in some internal point $l_2^0 = l_{e2}$.

Figure 3.9 and Figure 3.10 illustrate time–temperature history for the case of final distribution shown in Figure 3.8. Dashed curves correspond to the condition in Equation (3.13).

Instead of Equation (3.7) and Equation (3.8) and taking into consideration Figure 3.8, it is possible to write the following set of four equations:

$$\begin{cases} \theta(0, \Delta_1^0, \Delta_2^0) - \theta^* = -\varepsilon_{\min}^{(2)}; \\ \theta(l_{e2}, \Delta_1^0, \Delta_2^0) - \theta^* = +\varepsilon_{\min}^{(2)}; \\ \theta(1, \Delta_1^0, \Delta_2^0) - \theta^* = -\varepsilon_{\min}^{(2)}; \\ \dfrac{\partial \theta(l_{e2}, \Delta_1^0, \Delta_2^0)}{\partial l} = 0. \end{cases} \tag{3.16}$$

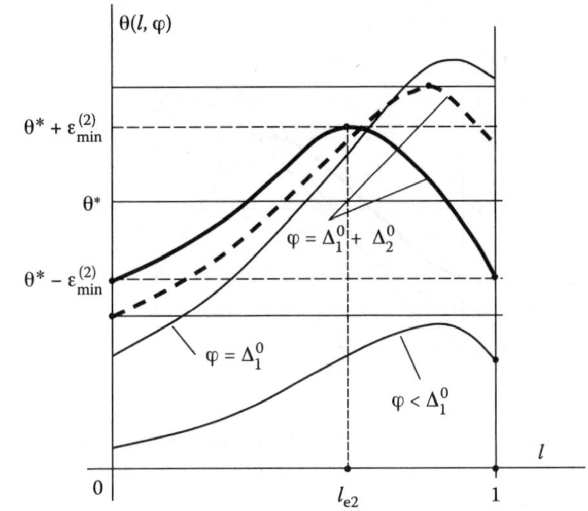

FIGURE 3.10 Temperature distribution across workpiece thickness/radius at different time points of two-stage optimal heating cycle: $\varepsilon = \varepsilon_{\min}^{(2)}$ (solid curves) and $\varepsilon \in (\varepsilon_{\min}^{(2)}, \varepsilon_{\min}^{(1)})$ (dashed curve).

As one can see, this set of equations includes four unknown variables: Δ_1^0, Δ_2^0, $\varepsilon_{\min}^{(2)}$, and l_{e2}.

Therefore, for any value of ε: $\varepsilon_{\min}^{(2)} \leq \varepsilon \leq \varepsilon_{\min}^{(1)}$, the shapes of temperature curve $\theta(l,\Delta^0)$ (Figure 3.11) and appropriate computational sets of Equation (3.12), Equation (3.15), and Equation (3.16) are well determined. After substitution of $\theta(l,\Delta^0)$ in the form of Equation (2.28) and Equation (2.29) into equation sets and using known numerical methods, these sets can be solved with respect to all sought parameters. Obtained optimal value Δ^0 represents the solution of the initial time-optimal problem in the form of Equation (2.34).

It should be emphasized that an ability to define minimum attainable minimax values $\varepsilon_{\min}^{(1)}$ and $\varepsilon_{\min}^{(2)}$ is of special interest. This follows from solution of the computational systems where values $\varepsilon_{\min}^{(1)}$ and $\varepsilon_{\min}^{(2)}$ are considered as unknown parameters of an optimal control process.

If the process demands better heating accuracy, i.e., the prescribed deviation ε should be *less than* $\varepsilon_{\min}^{(2)}$, then values of ε will belong to the following range:

$$\varepsilon_{\min}^{(3)} \leq \varepsilon < \varepsilon_{\min}^{(2)}, \tag{3.17}$$

and the problem will be significantly complicated.

It is feasible to determine only the set of possible variants of temperature profiles $\theta(l,\Delta^0)$. Every variant corresponds only to certain, but a priori unknown, initial data. It primarily depends on the value of heat losses during the heating process. Generally speaking, it is not possible in advance to choose a single

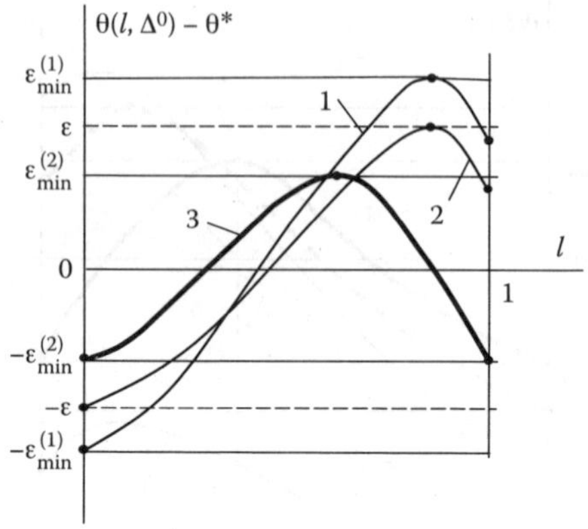

FIGURE 3.11 Temperature distributions across workpiece thickness/radius at the end of time-optimal heating process. 1: For $\varepsilon = \varepsilon_{\min}^{(1)}$; 2: for $\varepsilon \in (\varepsilon_{\min}^{(2)}, \varepsilon_{\min}^{(1)})$; 3: for $\varepsilon = \varepsilon_{\min}^{(2)}$.

variant from this set for each specific situation in contrast to the case of $\varepsilon_{\min}^{(2)} \le \varepsilon \le \varepsilon_{\min}^{(1)}$.

The proper choice of particular temperature profile can be performed in the course of a special computational procedure based on the alternance method.[2,5]

Potential difficulties can appear because, in the basic relationships in Equation (3.9) under conditions in Equation (3.17), the possible number of points l_i^0 can be greater than their minimum required number R. This means that various combinations of extremum points occur.

If value of ε belongs to the range:

$$\varepsilon_{\min}^{(3)} < \varepsilon < \varepsilon_{\min}^{(2)}, \tag{3.18}$$

then, according to Equation (3.3), Equation (3.8), and Equation (3.10), the following relationships could be written:

$$N = 3,\ R = N = 3,\ M_{\max} = N + 1 = 4\ . \tag{3.19}$$

All extremum points of curve $\theta(l,\Delta^0) - \theta^*$ (including $l = 0$) and the point $l = 1$ on the surface of a heated body can be considered as points l_j^0. That is why the maximum number of such points becomes equal to $M_{\max} + 1 = 5 > R = 3$. Consequently, which three out of five possible points represent points l_j^0 for each specific temperature profile remains unclear. Having no answer to this question, we cannot write a system of equations somewhat similar to Equation (3.12), Equation (3.15), or Equation (3.16). Therefore, it will not be possible to solve the optimal control problem using the suggested method.

The case of $\varepsilon = \varepsilon_{\min}^{(3)}$ is characterized by similar difficulties. Here, instead of Equation (3.19), we can write:

$$N = 3,\ R = N + 1 = 4,\ M_{max} = N + 1 = 4;\ M_{max} + 1 = 5 > R. \quad (3.20)$$

However, it is necessary to choose four points l_j^0 from the set of five possible points, even though the number of candidate variants decreases.

In actual induction heating processes, the limiting attainable heating accuracy $\varepsilon = \varepsilon_{inf}$ in Equation (3.1) (see Section 3.1) is usually greater than zero and equal to $\varepsilon_{min}^{(3)}$. The value of ε_{inf} is primarily defined by the level of heat losses from the billet surface. In many cases, this value proves to be unattainable when heat losses are large enough (required process time grows with no limit as ε approaches $\varepsilon_{min}^{(3)}$).[2,5] In practice, it is feasible to solve a time-optimal problem only if required accuracy ε in Equation (3.18) is not too close to $\varepsilon_{min}^{(3)} = \varepsilon_{inf}$.

The limited set of all possible combinations of points l_j^0 (satisfying condition of sign alternating in Equation 3.9) can be revealed on the basis of the usual physical concepts about final spatial temperature distribution.[5] It is also possible to find sets of points l_j^0 corresponding to values $\varepsilon_{min}^{(3)} = \varepsilon_{inf}$ and $\varepsilon_{min}^{(3)} > \varepsilon_{inf}$. Heating accuracy $\varepsilon = \varepsilon_{min}^{(3)}$ is always attainable if $\varepsilon_{min}^{(3)} > \varepsilon_{inf}$.

It is possible to further increase the heating accuracy for $\varepsilon_{min}^{(4)} \le \varepsilon < \varepsilon_{min}^{(3)}$ if control functions have four intervals of constancy. Similarly to the case of $\varepsilon_{min}^{(2)} \le \varepsilon < \varepsilon_{min}^{(1)}$, the temperature profile in this case can be uniquely defined.[5] However, as was mentioned earlier, the case when $\varepsilon_{min}^{(4)} \le \varepsilon < \varepsilon_{min}^{(3)}$ is not typical for induction heating of metals.

As one can see, conditions in Equation (3.18) lead to a quite complicated time-optimal control problem. Nevertheless, for conventional values of the heat losses $q(\varphi)$ in Equation (1.32), it is possible to simplify the problem, taking into account typical technological requirements with respect to values ε.

Most often, one of two practicable cases arises.[5] For two-stage control, the minimum admissible value $\varepsilon = \varepsilon_{min}^{(2)}$ decreases monotonically as heat losses diminish. When heat losses are negligible, $q(\varphi) = 0$, then the value $\varepsilon_{min}^{(2)}$ becomes equal to zero as well (Figure. 3.12). If heat losses are substantially small (for example,

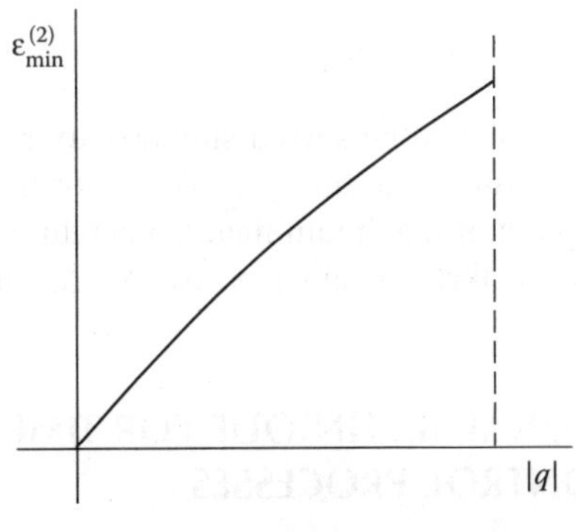

FIGURE 3.12 The value of $\varepsilon_{min}^{(2)}$ as a function of heat losses for two-stage control.

during heating of aluminum alloy ingots prior to hot forming), then the value $\varepsilon_{\min}^{(2)}$ becomes appreciably small and required values ε meet inequalities $\varepsilon_{\min}^{(2)} \leq \varepsilon < \varepsilon_{\min}^{(1)}$. Therefore, a single- or two-stage control can provide required heating accuracy in this case and a complicated control algorithm under $N \geq 3$ is not required.

If surface heat losses are appreciably large (for example, during heating of titanium and steel alloy ingots before hot forming), then, quite the contrary, heat losses increase up to the level when inequality $\varepsilon < \varepsilon_{\min}^{(2)}$ becomes true. Here, minimax $\varepsilon_{\min}^{(3)}$ coincides with $\varepsilon_{\inf} > 0$ and becomes unattainable. In this case, it is necessary to solve the time-optimal problem for the control functions with three intervals under conditions in Equation (3.18).

However, a high level of heat losses allows one to define unambiguously the simplest possible variant of curve $\theta(l,\Delta^0) - \theta^*$.[5] This variant differs by the fact that, due to the negative temperature gradient in billet surface layers (see Figure 3.5), the temperature "spike" does not appear during the third interval. In other words, during the third interval, due to shortage of heating power and/or due to relatively short heating duration, a final temperature distribution (typical for a two-stage control) is not distorted. In this case, the number of extremum points is equal to two (dotted curve in Figure 3.5) and becomes less than $M_{\max} = 4$.

As a result, the shape of curve $\theta(l,\Delta_1^0,\Delta_2^0,\Delta_3^0) - \theta^*$ for $\varepsilon_{\min}^{(3)} < \varepsilon < \varepsilon_{\min}^{(2)}$ replicates the shape of curve $\theta(l,\Delta_1^0,\Delta_2^0) - \theta^*$ for $\varepsilon = \varepsilon_{\min}^{(2)}$ (Figure 3.13).

If a prescribed value of ε satisfies the following condition, $\varepsilon < \varepsilon_{\min}^{(2)}$, then the appropriate set of four equations can be written as:

$$\begin{cases} \theta(0,\ \Delta_1^0,\ \Delta_2^0, \Delta_3^0) - \theta^* = -\varepsilon; \\ \theta(l_{e2},\ \Delta_1^0,\ \Delta_2^0,\ \Delta_3^0) - \theta^* = +\varepsilon; \\ \theta(1,\ \Delta_1^0,\ \Delta_2^0,\ \Delta_3^0) - \theta^* = -\varepsilon; \\ \dfrac{\partial\theta(l_{e2},\ \Delta_1^0,\ \Delta_2^0,\ \Delta_3^0)}{\partial l} = 0. \end{cases} \tag{3.21}$$

This set of equations should be solved similarly to those in Equation (3.16) with respect to sought optimal durations, Δ_1^0, Δ_2^0, Δ_3^0, of three control stages and coordinate, l_{e2}, of the point of maximum final temperature. Here, $\theta(l,\Delta_1,\Delta_2,\Delta_3)$ is a complicated function of three parameters, Δ_1, Δ_2, Δ_3, of optimal control and spatial coordinate l.

3.4 COMPUTATIONAL TECHNIQUE FOR TIME-OPTIMAL CONTROL PROCESSES

As has been shown in Section 3.3, for the most practical cases, the set of equations in Equation (3.12), Equation (3.15), Equation (3.16), or Equation (3.21) should be solved with respect to parameters of a time-optimal induction heating process.

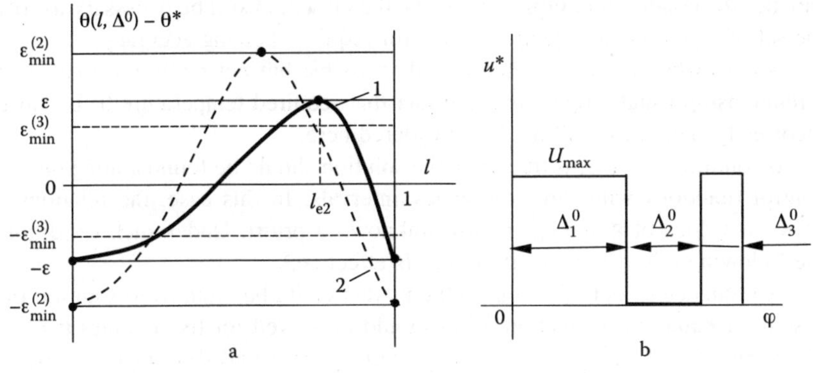

FIGURE 3.13 Final temperature profiles (a) and optimal control $u^*(\varphi)$ (b). 1: For $\varepsilon \in (\varepsilon_{min}^{(3)}, \varepsilon_{min}^{(2)})$ at high levels of heat losses; 2: for $\varepsilon = \varepsilon_{min}^{(2)}$.

For a particular application, the appropriate set of equations should be chosen and solved. The choice of set of equations can only be made if the location of value ε in the subsequence of inequalities (Equation 3.1) is known. This location is defined by Equation (3.3).

There are two quite different possible variants of value assignment for required accuracy ε:

1. The fixed numerical value of ε can be prescribed a priori according to specific technological requirements. (For example, it could be necessary that the final temperature in any point of billet would not deviate greater than 25°C from a desired temperature distribution.)
2. It can be assumed that value ε is equal to one of the minimax values in Equation (3.1). This means that, at the end of the optimal heating process, the deviation of final temperature distribution from the required one should be as small as possible for the chosen class of control inputs.

In both cases, the minimax values $\varepsilon_{min}^{(k)}$ in Equation (3.1) are not defined a priori.

Under given conditions, the following simple computational procedure is suggested for determining algorithms of optimal control for induction heating processes.

At the beginning, the sets of equations in Equation (3.12) and Equation (3.16) can be solved assuming that required accuracy $\varepsilon = \varepsilon_{min}^{(1)}$ and $\varepsilon = \varepsilon_{min}^{(2)}$ in Equation (2.34). This leads to definition of minimax values $\varepsilon_{min}^{(1)}$ and $\varepsilon_{min}^{(2)}$ as well as corresponding vectors $\Delta^0(\varepsilon_{min}^{(1)})$ and $\Delta^0(\varepsilon_{min}^{(2)})$ determining optimal control intervals. Therefore, the considered problem is already solved at this stage for case 2.

The optimal control problem is also solved for case 1 if the prescribed numerical value of ε coincides with one of the obtained numerical values of minimaxes $\varepsilon_{min}^{(1)}$ or $\varepsilon_{min}^{(2)}$. Otherwise, it is necessary to compare the prescribed numerical value of ε with obtained values of $\varepsilon_{min}^{(1)}$ and $\varepsilon_{min}^{(2)}$. If this value of ε satisfies the condition in Equation (3.13), then it would still be necessary to solve the set of equations in Equation (3.15) for required heating accuracy.

A case when $\varepsilon > \varepsilon_{\min}^{(1)}$ is theoretically possible but not really practical. Such a relationship usually underlines the fact that required temperature θ^* is chosen incorrectly. This case will not be considered here.

If condition $\varepsilon < \varepsilon_{\min}^{(2)}$ is true, then the solution should be found among optimal control functions with three constancy intervals. In this case, the relationship between values of ε and $\varepsilon_{\min}^{(3)}$ is also unknown a priori. Under such conditions, the following computational algorithm is effective.[5]

A sequence of optimization problems that could be reduced to solution of a system of equations (Equation 3.21) should be solved for fixed values $\varepsilon = \varepsilon' = \varepsilon_{\min}^{(2)} - kd\varepsilon$, $k = 1, 2, 3 \ldots$ decreasing from $\varepsilon_{\min}^{(2)}$ by small decrement $d\varepsilon$. At the beginning of this computational procedure, the value of ε' does not appreciably deviate from value $\varepsilon_{\min}^{(2)}$ and strict inequalities (Equation 3.18) certainly hold true for $\varepsilon = \varepsilon'$. Initial estimate for solution of such a set of equations can be chosen using results of the previous computation step. As value ε' decreases, there might be one of two possible cases.

1. Value ε' becomes equal to prescribed value ε in Equation (2.34) within admissible error. In this case, the initial optimal control problem is solved at this step of the described computation procedure. It should be remembered here that cases with ambiguous variations of possible spatial temperature distribution for three-stage controls are atypical for induction heating processes (see Section 3.3). Such cases lead to the more complicated computational algorithms described in reference 5 and further discussion regarding these specific cases will not be provided here.
2. At certain value $\varepsilon' = \varepsilon'' > \varepsilon$, exceeding required value in Equation (2.34), the following equality is true:

$$\left| \frac{d\varphi_{\min}^{0}(\varepsilon'')}{d\varepsilon} \right| = A \tag{3.22}$$

for optimal process time $\varphi_{\min}^{0}(\varepsilon'') = \sum_{i=1}^{3} \Delta_i^0(\varepsilon'')$, where A is some sufficiently great number chosen a priori.

Taking into consideration condition $\varphi_{\min}^{0}(\varepsilon) \to \infty$ at $\varepsilon \to \varepsilon_{\inf}$, we can assume that $\varepsilon'' \approx \varepsilon_{\inf}$ with an accuracy defined by value A (Figure 3.14). In this case, the initial problem is unsolvable as far as required accuracy is found to be higher than extremely attainable accuracy. This means that, under given initial data, it is not possible to provide the required uniformity of final temperature distribution using admissible control inputs. Therefore, some parameters of the control process should be changed to obtain required temperature distribution.

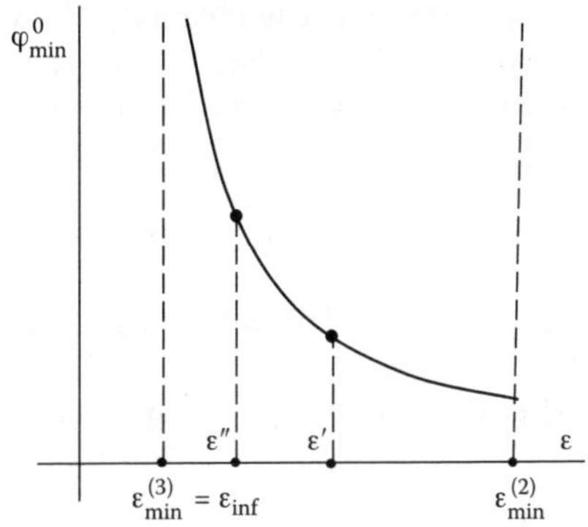

FIGURE 3.14 Minimum heating time as a function of heating accuracy.

If an induction heater has improved thermal insulation (refractory) that allows drastic reduction of surface heat losses leading to further reduction of the value ε_{inf} and inequality $\varepsilon_{inf} < \varepsilon$ in Equation (2.34) becomes true, then an original optimal control problem will be reduced to case 1. A similar effect can be obtained using higher frequency heating or increased power of internal heat sources.

All sets of equations can be solved by typical numerical methods, including the well-known iteration algorithm.[6] According to this algorithm, each system is divided into two blocks. The first block includes one equation, which represents the equality of temperature gradient in the point l_{e2} to zero (last equation in systems in Equation 3.12, Equation 3.15, Equation 3.16, and Equation 3.21). The second block includes all remaining equations of the set.

For each kth approximation $\Delta^{[k]}$ of sought vector of parameters Δ^0, the appropriate approximation $l_{e2}^{[k]}$ of coordinate $l_2^0 = l_{e2}$ can be determined as a solution of the first equation that forces temperature gradients to be equal to zeroes:

$$\frac{\partial\theta\left(l_{e2}^{[k]}, \Delta^{[k]}\right)}{\partial l} = 0, \; k = 0, 1, 2, \ldots . \tag{3.23}$$

The next approximation, $\Delta^{[k+1]}$, is defined as a solution of the set of equations of the second block, where it is assumed that $l_{e2} = l_{e2}^{[k]}$. For example, in the case of Equation (3.16), this block consists of three equations:

$$\begin{cases} \theta(0, \Delta_1^{[k+1]}, \Delta_2^{[k+1]}) - \theta^* = -\varepsilon_{min}^{(2)}; \\ \theta(l_{e2}^{[k]}, \Delta_1^{[k+1]}, \Delta_2^{[k+1]}) - \theta^* = +\varepsilon_{min}^{(2)}; \\ \theta(1, \Delta_1^{[k+1]}, \Delta_2^{[k+1]}) - \theta^* = -\varepsilon_{min}^{(2)}. \end{cases} \tag{3.24}$$

After solving this set of equations, one can obtain $\Delta_1^{[k+1]}$, $\Delta_2^{[k+1]}$, and $\varepsilon_{\min}^{(2)}$ on each $(k + 1)$th step of the computational procedure ($k = 0, 1, 2, ...$).

The following set of three equations can be written regarding Equation (3.21):

$$\begin{cases} \theta(0, \Delta_1^{[k+1]}, \Delta_2^{[k+1]}, \Delta_3^{[k+1]}) - \theta^* = -\varepsilon; \\ \theta(l_{e2}^{[k]}, \Delta_1^{[k+1]}, \Delta_2^{[k+1]}, \Delta_3^{[k+1]}) - \theta^* = +\varepsilon; \\ \theta(1, \Delta_1^{[k+1]}, \Delta_2^{[k+1]}, \Delta_3^{[k+1]}) - \theta^* = -\varepsilon. \end{cases} \tag{3.25}$$

Here, $\Delta_1^{[k+1]}$, $\Delta_2^{[k+1]}$, and $\Delta_3^{[k+1]}$ are unknown parameters; however, the value of ε is prescribed. The sets of Equation (3.12) and Equation (3.15) are solved in a similar manner.

The described iteration procedure is repeated for predetermined initial value of $\Delta^{[0]}$ until value $\Delta^{[r]}$ coincides satisfactorily with $\Delta^{[r+1]}$ on some step under $k = r$. Convergence of this procedure is typically guaranteed only if the initial value of $\Delta^{[0]}$ is chosen properly. Values $\Delta^{[0]}$ can be determined with fair accuracy using the following algorithms.

If $\varepsilon = \varepsilon_{\min}^{(1)}$, then the value $\Delta^{[0]} = \Delta_1^{[0]}$ can be found as a duration of the heating interval with constant power $u(\varphi) \equiv U_{\max}$ that would be necessary to provide billet final average temperature equal to required average temperature θ^*. For $\varepsilon = \varepsilon_{\min}^{(2)}$, it is possible to assume that duration $\Delta_1^{[0]}$ of the first interval of two-stage process control is equal to the value $\Delta_1^{[0]}(\varepsilon_{\min}^{(1)}) \cong \Delta_1^{[0]}(\varepsilon_{\min}^{(2)})$ for one-stage control. The second time interval can be calculated approximately as[5]:

$$\Delta_2^{[0]} \cong -\frac{1}{\mu_1^2} \ln \frac{\mu_1^2 K^2(\mu_1) q_0}{4W_1(\xi)(1 - K(\mu_1))}, \tag{3.26}$$

where q_0 is time-average value of heat losses in Equation (1.32), which can be calculated, according to Equation (1.34); μ_1 is the first root of Equation (1.43). All other denotations are discussed in Chapter 1.

For $\varepsilon = \varepsilon_{\min}^{(2)}$, there is a simple exponential relationship between the value $\Delta_2^{[0]}$ and value of heat losses. As shown in Rapoport,[5] the following expression takes place:

$$l_{e2}^{[0]} \approx \frac{2}{3}, \tag{3.27}$$

i.e., the coordinate of point of maximum temperature does not depend on optimal process parameters.

The values $(\Delta_1^{[0]}, \Delta_2^{[0]})$ are defined as a solution of the set of equations in Equation (3.16). As a rule, these values can be used as initial approximation for solving the set of equations in Equation (3.15). In this case, values ε should occupy the following range: $\varepsilon_{\min}^{(2)} < \varepsilon < \varepsilon_{\min}^{(1)}$.

Every set of equations (Equation 3.21) is solved for decreasing sequence of values $\varepsilon = \varepsilon'$. These values are obtained by utilizing a three-stage control. Initial estimates $\Delta^{[0]}$ used for solution of such systems can be chosen by applying results obtained from a previous step of computation. During the first step of this computational procedure, the values $\Delta_1^{[0]}(\varepsilon_{\min}^{(2)})$, $\Delta_2^{[0]}(\varepsilon_{\min}^{(2)})$ are used as initial estimates of $\Delta_1^{[0]}, \Delta_2^{[0]}$ for $\varepsilon = \varepsilon' = \varepsilon_{\min}^{(2)} - d\varepsilon$.

Unknown values of minimaxes $\varepsilon_{\min}^{(1)}$ and $\varepsilon_{\min}^{(2)}$ could be excluded from calculational sets of equations in Equation (3.12) and Equation (3.16), replacing initial equations by their sums or differences. This allows reducing the system dimension, decreasing the number of sought unknowns, and simplifying calculations. Therefore, by combining the first two equations in Equation (3.12), it is possible to obtain the following set of two equations with respect to two unknowns — Δ_1^0 and l_{e2}:

$$\begin{cases} \theta(0,\ \Delta_1^0) + \theta(l_{e2},\ \Delta_1^0) - 2\theta^* = 0; \\ \dfrac{\partial\theta(l_{e2},\ \Delta_1^0)}{\partial l} = 0. \end{cases} \tag{3.28}$$

Combining the first equation with the second and the second equation with the third in Equation (3.16), it is possible to obtain the set of three equations with respect to three unknowns — Δ_1^0, Δ_2^0, and l_{e2} (instead of four equations in Equation 3.16):

$$\begin{cases} \theta(0,\ \Delta_1^0,\ \Delta_2^0) + \theta(l_{e2},\ \Delta_1^0,\ \Delta_2^0) - 2\theta^* = 0; \\ \theta(l_{e2},\ \Delta_1^0,\ \Delta_2^0) + \theta(1,\ \Delta_1^0,\ \Delta_2^0) - 2\theta^* = 0; \\ \dfrac{\partial\theta(l_{e2},\ \Delta_1^0,\ \Delta_2^0)}{\partial l} = 0. \end{cases} \tag{3.29}$$

Minimax values are calculated from appropriate equalities in Equation (3.12) and Equation (3.16) after solving sets of equations in Equation (3.28) and Equation (3.29). Therefore,

$$\varepsilon_{\min}^{(1)} = \theta(l_{e2},\ \Delta_1^0) - \theta^* \quad \text{or } \varepsilon_{\min}^{(1)} = \theta^* - \theta(0,\ \Delta_1^0)$$

and

$$\varepsilon_{\min}^{(2)} = \theta(l_{e2},\ \Delta_1^0,\ \Delta_2^0) - \theta^*\ ;\ \varepsilon_{\min}^{(2)} = \theta^* - \theta(0,\ \Delta_1^0,\ \Delta_2^0)\ , \text{or } \varepsilon_{\min}^{(2)} = \theta^* - \theta(1,\ \Delta_1^0,\ \Delta_2^0)\ .$$

Using the previously mentioned linear dependence of a set of equations from ε, it is possible to significantly simplify the computational procedure for three-stage optimal control when $\varepsilon < \varepsilon_{\min}^{(2)}$. For this purpose, it is necessary to "swap the roles" of values ε and Δ_3^0 in Equation (3.21). It should be assumed that value of ε is unknown instead of value Δ_3^0 (which is considered as fixed third constancy interval of optimal control function). Then, this new unknown can be excluded using linear combinations of equations in Equation (3.21) like Equation (3.28) and Equation (3.29). As a result, the set in Equation (3.21) can be rewritten with respect to three unknowns Δ_1^0, Δ_2^0 l_{e2} for preassigned value Δ_3^0:

$$\begin{cases} \theta(0, \Delta_1^0, \Delta_2^0, \Delta_3^0) + \theta(l_{e2}, \Delta_1^0, \Delta_2^0, \Delta_3^0) - 2\theta^* = 0; \\ \theta(l_{e2}, \Delta_1^0, \Delta_2^0, \Delta_3^0) + \theta(1, \Delta_1^0, \Delta_2^0, \Delta_3^0) - 2\theta^* = 0; \\ \dfrac{\partial\theta(l_{e2}, \Delta_1^0, \Delta_2^0, \Delta_3^0)}{\partial l} = 0. \end{cases} \tag{3.30}$$

Computation results[5] show that the value Δ_3^0 decreases with ε on the interval $[\varepsilon_{\min}^{(3)}; \varepsilon_{\min}^{(2)}]$ until $\Delta_3^0 = 0$ at $\varepsilon = \varepsilon_{\min}^{(2)}$ (Figure 3.15). In this case, instead of solving a sequence of sets of four equations (Equation 3.21) using small step $d\varepsilon$, it would be possible to transform an initial optimal control problem into a much simpler problem that would require solving the sequence of sets of equations in Equation (3.30) for sequence of values Δ_3^0 increasing from zero using appreciably small step $d\Delta_3^0$. The value ε can be calculated for the appropriate step from any of the first three equations in Equation (3.21). Obtained values of ε diminish with increasing sequence of prescribed values of Δ_3^0. Here, the approach similar to solution of the set in Equation (3.21) is used.

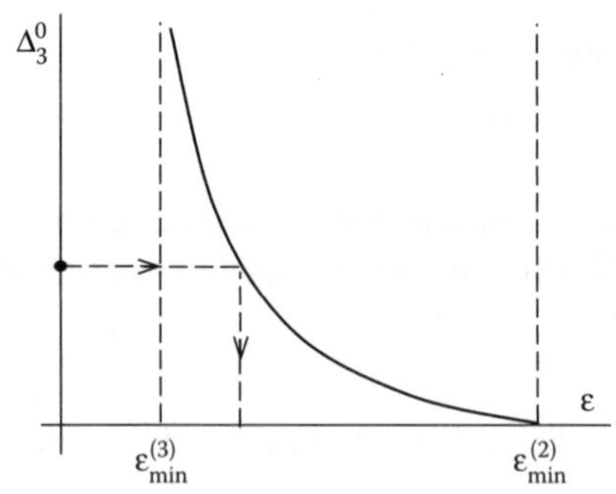

FIGURE 3.15 Duration of the third stage of heating cycle as a function of required heating accuracy.

The finite number of terms of series in Equation (2.29) is always used in governing equation sets. An important question arises at this point with regard to an accuracy of obtained results that depends on computational accuracy of infinite series.

Let us assume that $\tilde{\Delta}^0$ is the exact solution of a set of equations obtained for finite number $S < \infty$ of the series terms in Equation (2.29). Let us also assume that an error of temperature field computation, δ, is caused by truncation. Then, instead of basic equations in Equation (3.7), the following expressions hold true at points l_j^0:

$$\left| \left| \theta(l_j^0, \tilde{\Delta}^0) - \theta^* \right| - \varepsilon \right| \leq \delta, \; j = 1, 2, ..., R \; . \qquad (3.31)$$

It can be shown[5] that, if the value δ is appreciably small, then the variation of the exact solution Δ^0 of the set of equations that is obtained with all terms of the series in Equation (2.29) would differ from $\tilde{\Delta}^0$ for the value that is no greater than 2δ:

$$\left| \Delta_i^0 - \tilde{\Delta}_i^0 \right| \leq 2\delta \; for \; all \; i = \overline{1, N} \; . \qquad (3.32)$$

For convergent Fourier series in Equation (2.29), $\delta \to 0$ under $S \to \infty$. Therefore, $\tilde{\Delta}^0$ converges to Δ^0 at $S \to \infty$ according to Equation (3.32). Experience of practical computation shows[5] that for $S \geq 10 \ldots 20$, a quite high accuracy is achieved for computation of temperature distribution. Maximum error does not exceed 0.5%. Therefore, Δ^0 can be calculated with the same accuracy in all practicable cases. Without considerable difficulties, the required number of series terms in the set of equations can be utilized using modern computers.

3.5 APPLICATION OF THE SUGGESTED METHOD TO MODEL PROBLEMS BASED ON TYPICAL COST FUNCTIONS

An engineering technique for solving general time-optimal control problems for induction heating of metals has been discussed in previous chapters. It has been shown that, with respect to the mathematical model (Equation 1.27 through Equation 1.35), time-optimal control problems can be reduced to the nonstandard problem of mathematical programming (Equation 2.34). Naturally, one might be interested whether it is possible to use the suggested method for other types of different typical optimization problems discussed in Section 2.9.

First of all, let us notice that the previously described computational technique remains unchanged as applied to the optimization problem (Equation 2.33) for steady-state continuous heating. Such heating processes are described by the same mathematical model (Equation 1.27 through Equation 1.35) with substitution of the variable (Equation 2.6). However, lengths of inductor sections, Δ_i^*, should be considered as sought parameters instead of time intervals, Δ_i, of optimal control for static induction heating. Moreover, all properties of spatial temperature distributions at the end of optimal static heating relate to temperature distributions over billet cross-section at the exit of a continuous heater. Therefore, all expressions for $\theta(l,\Delta)$ in the sets of equations in Equation (3.12), Equation (3.15), Equation (3.16), and Equation (3.21) should be changed for $\theta(l,\Delta^*/V)$ by simple substitution $\Delta_i = \Delta_i^* / V,\ i = \overline{1,N}$.

The method can be applied to the optimization problems for heating when maximum accuracy is required. Such problems are formulated using Equation (2.35), Equation (2.36), and Equation (2.37). As shown in Section 2.9.1, they can be reduced to the appropriate time-optimal problems under the absence of excessive heating time.

A multicoil continuous heater consists of a number of individual coils. Instead of using the lengths of coils of the induction heater, a voltage applied to each coil can be used as control input in Equation (2.38). In this case, expression $\theta(l,U)$ should be used as a representation of radial temperature profile at the exit end of the multicoil induction heater instead of $\theta(l,\Delta^*/V)$ in Equation (2.37). Nevertheless, as investigations and computer modeling evaluation show, all properties of final temperature distribution, governing sets of equations, and computational procedures described in Section 3.1 through Section 3.4 are kept identical to the time-optimal problem for static heating processes. It will only be required to substitute analytical expressions for $\theta(l,U^0)$ instead of $\theta(l,\Delta^0)$ (see Section 2.9.1) in governing sets of equations. In addition, unknowns $\Delta_1^0, \Delta_2^0, \ldots, \Delta_N^0$ should be changed to $U_1^0, U_2^0, \ldots, U_N^0$.

For the problem described by Equation (2.42) that represents optimization of minimum power consumption during induction heating, the described method will have various forms (see Section 2.9.2). In this case, the general optimal control algorithm is the same and time lengths of alternating intervals of heating and soaking remain unknown. Therefore, the vector Δ should be defined. In addition, basic properties in Equation (3.7) and Equation (3.8) remain valid for the temperature distribution at the end of the optimal process. However, alternance property (Equation 3.9) is violated for such distribution, and the rule in Equation (3.3) will not be appropriate for the definition of number of constancy intervals N.[2,5]

Therefore, if energy saving is a goal of the optimal control algorithm, then it will not be possible to write the set of equations by simply applying conclusions regarding temperature profile $\theta(l,\Delta^0)$ (appropriate validation was made in Section 3.3). The difference between minimum power consuming and time-optimal control algorithms (assuming similar required heating accuracy) has been discussed in Section 2.9.2. Thus, in a case when $\varepsilon = \varepsilon_{\min}^{(1)}$, the time-optimal control algorithm would not be optimal for power cost function. Energy consumption can be calculated as:

$$I_E = U_{\max}\Delta_1. \tag{3.33}$$

It is possible to reduce I_E by decreasing the first (and in this case single) control interval in comparison with duration of time-optimal control. To obtain required heating accuracy, it is necessary to apply the second interval that represents temperature soaking. During soaking stage, there would be no energy applied to induction coil: $u(\varphi) \equiv 0$. There is a unique correlation between the value of $\varepsilon_{\min}^{(1)}$ and maximum value Δ_2. If that value would be exceeded, then the inverse effect would take place: temperature nonuniformity will worsen due to surface heat losses. This situation can take place because, during the soaking stage, the surface temperature falls lower than minimum admissible level. In this particular case, the following expressions could be written for $N = 2$ according to basic rule $R = N$ in Equation (3.7) and Equation (3.10):

$$R = N = 2 \ for \ \varepsilon = \varepsilon_{\min}^{(1)} > \varepsilon_{\min}^{(2)};\ M_{\max} = N = 2\ .$$

As one can see, if $\varepsilon = \varepsilon_{\min}^{(1)}$ and $N = 2$, there is only one possible final temperature distribution for a minimum power consuming process. Figure 3.16 shows this temperature profile. As this takes place, because additional energy is consumed during following odd intervals, a two-stage control function would be optimal with respect to criterion I_E.

The control that optimizes energy consumption of an induction heater appreciably differs from time-optimal control for $\varepsilon = \varepsilon_{\min}^{(1)}$. A minimum power consumption process consists of two stages (compared to a single-stage control for a time-optimal problem). Therefore, the condition in (Equation 3.3) would not be valid in this case. The final temperature distribution is characterized by reaching minimum admissible temperatures in the billet center and on the billet surface. The final maximum temperature does not reach its utmost admissible value (see Figure 3.16). Thus, the alternance condition in Equation (3.9) that has been established for time-optimal processes is violated.

It is proper to remember at this point that, at the end of a time-optimal process, the maximum temperature deviations of final temperature from the required one are reached with different signs at the points $l = 0$ and $l = l_{e2}$ (see Figure 3.6).

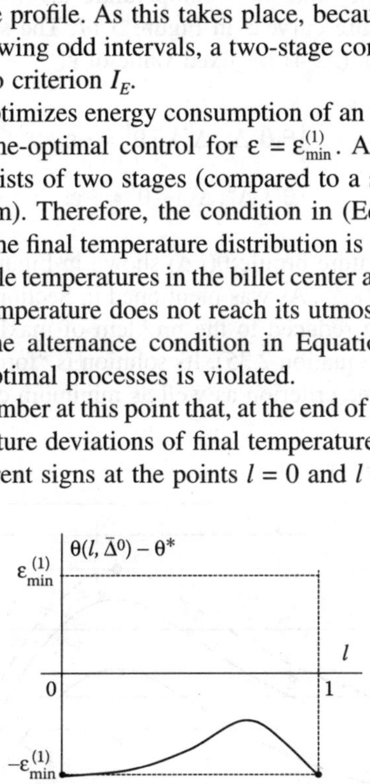

FIGURE 3.16 Temperature profile at the end of minimum power consumption process for $\varepsilon = \varepsilon_{\min}^{(1)}$.

As a result, according to the temperature curve in Figure 3.16, for $\varepsilon = \varepsilon_{\min}^{(1)}$, the governing set of equations can be written as:

$$\begin{cases} \theta(0, \bar{\Delta}_1^0, \bar{\Delta}_2^0) - \theta^* = -\varepsilon_{\min}^{(1)}; \\ \theta(1, \bar{\Delta}_1^0, \bar{\Delta}_2^0) - \theta^* = -\varepsilon_{\min}^{(1)}. \end{cases} \tag{3.34}$$

Here, time intervals of power cost optimal control are denoted as $\bar{\Delta}_1^0$ and $\bar{\Delta}_2^0$ in contrast to parameters of the time-optimal process.

The given set of two equations can be solved with respect to two unknowns, $\bar{\Delta}_1^0, \bar{\Delta}_2^0$, under prescribed value of $\varepsilon_{\min}^{(1)}$ by substitution of similar expressions $\theta(l, \bar{\Delta}_1, \bar{\Delta}_2)$. Computational techniques described in Section 3.4 can be used here.

For prescribed values ε, such as $\varepsilon_{\min}^{(2)} < \varepsilon < \varepsilon_{\min}^{(1)}$, it is also possible to pass from time-optimal control to power cost optimal control. The results will be similar in both cases. For a particular value of ε, the optimal control function consists of two constancy intervals. Temperature distribution $\theta(l, \bar{\Delta}_1^0, \bar{\Delta}_2^0)$ corresponds to the temperature curve 2 in Figure 3.17. The set of equations will be the same as in Equation (3.34) for fixed value of ε:

$$\begin{cases} \theta(0, \bar{\Delta}_1^0, \bar{\Delta}_2^0) - \theta^* = -\varepsilon; \\ \theta(1, \bar{\Delta}_1^0, \bar{\Delta}_2^0) - \theta^* = -\varepsilon. \end{cases} \tag{3.35}$$

The unique temperature profile $\theta(l, \Delta)$ shown in Figure 3.8a (curve 2) corresponds to the value $\varepsilon = \varepsilon_{\min}^{(2)}$. As was mentioned in Section 3.1, for this particular case, a problem can be reduced to the problem of maximum heating accuracy for two-stage control (Equation 2.35). Its solution is "forced" to be optimal with respect to minimum time criterion as well as minimum energy consumption.

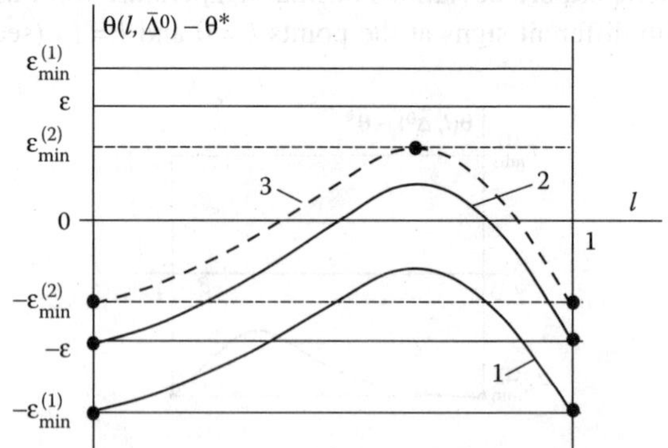

FIGURE 3.17 Temperature profiles at the end of minimum power consumption process. 1: For $\varepsilon = \varepsilon_{\min}^{(1)}$; 2: for $\varepsilon_{\min}^{(2)} < \varepsilon < \varepsilon_{\min}^{(1)}$; 3: for $\varepsilon = \varepsilon_{\min}^{(2)}$.

For the case $\varepsilon_{\min}^{(3)} < \varepsilon < \varepsilon_{\min}^{(2)}$, we would restrict the discussion to the temperature profiles that are typical for induction heating and correspond to the two-stage control (Section 3.3). There is a general analogy to the previously described problem of time-optimal control (Figure 3.13). Maximum deviations of final temperature from desired one can be reached in three possible points on the coordinate axis. These points are $l = 0$, $l = 1$, and $l = l_{e2}$ — i.e., $R \le 3$ in Equation (3.7). It follows immediately that $N = 3$.

On the one hand, $N \ge 3$ under condition $\varepsilon_{\min}^{(3)} < \varepsilon < \varepsilon_{\min}^{(2)}$. However, on the other hand, if $R \le 3$, then $N < 4$ according to the basic condition $R = N$ and therefore $R = N = 3$. The only possible shape of final temperature distribution in this case is shown in Figure 3.13a (curve 1). The set of governing equations can be written as Equation (3.21). The solution of this system is $\Delta^0 = \bar{\Delta}^0$. This solution uniquely defines the optimal control with respect to minimum time criterion and minimum energy consumption.

Therefore, obtained control algorithms are optimal respectively to both considered criteria when value ε satisfies the following requirement: $\varepsilon_{\min}^{(3)} < \varepsilon \le \varepsilon_{\min}^{(2)}$. The solution of the maximum heating accuracy problem for $\varepsilon = \varepsilon_{\min}^{(3)}$ was analyzed in Section 3.3 and Section 3.4.

3.6 EXAMPLES

3.6.1 Solution of Time-Optimal Control Problem

As an example, let us consider a typical problem of time-optimal control for induction heating of cylindrical billets. This process is described by a heterogeneous one-dimensional equation of heat conductivity (Equation 1.27 through Equation 1.35) for radial axisymmetric temperature field $\theta(l,\varphi)$ (according to condition $\Gamma = 1$).

Applying the previously described method, computations of optimal heating modes have been conducted with respect to typical relative values of final temperature θ^* and parameter ξ under $\theta_0 = 0$ in Equation (2.28) and Equation (2.29) for required heating accuracy ε and surface heat losses, estimated by value of *Bi* criterion in Equation (1.34). Equations for final temperature distribution $\theta(l,\Delta)$ in the form of Equation (2.28) and Equation (2.29) were used in all sets of equations. For $\Gamma = 1$, Bessel zero order functions $J_0(\mu_n l)$ should be taken as $K(\mu_n l)$ (see Chapter 1).

As was shown earlier, if surface heat losses are appreciable, then shape of the radial temperature distribution $\theta(l,\Delta^0)$ at the end of an optimal process is not changed, regardless of that it was obtained when passing from two-stage control for $\varepsilon = \varepsilon_{\min}^{(2)}$ to three-stage control for $\varepsilon_{\min}^{(3)} \le \varepsilon < \varepsilon_{\min}^{(2)}$ (see Section 3.3 and Section 3.4). In this case, a computational algorithm considered earlier can be used for all ε that satisfy condition $\varepsilon_{\min}^{(3)} \le \varepsilon \le \varepsilon_{\min}^{(1)}$ (see Section 3.4).

When surface thermal losses are not as appreciable as was considered in the preceding case, computations become complicated but only if it is necessary to provide the heating accuracy $\varepsilon < \varepsilon_{\min}^{(2)}$. For such values of ε, several variants of

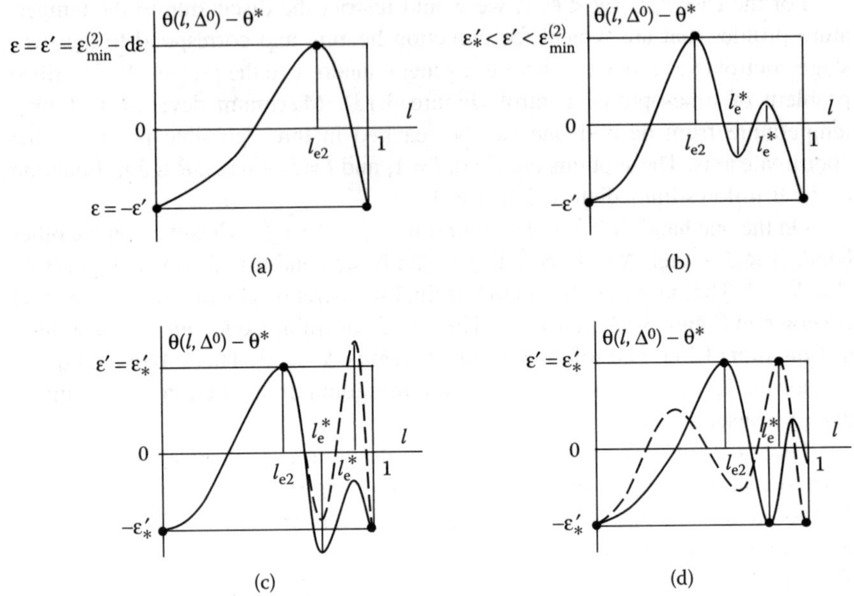

FIGURE 3.18 Variants of final temperature profiles for different values of heating accuracy (under low level of heat losses).

temperature profile $\theta(l,\Delta^0)$ exist. However, in each particular case, it is not too difficult to reveal a unique profile variant from all possible ones. The choice of particular temperature profile is performed in the course of computational procedure (Section 3.4) on the basis of the continuous nature of dependencies Δ^0 and $\theta(l,\Delta^0)$ on ε.

Let us assume that values $\varepsilon = \varepsilon' < \varepsilon_{\min}^{(2)}$ are slightly different from $\varepsilon_{\min}^{(2)}$. Due to previously mentioned continuity for any values of $Bi > 0$ (see Figure 3.18a), the shape of the curve $\theta(l,\Delta^0(\varepsilon'))$ is unchanged in comparison with $\theta(l,\Delta^0(\varepsilon_{\min}^{(2)}))$. Then, for such values of ε', the computational algorithm is always reduced to the solution of sequence of sets of equations (Equation 3.21) for values $\varepsilon = \varepsilon' = \varepsilon_{\min}^{(2)} - kd\varepsilon$, $k = 1, 2, \ldots$ decreasing by small step $d\varepsilon$.

In a case of sufficiently large surface thermal losses, the shape of the curve $\theta(l,\Delta^0(\varepsilon))$ and the equations in Equation (3.21) are unchanged within the whole interval $\varepsilon_{\min}^{(3)} \le \varepsilon \le \varepsilon_{\min}^{(2)}$.

For small values of Bi, there are possible variations of the shape of the curve $\theta(l,\Delta^0(\varepsilon'))$ due to appearance (according to a rule in Equation 3.10) of additional extremum points (Figure 3.18b). In this case, the system (Equation 3.21) remains unchanged while ε' decreases until inadmissible inequality $|\theta(l_e^*, \Delta^0(\varepsilon_*')) - \theta^*| > \varepsilon_*'$ becomes true for some value $\varepsilon = \varepsilon_*' < \varepsilon_{\min}^{(2)}$, at least for one of the new appearing extremum points $l_e^* \in (0, 1)$. Naturally, this point does not belong to the set of points $l_1^0 = 0$, $l_2^0 = l_{e2}$, $l_3^0 = 1$, defined by the system in Equation (3.21) (Figure 3.18c). Final temperature profile varies while ε' decreases. Two examples of these variations are shown by continuous and dotted lines in Figure 3.18c through Figure 3.18d.

The new temperature profile $\theta(l,\Delta^0)$ corresponds to value $\varepsilon = \varepsilon'_*$. This curve should be identical to profile $\theta(l,\Delta^0(\varepsilon'_* + d\varepsilon))$ on the previous step. In addition, it should still satisfy the basic relationship in Equation (3.9) at $R = N = 3$. Also, it should correspond to the choice l_e^* as one of new points, l_i^0, in Equation (3.9).

Instead of Equation (3.21), the following set of equations corresponds to the temperature profile shown in Figure 3.18c (solid line):

$$\begin{cases} \theta(0,\Delta_1^0,\Delta_2^0,\Delta_3^0) - \theta^* = -\varepsilon'_*; \\ \theta(l_{e2},\Delta_1^0,\Delta_2^0,\Delta_3^0) - \theta^* = +\varepsilon'_*; \\ \theta(l_e^*,\Delta_1^0,\Delta_2^0,\Delta_3^0) - \theta^* = -\varepsilon'_*; \\ \dfrac{\partial\theta(l_{e2},\Delta_1^0,\Delta_2^0,\Delta_3^0)}{\partial l} = \dfrac{\partial\theta(l_e^*,\Delta_1^0,\Delta_2^0,\Delta_3^0)}{\partial l} = 0. \end{cases} \tag{3.36}$$

Then, final optimal temperature distribution will take a shape shown in Figure 3.18d (solid line).

In case of the temperature profile indicated by a dashed curve in Figure 3.18c, it is possible to write the following set of equations:

$$\begin{cases} \theta(0,\Delta_1^0,\Delta_2^0,\Delta_3^0) - \theta^* = -\varepsilon'_*; \\ \theta(l_e^*,\Delta_1^0,\Delta_2^0,\Delta_3^0) - \theta^* = +\varepsilon'_*; \\ \theta(1,\Delta_1^0,\Delta_2^0,\Delta_3^0) - \theta^* = -\varepsilon'_*; \\ \dfrac{\partial\theta(l_e^*,\Delta_1^0,\Delta_2^0,\Delta_3^0)}{\partial l} = 0. \end{cases} \tag{3.37}$$

The final optimal profile will take a shape shown in Figure 3.18d (dashed line).

Updating systems of equations can be repeated several times in the process of further ε' decreasing from $\varepsilon_{\min}^{(2)}$ until $\varepsilon_{\min}^{(3)}$. Thus, basic rules in Equation (3.9) and Equation (3.10) can be used with respect to each step. The shape of the curve $\theta(l,\Delta^0)$ and appropriate set of equations can be explicitly defined in computational procedure for all values $\varepsilon \in (\varepsilon_{\min}^{(3)}, \varepsilon_{\min}^{(2)})$ at any $Bi > 0$. As examples, some computation results can be found in Figure 3.19 through Figure 3.22.

For prescribed values of parameter ξ and final temperature θ^*, the shape of temperature curve $\theta(l,\Delta^0(\varepsilon))$ is defined uniquely by the value of criterion *Bi* in Equation (1.34). The values of maximum achievable heating accuracy for considered control inputs also depend strongly on the value of *Bi* (see Figure 3.19 and Figure 3.20).

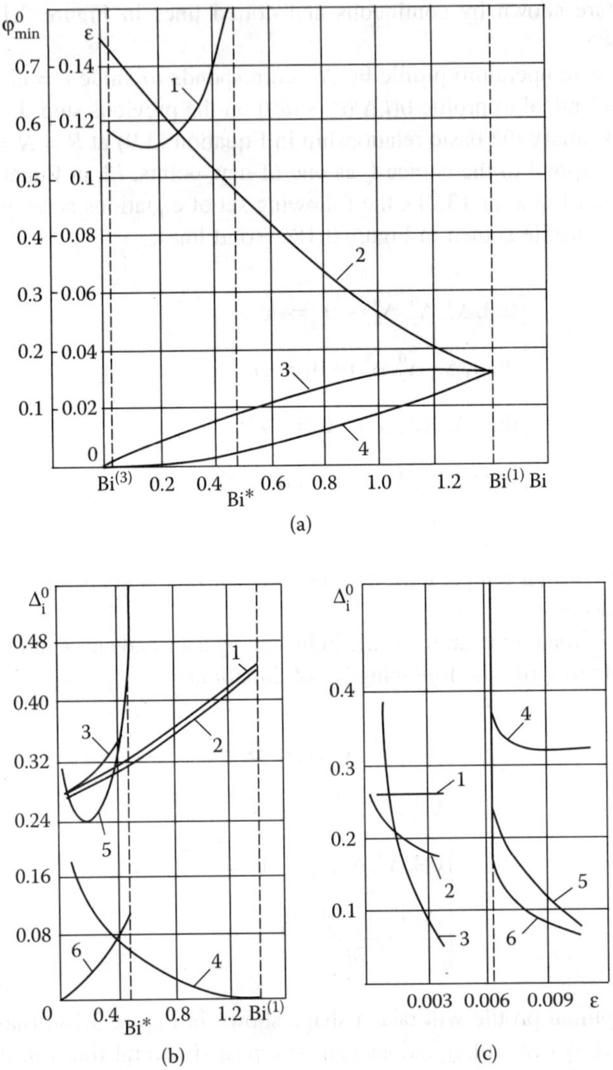

FIGURE 3.19 Computation results for optimal control process parameters. a: Dependencies of maximum achievable heating accuracy $\varepsilon_{\min}^{(N)}$ and optimal heating time $\varphi_{\min}^{0}$ on value Bi; 1: $\varphi_{\min}^{0}$ $(\varepsilon_{\min}^{(3)})$; 2: $\varepsilon_{\min}^{(1)}$; 3: $\varepsilon_{\min}^{(2)}$; 4: $\varepsilon_{\min}^{(3)}$ b: Dependencies of optimal process parameters Δ_i^0 on value Bi; 1: $\Delta_1^0(\varepsilon_{\min}^{(1)})$; 2: $\Delta_1^0(\varepsilon_{\min}^{(2)})$; 3: $\Delta_1^0(\varepsilon_{\min}^{(3)})$; 4: Δ_2^0 for $\varepsilon_{\min}^{(2)}$; 5: Δ_2^0 for $\varepsilon_{\min}^{(3)}$; 6: $\Delta_3^0 \times 5$ for $\varepsilon_{\min}^{(3)}$. c: Dependencies of optimal process parameters Δ_i^0 on heating accuracy $\varepsilon \in (\varepsilon_{\min}^{(3)}, \varepsilon_{\min}^{(2)})$; 1: for Δ_1^0, $Bi = 0.1$; 2: for Δ_2^0, $Bi = 0.1$; 3: for $\Delta_3^0 \times 10^2$, $Bi = 0.1$; 4: for Δ_1^0, $Bi = 0.5$; 5: $\Delta_3^0 \times 10$, $Bi = 0.5$; 6: Δ_2^0, $Bi = 0.5$.

All variation ranges of Bi can be divided into several intervals by its characteristic values $Bi^{(3)}$, Bi^*, and $Bi^{(1)}$. These values depend only on ξ and θ^*. It should be pointed out that $0 < Bi^{(3)} < Bi^* < Bi^{(1)}$. Each of these values corresponds to a

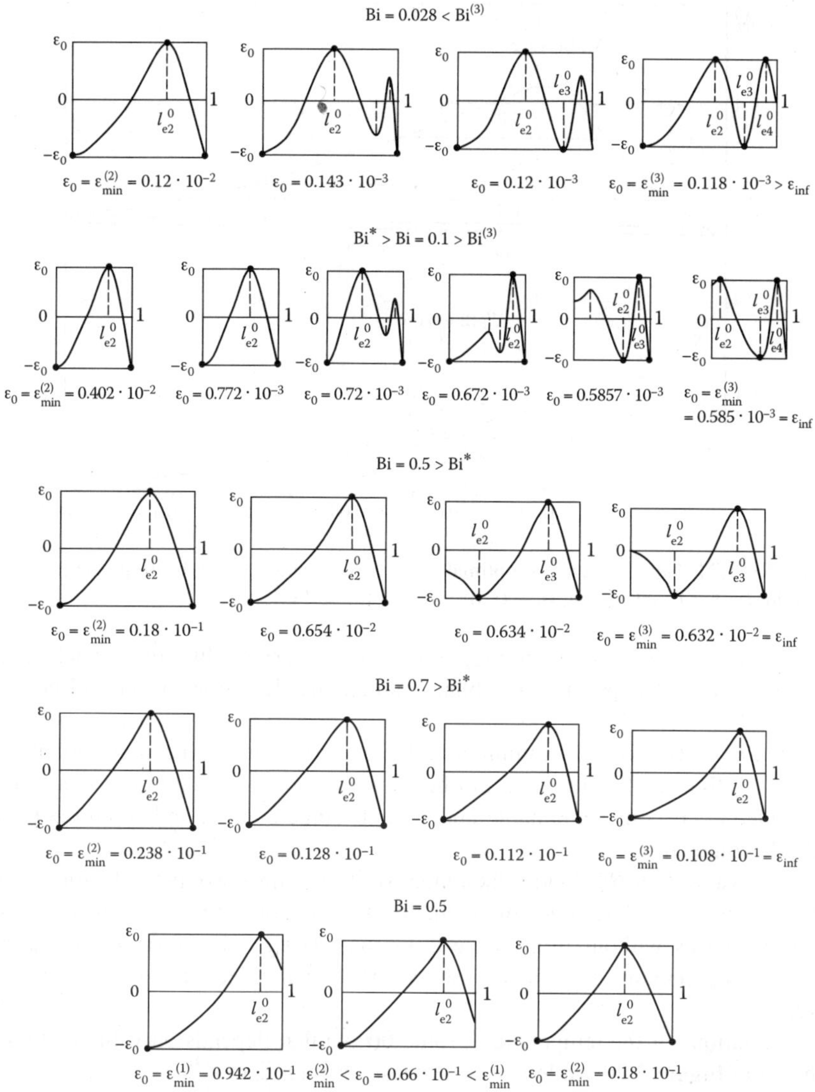

FIGURE 3.20 Temperature distribution $\theta(l,\Delta^0) - \theta^*$ across workpiece thickness/radius at the end of time-optimal heating process for $\Gamma = 1$, $\xi = 4$, $\theta_0 = 0$, and $\theta^* = 0.5$.

certain shape of a curve $\theta(l,\Delta^0)$ for utmost heating accuracy ε_{inf} in Equation (3.1) and to an appropriate set of equations. Characteristic values $Bi^{(3)}$, Bi^*, and $Bi^{(1)}$ can be considered as additional unknown variables in a set of equations. These variables can be determined during the solving process.[5]

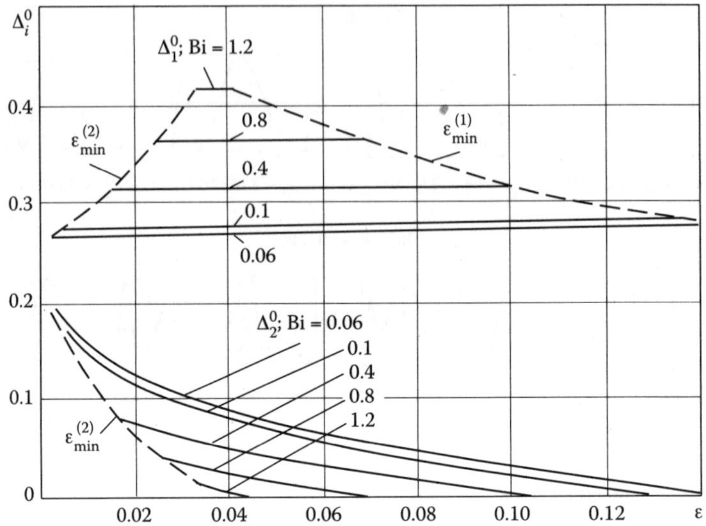

FIGURE 3.21 Dependencies of optimal process parameters Δ_i^0 on heating accuracy and heat losses for $\Gamma = 1$, $\xi = 4$, $\theta_0 = 0$, $\theta^* = 0.5$, and $\varepsilon \in [\varepsilon_{min}^{(2)}, \varepsilon_{min}^{(1)}]$.

For $Bi \in [0, Bi^{(3)})$, condition $\varepsilon_{min}^{(3)} > \varepsilon_{inf}$ is correct. For values of thermal losses less than $Bi^{(3)}$, it is possible to obtain temperature deviation that would be less than $\varepsilon_{min}^{(3)}$.

For $Bi \in [Bi^{(3)}, Bi^{(1)})$, the minimax value $\varepsilon_{min}^{(3)}$ coincides with extreme attainable value ε_{inf}. Starting from the level of heat losses, corresponding to the value $Bi^{(1)}$, the value ε_{inf} becomes attainable under single-stage control, and condition $\varepsilon_{inf} = \varepsilon_{min}^{(1)}$ becomes true.

The value $Bi = Bi^*$ divides the range $[Bi^{(3)}, Bi^{(1)})$ into two intervals: for values $Bi < Bi^*$, the utmost accuracy of heating ε_{inf} is attainable. If $Bi > Bi^*$, then process time increases without limit while ε approaches ε_{inf}. Therefore, limit heating accuracy ε_{inf} remains unattainable (see Section 3.3 and Section 3.4, and Equation 3.22).

Variations of the temperature profile $\theta(l, \Delta^0)$ that depends on value of Bi are shown in Figure 3.20 while $\varepsilon = \varepsilon_0$ decreases from $\varepsilon_{min}^{(2)}$ to $\varepsilon_{min}^{(3)}$.

Temperature distribution $\theta(l, \Delta^0)$ that corresponds to $Bi^{(3)}$ has one additional critical point, l_j^0, in comparison with a minimal number of points required according to the rule in Equation (3.8). This means that the number of these points, R, becomes greater by one as it is for $Bi^{(1)}$ (Figure 3.23b). Therefore, the additional equation should be incorporated into the set of equations. The values of $Bi^{(3)}$ or $Bi^{(1)}$ can be considered as corresponding additional unknowns. The value of Bi^* can be defined with desired precision as maximum value of all Bi for which a temperature distribution $\theta(l, \Delta^0)$ has the shape of a curve 1 (Figure 3.23b) (assuming condition $\varepsilon = \varepsilon_{min}^{(3)} = \varepsilon_{inf}$). The described computational algorithm allows defining values $Bi^{(3)}$, Bi^*, and $Bi^{(1)}$ and all unknown

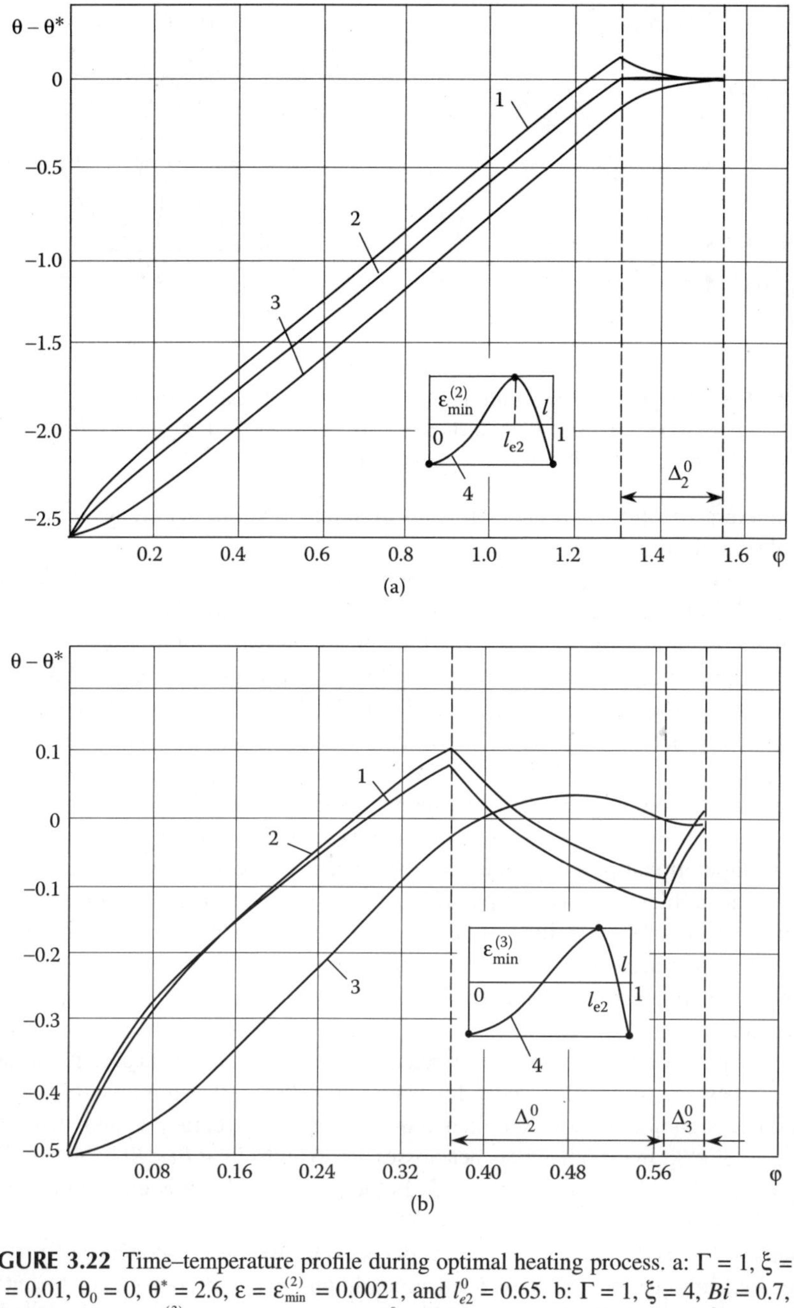

FIGURE 3.22 Time–temperature profile during optimal heating process. a: $\Gamma = 1$, $\xi = 5$, $Bi = 0.01$, $\theta_0 = 0$, $\theta^* = 2.6$, $\varepsilon = \varepsilon_{min}^{(2)} = 0.0021$, and $l_{e2}^0 = 0.65$. b: $\Gamma = 1$, $\xi = 4$, $Bi = 0.7$, $\theta_0 = 0$, $\theta^* = 0.5$, $\varepsilon = \varepsilon_{min}^{(3)} = \varepsilon_{inf} \cong 0.011$, and $l_{e2}^0 = 0.84$; 1: for $\theta(1,\varphi)$; 2: for $\theta(l_{e2}^0,\varphi)$; 3: for $\theta(0,\varphi)$; 4: for $\theta(l,\Delta^0) - \theta^*$.

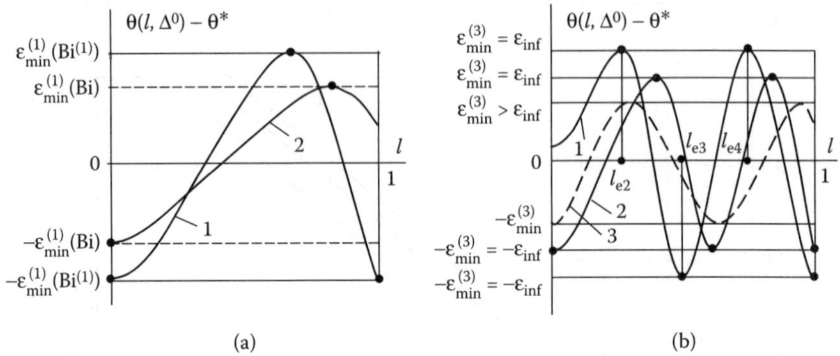

FIGURE 3.23 Optimal temperature profiles at the end of heating for $\varepsilon = \varepsilon_{min}^{(1)}$ and $\varepsilon = \varepsilon_{min}^{(3)}$. a: For $\varepsilon = \varepsilon_{min}^{(1)}$; 1: $Bi = Bi^{(1)}$; 2: $Bi < Bi^{(1)}$. b: For $\varepsilon = \varepsilon_{min}^{(3)}$; 1: $\varepsilon_{min}^{(3)} = \varepsilon_{inf}$; $Bi^{(3)} < Bi < Bi^{(1)}$; 2: $\varepsilon_{min}^{(3)} = \varepsilon_{inf}$; $Bi = Bi^{(3)}$; 3: $\varepsilon_{min}^{(3)} > \varepsilon_{inf}$; $Bi < Bi^{(3)}$.

parameters of optimal process for any required heating accuracy within the whole range of surface heat losses.

Dependence of optimal heating time on Bi for $\varepsilon = \varepsilon_{min}^{(3)}$ (Figure 3.19a, curve 1) clearly demonstrates the fact that it is impossible to obtain limit heating accuracy $\varepsilon_{inf} = \varepsilon_{min}^{(3)}$ for $Bi > Bi^*$. The relationship between minimaxes $\varepsilon_{min}^{(1)}$, $\varepsilon_{min}^{(2)}$, and $\varepsilon_{min}^{(3)}$ as a complex function of thermal losses is illustrated by curves 2 through 4 in Figure 3.19a. Figure 3.19b, Figure 3.19c, and Figure 3.21 show dependencies of optimal process parameters Δ_i^o on value Bi for various heating accuracy values $\varepsilon \in [\varepsilon_{min}^{(3)}, \varepsilon_{min}^{(1)}]$. Figure 3.20 shows variations of the temperature profiles while ε decreases from $\varepsilon_{min}^{(2)}$ to $\varepsilon_{min}^{(3)}$ for different values of Bi. The temperature field evolving over optimal heating process time is shown in Figure 3.22.

Computational procedure becomes simpler if it is necessary to solve only a particular problem — for example, if it is necessary to calculate the time-optimal heating process for prescribed value ε and fixed value Bi. In this case, calculations are performed according to the procedure described earlier. As has been mentioned, the simplest possible cases are at the same time the most widespread in practice.

Two polar cases have been described in Section 3.3. In the first case, the surface heat losses are so small (usually for values $Bi \le Bi^{(3)}$), that minimax $\varepsilon_{min}^{(3)} \ge \varepsilon_{inf}$ becomes significantly smaller than required heating accuracy ε and therefore $\varepsilon_{min}^{(2)} \le \varepsilon \le \varepsilon_{min}^{(1)}$. In the other case, heat losses are so high ($Bi^* < Bi < Bi^{(1)}$), that $\varepsilon_{min}^{(3)} = \varepsilon_{inf}$ and $\varepsilon_{min}^{(3)} \le \varepsilon \le \varepsilon_{min}^{(2)}$ or even $\varepsilon < \varepsilon_{min}^{(3)}$. In the first case, the problem is reduced to solving sets of equations in Equation (3.12) and Equation (3.16) and, if necessary, Equation (3.15). In the second case, it is necessary to solve a sequence of sets of equations in Equation (3.21). As soon as condition $\varepsilon < \varepsilon_{min}^{(3)}$ becomes true, the initial problem is unsolvable.

As an example, let us consider a time-optimal process of induction heating of titanic alloy cylindrical billets before hot forming. Diameter of the billet is 0.54 m, applied frequency is 50 Hz, and required final temperature is 1050°C.

Let us assume that the maximal surface density of heating power is about 106 kW/m². The average values of electromagnetic and thermophysical parameters are:

$$\rho = 180 \cdot 10^{-8} \text{Ohm} \cdot m \ , \ \lambda = 14 \frac{W}{m \cdot {}^0C} \ , \ a = 4.34 \cdot 10^{-6} \frac{m^2}{s} \ ,$$

where

ρ is electrical resistivity.
λ is thermal conductivity of the metal.
a is temperature conductivity of heated material.

Let us assume that the average value of convection heat transfer coefficient during the heating is equal to 35 to 40 W/(m² · °C), which corresponds to the value of $Bi \cong 0.7$ for surface thermal losses. The heating process is described precisely enough by model (Equation 1.27 through Equation 1.35). Using the system of relative units discussed in Chapter 1, we can obtain $\theta^* = 0.5$ and $\xi = 4$ for the case when initial temperature equals ambient temperature of 30°C.

For considered initial data and calculated values of θ^* and ξ, the sought parameters of a time-optimal heating process can be obtained using computation results presented in Figure 3.19 through Figure 3.21. Computational procedure has been performed according to the algorithm described previously. For values $\xi = 4$, $\theta^* = 0.5$, and $Bi = 0.7$ (Figure 3.19), the following results can be obtained:

$$\varepsilon_{\min}^{(1)} = 0.075;\ \varepsilon_{\min}^{(2)} = 0.0246;\ \varepsilon_{\min}^{(3)} = \varepsilon_{\inf} \cong 0.01;\ Bi^{(3)} = 0.031;$$
$$Bi^* \cong 0.460;\ Bi^{(1)} = 1.400.$$

Therefore, $Bi^* < Bi = 0.7 < Bi^{(1)}$.

Results can be rewritten in absolute units as:

$$\varepsilon_{\min}^{(1)} = 153°C;\ \varepsilon_{\min}^{(2)} = 50.3°C;\ \varepsilon_{\min}^{(3)} = \varepsilon_{\inf} \cong 21°C.$$

Corresponding values of optimal process intervals of heating and temperature soaking are presented in Table 3.1 for minimax errors. The minimax $\varepsilon_{\min}^{(1)} = 153°C$ defines nonuniformity of heating, which is typically unacceptable for the majority of hot forming applications where the higher accuracy is usually required — i.e., $\varepsilon < \varepsilon_{\min}^{(1)}$.

If $\varepsilon_{\min}^{(2)} = 50.3°C \le \varepsilon < 153°C$, then it is necessary to provide two-stage optimal control, i.e., "heating–temperature soaking" mode. However, in the case of $\varepsilon_{\inf} = \varepsilon_{\min}^{(3)} = 21°C \le \varepsilon < \varepsilon_{\min}^{(2)} = 50.3°C$, it would be necessary to apply a three-stage control process that would include heating, soaking, and reheating process stages.

Utmost heating accuracy coincides with minimax $\varepsilon_{\min}^{(3)}$ and it appears unattainable under conditions $Bi = 0.7 > Bi^* = 0.46$. Therefore, for $\varepsilon \le 21°C$, the

TABLE 3.1
Results of Time-Optimal Heating for Titanic Alloy Billets

N	$\varepsilon_{min}^{(N)}$, °C	Δ_1^0, min	Δ_2^0, min	Δ_3^0, min
1	153.0	104.0	—	—
2	50.3	105.0	11.6	—
3	23.0	102.0	23.5	8.0

problem of heating with required accuracy is unsolvable. To obtain smaller deviations from θ^* (for example, less than 21°C), it would be necessary to change parameters of induction installation. For example, refractory would need to be installed as a thermal insulation that would provide smaller values of Bi. Typical technological requirements define the value of ε higher than $\varepsilon_{inf} \cong 21$°C, but quite close to it. Then, an optimal process can be implemented using three-stage control (Figure 3.22b); the temperature distribution for this case is shown in Figure 3.20.

As the next example, let us consider the process of heating of aluminum alloy cylindrical billets prior to hot forming. Input data: diameter of the billet is 0.48 m; applied frequency is 50 Hz; required temperature is 460°C; and the maximum surface density of heating power is 130 kW/m^2. For low thermal losses, the characteristic process parameters will be as: $\xi = 18$; $\theta^* = 2.0$; $Bi = 0.05$.[5] Solution of this time-optimal problem leads to the following results[5]:

$$\varepsilon_{min}^{(1)} = 38°\text{C};\ \varepsilon_{min}{}^{(2)} = 1.5°\text{C};\ Bi^{(3)} = 0.15;\ Bi^* = 0.35;\ Bi^{(1)} = 0.42.$$

Therefore, $Bi = 0.05 < Bi^{(3)}$; $\varepsilon_{min}^{(3)} > \varepsilon_{inf}$.

Condition $\varepsilon_{min}^{(2)} < \varepsilon < \varepsilon_{min}^{(3)}$ usually satisfies the required heating accuracy for hot forming of aluminum alloys. Therefore, two-stage control is typically sufficient (Figure 3.22a).

3.6.2 Solution of Minimum Power Consumption Problem

Figure 3.24 shows some comparative results for computation of optimal control when heating thick cylinders. Criteria of minimum time and power cost were taken into consideration. Required heating accuracy ε varies from $\varepsilon_{min}^{(2)}$ up to $\varepsilon_{min}^{(1)}$. As shown in Section 3.5, for such values of required accuracy ε, optimal control algorithms are substantially different. They coincide only for the boundary value $\varepsilon = \varepsilon_{min}^{(2)}$.

Calculations of heating and soaking time intervals that are optimum with respect to energy consumption are reduced to solving a set of equations (Equation 3.35) for all values of ε that belong to the range of $\varepsilon_{min}^{(2)} \le \varepsilon \le \varepsilon_{min}^{(1)}$. Obtained dependencies demonstrate that energy consuming optimal algorithms lead to a gain of power consumption profit and to production rate loss when heating accuracy decreases.

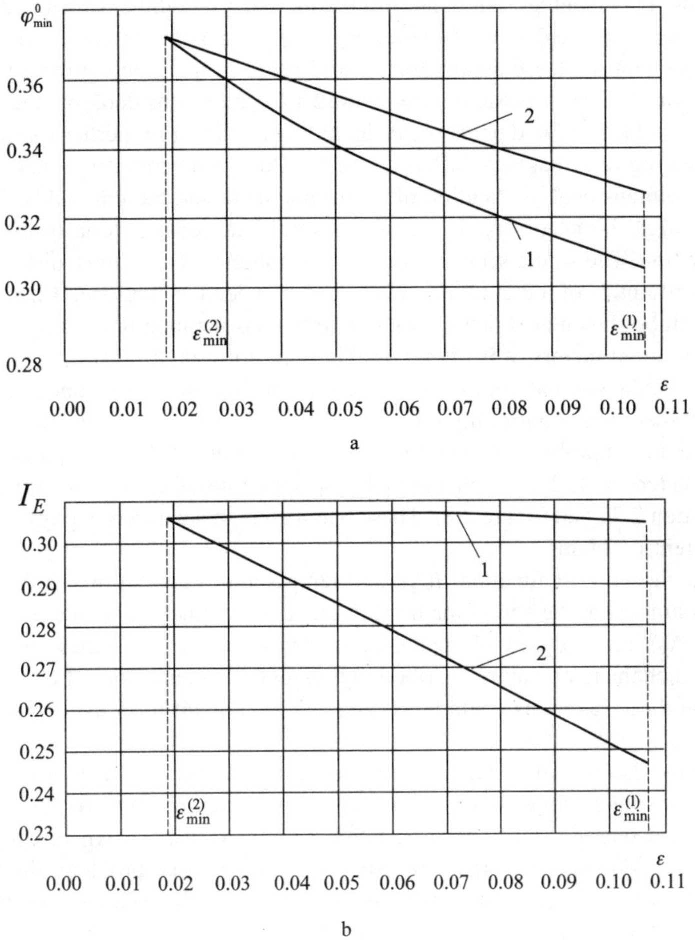

FIGURE 3.24 Optimal heating time φ° (a) and energy consumption Io (b) as functions of required heating accuracy ε: $\varepsilon^{(2)}_{min} \le \varepsilon \le \varepsilon^{(1)}_{min}$ for Bi = 0.5; ξ = 0.4; θ_0 = –0.5: 1 – time-optimal process; 2 – optimal energy consumption process. (The computations were performed by A.N. Diligenskaya.)

3.7 GENERAL PROBLEM OF PARAMETRICAL OPTIMIZATION OF INDUCTION HEATING PROCESSES

Up to this point, several simplified problems of optimal control of induction heating prior to metal hot working were considered. Only extremely simplified linear models of spatial–one-dimensional temperature fields were investigated. Some basic assumptions have been made with respect to those simplified problems. This includes, but is not limited to, using one-dimensional mathematical models and neglecting nonlinearity of physical properties of the heated metals. Obviously, these assumptions could result in noticeable error for certain applications. More precise models should take into consideration the several complex and interrelated phenomena, including nonlinear properties (particularly when heating ferromagnetic billets above the Curie temperature) as well as two- or three-dimensional distribution of electromagnetic and thermal fields. In addition, it might be necessary to consider several additional criteria or even their combination. The whole set of additional technological constraints, disturbances, and uncertainties of a controlled system can be added as well (see Chapter 2).

It will be shown next that even taking into consideration practically all of the most important factors, it is often possible to perform preliminary parameterization of sought optimal control inputs. Parameterization of the control problem requires specifying control input function by one or more variables. In cases like these, the initial problem is reduced to searching for number N and optimal values of parameters Δ_i, 1, 2, …, N that uniquely specify optimal control input (similarly to Equation 2.27 and Figure 2.8). These parameters make different physical sense in different problems.

The chosen cost function, $I(\Delta)$, and temperature distribution, $t(x,\Delta)$, within billet volume Ω at the end of the heating process are functions of the set $\Delta = (\Delta_1, \Delta_2, \ldots, \Delta_N)$ and vector x of spatial coordinates. Vector x includes coordinates along which the temperature distribution is essentially nonuniform. Let us assume now that these functions could be obtained in an explicit form using any known method.

Let us also assume that, at the end of a heating process, it is necessary to provide absolute temperature deviation ε of final temperature $t(x,\Delta)$ from the preassigned one, t^*, similarly to Equation (2.30). As was shown in Chapter 2, similarly to Equation (2.34) for the simplest time-optimal problem, the task of vector $\Delta^0 = (\Delta_1^0, \Delta_2^0, \ldots, \Delta_N^0)$ search can be reduced to the following problem of mathematical programming:

$$\begin{cases} I(\Delta) \to \min\limits_{\Delta}; \\ \Phi(\Delta) = \max\limits_{x \in \Omega} \left| t(x, \Delta) - t^* \right| \le \varepsilon. \end{cases} \tag{3.38}$$

The problem in Equation (3.38) proves to be more difficult than those in Equation (2.34) or Equation (2.42). This is due to the fact that the cost function is more complicated in comparison with the simpler ones that correspond to Equation (2.31) and Equation (2.41). In addition, the basic difficulties arise here because the spatial distribution of final temperature $t(x,\Delta)$ could be noticeably different compared to cases described earlier. These features primarily deal with nonlinear and multidimensional models of temperature fields.

Nevertheless, it is possible to show[5] that, if some preliminary conditions are met, then the optimal control method offered in the present text could be applied in its basic aspects to the general case (Equation 3.38). Of course, not all optimal control results discussed earlier for the simplified IHP models can be easily transferred for these more complex cases.

High generality of the basic property of the temperature distribution at the optimal process end allows one to extend the area of application of the considered method onto the wide problem variety. Similarly to Equation (3.7) and Equation (3.8), the number, R, of points within billet where maximum admissible deviations are reached, is not less than number N of parameters of sought optimal control. This rule does not depend on types of IHP models, cost functions, technological constraints, control inputs, and other factors.[5]

The analysis similar to the one provided in Section 3.1 for time-optimal IHP allows writing alternance property in the form similar to relations in Equation (3.7) and Equation (3.8):

$$\begin{cases} \left| t(x_j^0, \Delta^0) - t^* \right| = \varepsilon, \ j = 1, 2, ..., R; \\ R = \begin{cases} N, \ if \ \varepsilon > \varepsilon_{\min}^{(N)}; \\ N+1, \ if \ \varepsilon = \varepsilon_{\min}^{(N)}, \end{cases} \end{cases} \qquad (3.39, 3.40)$$

where $x_j^0 \in \Omega, \ j = \overline{1, R}$ are coordinates of points within the billet, where maximum deviations of the final temperature from the prescribed one are reached.

These deviations are equal to ε. In contrast to Equation (3.7), in cases of multidimensional thermal models, we consider x_i^0 as a set of jth point coordinates along a corresponding axis. As stated before, $\varepsilon_{\min}^{(N)}$ in Equation (3.40) represents the maximal possible value of the utmost heating accuracy — i.e., minimal possible deviation from t^*. Such accuracy can be achieved by applying control actions explicitly characterized by a set of N parameters vector $\Delta = (\Delta_1, \Delta_2, \ldots, \Delta_N)$.

In the general case (Equation 3.38) and similarly to time-optimal control inputs, it is assumed that control inputs could be defined by this set of N parameters, Δ_l, for which the sequence of inequalities of the type in Equation (3.1) holds true:

$$\varepsilon_{\min}^{(1)} > \varepsilon_{\min}^{(2)} > ... > \varepsilon_{\min}^{(N)} > \varepsilon_{\min}^{(N+1)} > ... > \varepsilon_{\min}^{(N^*)} = \varepsilon_{\inf} > 0 \ . \qquad (3.41)$$

Let us notice that, in the general case (Equation 3.38), the condition in Equation (3.3) *is no longer true*. This condition directly links the number, N, with the place of preset value, ε, in the series of inequalities in Equation (3.41). This fact was discussed earlier, considering as an example an optimization of consumed energy of induction heater (Section 3.5).

Generally speaking, Equation (3.40) is only true under assumption that a number, N, is known and does not necessarily correspond to a condition (Equation 3.3).[2] This situation differs from the case described earlier by Equation (3.8), where the number of optimal control parameters N directly depends on ε.

Properties in Equation (3.39) and Equation (3.40) mean that, for known number N that corresponds to required accuracy ε, such R (number of the points x_i^0 within billet) will be necessarily found, which is related with N by Equation (3.40). Thus, in reality the actual number of these points can be greater than R, and the problem of defining the value N should be solved separately for each specific case.

Properties in Equation (3.39) and Equation (3.40) are correct, taking into consideration preliminary conditions that prove to be formal enough and not restrictive. As a rule, they are met in practice for problems of actual IHP optimization.[2,5] The only exception deals with simplified and uncommon cases from a practical standpoint when requirements to heating accuracy are insignificant and "automatically" satisfied for the optimal control solution $\Delta = \Delta^0$. This case takes place if Δ^0 is a minimum point of $I(\Delta)$, and therefore

$$\frac{\partial I(\Delta^0)}{\partial \Delta_i} = 0, \; i = \overline{1,N}. \tag{3.42}$$

Vector Δ^0 could be found from these N equations. Solution of the problem in Equation (3.38) can be found relatively easily by considering the set of N equations that should be solved with respect to unknowns Δ^0 for given $I(\Delta)$.

We shall assume further that, considering only the main variant with an "active" constraint role in Equation (3.38), Δ^0 is not an extremum point of $I(\Delta)$. Similarly to techniques described earlier in this chapter, relations in Equation (3.39) and Equation (3.40) create potential possibilities for their transformation into corresponding systems of equations with respect to sought parameters of the optimal process. In comparison with simple IHP optimization, several important peculiarities take place in this case.

One of them deals with the nature of final spatial temperature distribution. Its accurate computation represents a much more difficult problem than a simple definition of curve $\theta(l,\Delta^0)$ obtained from considering a one-dimensional mathematical model in a time-optimal control problem. Actually, in cases like this, when general alternance equalities (Equation 3.9) are not met, variants of locations of points x_j^0 could not be unambiguously set a priori. Therefore, final temperature distribution and set of governing equations also could not be found easily. In addition, the definition of function $t(x,\Delta)$ in the explicit form represents the separate task for complex IHP models.

Nevertheless, the alternance method can be successfully applied for a wide variety of specific IHP optimization problems of the type in Equation (3.38).[2,5,7–21] This conclusion is based on the following general reasons. In the great majority of cases, $N = 1$ for $\varepsilon = \varepsilon^{(1)}_{\min}$ in Equation (3.38). This simplest case corresponds to the most typical modes of induction heating applied to various models and cost functions. It is relatively easy in this case to predict the time history and the shape of final temperature profile along the radius/thickness of the heated workpiece. If expression $t(x,\Delta) \equiv t(x,\Delta_1)$ is known, then conversion of relations in Equation

(3.39) and Equation (3.40) into the corresponding set of equations becomes straightforward (as well as in Section 3.3).

It would be possible to determine Δ_1 and $\varepsilon_{\min}^{(1)}$ using known numerical methods. Under typical conditions for a given heating accuracy $\varepsilon < \varepsilon_{\min}^{(1)}$, the sought optimal control (i.e., unknown vector Δ^0) could be found by solving the problem in Equation (3.38) for the series of fixed values $\varepsilon = \varepsilon'$ descending by sufficiently small step $d\varepsilon$ starting from initial value $\varepsilon_{\min}^{(1)}$. It can be done similarly to the procedure suggested in Section 3.4 and applied to a particular example discussed in Section 3.6.

The continuous nature of the final temperature distribution and all optimal process parameters as functions of ε allows one to establish (using basic properties in Equation 3.39 and Equation 3.40) optimal final temperature profile. It is possible to write a set of equations and to choose its initial approximation using results of a previous step. In the course of this algorithm implementation, the number N usually can be determined by some simple calculation experiments. This number grows with decreasing values of ε'.[5] This procedure often becomes simpler if known physical features are employed.

The following chapters will detail the way in which the alternance method is extended onto a wide range of IHP and IHI optimization problems for static and continuous heaters as well as onto more general control problems for production complexes that involve interrelated heating and hot forming processes. Computational algorithms suggested later noticeably extend the optimization technique, allowing its usage directly for optimal operational modes and optimal IHI design for the variety of technological processes.

The supposed method has some noticeable advantages in comparison with complicated and time-consuming numerical techniques developed for solution of mathematical programming problems similar to those in Equation (3.38).[22–24] This method is based upon evident qualitative features of the physical nature of optimal induction heating processes. These features are used in the course of a computational procedure and also as effective means to check received results. In addition, these features allow one to find (depending on required heating accuracy) the number and location of the points of maximal and minimal temperatures at the end of an optimal process. All the foregoing allow a dramatic decrease in the required amount of calculations by reduction of the wide range of initial optimization problems to special, easy-to-solve sets of equations.

At the same time, the alternance method offers a way to calculate the maximal heating accuracy — i.e., values of minimaxes $\varepsilon_{\min}^{(i)}$, $i = \overline{1, N^*}$ in Equation (3.41), including maximum achievable accuracy $\varepsilon_{\inf} = \varepsilon_{\min}^{(N^*)}$, within a particular class of the control inputs. This opportunity represents a particular interest with respect to IHP technology. The alternance method also allows one to establish the outline of spatial final temperature distribution at the end of an optimal process for any given heating accuracy. This information could be used for creating technological charts and programs and also for synthesis of optimal control systems with temperature feedback.

REFERENCES

1. Collatz, L. and Krabs, W., *Approximations Theorie. Tschebyscheffsche Approximation mit Anwendungen*, B.G. Feubner, Stuttgart, 1973.
2. Rapoport, E.Ya., *Alternance Method for Solving Applied Optimization Problems*, Nauka, Moscow, 2000.
3. Butkovskii, A.G., *Optimal Control Theory for Systems with Distributed Parameters*, Nauka, Moscow, 1965.
4. Butkovskij, A.G., Malyj, S.A., and Andreev, Yu.N., *Optimal Control of Metals Heating*, Metallurgy, Moscow, 1972.
5. Rapoport, E.Ya., *Optimization of Induction Heating of Metals*, Metallurgy, Moscow, 1993.
6. Ortega, J.M. and Rheinboldt, W.C., *Iterative Solution of Nonlinear Equations in Several Variables*, Academic Press, New York and London, 1970.
7. Rapoport, E.Ya., Exact method in problems of optimization of nonstationary heat conductivity processes, *Izvestia* AN SSSR, *Energetika i transport*, 4, 137, 1978.
8. Rapoport, E.Ya., Moving control in optimization problems for induction heating of metals, in *Mobile Control of Distributed Systems*, Nauka, Moscow, 82, 1979.
9. Rapoport, E.Ya., Particular optimization problems for modes of metal heating prior to hot working, *Physica i himija obrabotki materialov*, 3, 54, 1984.
10. Rapoport, E.Ya. and Zimin, L.S., Optimal control of slabs induction heating prior to rolling, *Physica i himija obrabotki materialov*, 3, 21, 1986.
11. Rapoport, E.Ya., Method for computation of optimal processes of materials heat treatment, *Physica i himija obrabotki materialov*, 5, 5, 1987.
12. Pleshivtseva, Yu.E. et al., Spatial–time control of nonstationary heat transfer processes, *Vestnik SamGTU*, 1, 208, 1994.
13. Rapoport, E.Ya., Optimization of induction heating processes, *Proc. 40th Int. Wissenschaftliches Kolloquium*, Technische Universitaet Ilmenau, Thueringen, 4, 48, 1995.
14. Rapoport, E.Ya., Parametric optimization of coupled electromagnetic and temperature fields in induction heating, *Proc. Int. Symp. Electromagn. Fields Electr. Eng. (ISEF'95)*, Thessaloniki, 319, 1995.
15. Rapoport, E.Ya., Alternance properties of optimal temperature distribution and computational algorithms in problems of induction heating processes parametric optimization, *Proc. Int. Induction Heating Semin. (IHS-98)*, Padua, 443, 1998.
16. Rapoport, E.Ya., Pleshivtseva, Yu.E., and Livshits, M.Yu., Alternance method in problems of induction heating processes: basic principles and experience of applications, *ISEF'99, Proc. Int. Symp. Electromagn. Fields Electr. Eng.*, Pavia, 141, 1999.
17. Rapoport, E.Ya. and Pleshivtseva, Yu.E., Optimal control of electric heating by mobile internal sources, *Proc. Int. Semin. Heating Internal Sources (HIS-01)*, Padua, 667, 2001.
18. Pleshivtseva, Yu.E. et al., Potentials of optimal control techniques in induction through heating for forging, in *Proc. Int. Sci. Colloquium Modeling for Electromagn. Process.*, Hannover, 145, 2003.
19. Pleshivtseva, Yu.E. et al., Optimal control techniques in induction through heating for forging, in *Proc. Int. Symp. Heating Electromagn. Sources HES-04,* Padua, 97, 2004.

20. Rapoport, E.Ya., Livshits, M.Yu, and Pleshivtseva, Yu.E., Alternance method in optimization problems for processes of technological thermal physics: basics of theory, computational algorithms, experience of application, *Teplomassoobmen MMF-2000,* Minsk, 3, 298, 2000.
21. Rapoport, E.Ya. and Pleshivtseva, Yu.E., Special optimization methods in heat-conductivity inverse problems, *Izvestia RAN*, Energetica, Moscow, 5, 144, 2002.
22. Demjanov, V.F. and Malozemov, V.N., *Introduction in Minimax*, Nauka, Moscow, 1972.
23. Demjanov, V.F. and Vasiljev, L.V., *Nondifferentiable Optimization*, Nauka, Moscow, 1981.
24. Polak, E., Semi-infinite optimization in engineering design, *Lecture Notes in Economics and Mathematical Systems,* Springer–Verlag, New York, 215, 1981.

4 Optimal Control of Static Induction Heating Processes

This chapter discusses the wide range of optimal control problems for static induction heating of metals prior to hot forming. The suggested parametric optimization method will be used as a basis for solving optimization problems, taking into consideration real-life production requirements. The variety of cost functions applied to static induction heating processes will be considered in this chapter. An overview of the most typical process restraints will be provided for linear and nonlinear, one- and two-dimensional, and deterministic and bounded nondeterministic models of static induction heating.

4.1 TIME-OPTIMAL CONTROL FOR LINEAR ONE-DIMENSIONAL MODELS OF STATIC IHP WITH CONSIDERATION OF TECHNOLOGICAL RESTRAINTS

Detailed description of the time-optimal problem for static induction heating has been provided in Chapter 2 and Chapter 3 with respect to restraint on control input (see form of Equation 2.7). That restraint is defined by power supply limitations (Section 2.5.1).

Metal heating processes impose the two most general requirements with regard to the temperature distribution during the heating process. The first requirement demands that the maximum value $\theta_{\max}(\varphi)$ of the temperature within the heated workpiece should be below a certain admissible value θ_{adm}. If this value will be exceeded, then irreversible adverse changes in material structural properties (i.e., grain boundary liquation, burning, etc.) and even metal melting could take place.

Secondly, the temperature differences within the whole volume of the heated workpiece should be restricted during heating in such a way that the maximum value, $\sigma_{\max}(\varphi)$, of tensile thermal stresses $\sigma(l,\varphi)$ (due to thermal gradients) would not exceed prescribed admissible value σ_{adm}, which corresponds to ultimate stress limit of the heated material (see Section 2.5.2). Additional technological constraints add complexity to the OCP solution and appropriate computational procedures.[1] In the present section, we shall extend the alternance method to the solution of OCP with additional restraints applied to the linear one-dimensional model of static heating process.

4.1.1 General Overview of Optimal Heating Power Control

An internally generated heating power is the most significant process parameter that affects temperature distribution within the inductively heated workpiece. Proper heating process control involves appropriate choice of the internal heat sources (eddy current) power density represented by function $F(\xi,l,\varphi) = W(\xi,l)u(\varphi)$ in the right side of Equation (1.27). However, such control functions in many cases are too complex to be practically applied. Therefore, it is important to develop a set of particular control inputs that could be relatively easily applied in engineering practice. At first look, this may seem a relatively simple process that calls for the total consumed heating power $u(\varphi)$ as control input. In this respect, voltage applied to the induction coil can be considered as a required control function (see Section 2.4).

The time-optimal control problem (formulated in general form in Section 2.8) will be considered for static heating with prescribed accuracy. The mathematical model is described by Equation (1.27) through Equation (1.35). Time-dependent heating power $u(\varphi)$ is chosen as lumped control input restrained by Equation (2.7). However, in contrast to Section 2.8, the technological constraints in Equation (2.11) and Equation (2.13) now will be taken into account. Therefore, it is necessary to select such control function $u(\varphi) = u^*(\varphi)$ that provides steering workpiece initial temperature distribution to desired temperature θ^* with prescribed accuracy ε (according to Equation 2.25) in minimal optimal process time under conditions in Equation (2.11) and Equation (2.13).

It has been proved earlier (see Section 2.8) that the time-optimal control algorithm consists of alternating stages of heating with maximum power $u \equiv U_{max}$ (heat ON) and subsequent soaking/cooling under $u \equiv 0$ (heat OFF) cycles. Consequently, optimal control $u^*(\varphi)$ takes the form of Equation (2.27) (Figure 2.9). Let us now consider a solution of the optimal problem subject to additional restraints. Such a solution will look different compared to a solution of OCP discussed earlier in Section 2.8.

If the computations reveal that conditions in Equation (2.11) and Equation (2.13) are not violated under control input in the form of Equation (2.27) during the whole heating process, then the obtained OCP solution should be considered as optimal. Thus, a necessity in variation of optimal control mode arises only when at least one of inequalities in Equation (2.11) and Equation (2.13) does not hold true under the control input function (Equation 2.27). In this case, optimal time-varying heating power can be defined according to arrangement of curves $\theta_{max}(\varphi)$ and $\sigma_{max}(\varphi)$. Study of the shapes of curves $\theta_{max}(\varphi)$ and $\sigma_{max}(\varphi)$ could indicate in which process stage — heating under $u^*(\varphi) = U_{max}$ or soaking under $u^*(\varphi) = 0$ — the control input should be varied to satisfy the requirements in Equation (2.11) and Equation (2.13).

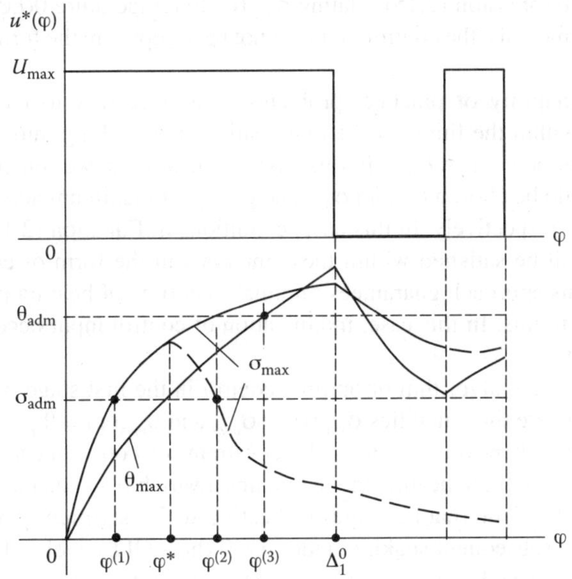

FIGURE 4.1 Shapes of curves $\theta_{max}(\varphi)$ and $\sigma_{max}(\varphi)$ during the time-optimal heating process without technological constraints.

In the case of uniform initial temperature distribution within heated workpiece, the shapes of curves $\theta_{max}(\varphi)$ and $\sigma_{max}(\varphi)$ during heating under $u^*(\varphi) = U_{max}$ can be found quite easily based on common sense and obvious technical conditions (Figure 4.1). In particular, the temperature maximum $\theta_{max}(\varphi)$ continuously increases during the heating process. The maximum of tensile thermal stresses is approximately proportional to the difference between the average temperature of the billet θ_{av} (φ) and temperature in the center $\theta(0,\varphi)$ (see Expression 2.15). During the initial heating stage, $\sigma_{max}(\varphi)$ increases as this difference grows. The further variation in $\sigma_{max}(\varphi)$ depends upon the heat losses $q(\varphi)$ from the surface of the heated billet.

If these losses are approximately constant in Equation (1.32), then $\sigma_{max}(\varphi)$ approaches monotonously to some constant level (Figure 4.1, solid curve) as rates of temperatures rise in the center and on the surface are equalized. Due to convection heat losses (see Equation 1.34), there will be a reduction of surface-to-core temperature gradients because heat losses from the surface grow when increasing surface temperature $\theta(1,\varphi)$. This can result in reducing σ_{max} after achievement of a certain maximum at $\varphi = \varphi^*$ (Figure 4.1, dotted curve).

Figure 4.1 shows that, during the first control stage, the condition in Equation (2.13) is violated on intervals $(\varphi^{(1)},\Delta_1^0)$ or $(\varphi^{(1)},\varphi^{(2)})$. This takes place because the permissible limit of tensile thermal stresses was exceeded. The inequality in Equation (2.11) is violated during the interval $(\varphi^{(3)},\Delta_1^0)$ due to inadmissible value of temperature maximum. It is assumed that dependencies $\sigma_{max}(\varphi)$ or $\theta_{max}(\varphi)$ can be computed from the known solutions of the equations for heating process according to Expression (2.15) relating $\sigma_{max}(\varphi)$ to temperature $\theta(l,\varphi)$. Therefore, during these intervals, the control input cannot be accepted in the form of Equation (2.27).

For the majority of practical problems, it is necessary to modify optimal modes only within the limits of the first heating stage, taking into consideration that maximum power $u \equiv U_{max}$. In this case, the heating power on corresponding intervals should be chosen to hold σ_{max} and/or θ_{max} at maximum admissible levels σ_{adm} and θ_{adm}, respectively. In this case, conditions in Equation (2.11) and Equation (2.13) will be satisfied within these intervals in the form of corresponding equalities. This approach guarantees minimal total time of heating process under additional restraints. In this case, modification of control input becomes as minimal as possible.

As a result, the algorithm of optimal control in the first stage is complicated by intervals where the equalities $\sigma_{max}(\varphi) \equiv \sigma_{adm}$ and $\theta_{max}(\varphi) \equiv \theta_{adm}$ hold true (so-called intervals of movement along the constraint or constraint motion). In the other stages, the optimal heating mode remains invariable — i.e., the time-optimal control consists of alternating stages of heating with maximum power $u \equiv U_{max}$ (heat ON) and subsequent soaking under $u \equiv 0$ (heat OFF) cycles. In the general case, the number and sequence order of intervals with $u^*(\varphi) \equiv U_{max}$, $\sigma_{max}(\varphi) \equiv \sigma_{adm}$, and $\theta_{max}(\varphi) \equiv \theta_{adm}$ can be different and depend on initial temperature distribution within the workpiece.

In the conventional case of uniform initial temperature distribution within the heated workpiece (Section 2.8), the analysis of possible variations of temperature profiles leads to the basic algorithm of optimal heating power shown in Figure 4.2. During the first interval, an accelerated heating takes place under maximum power $u^*(\varphi) \equiv U_{max}$ at continuously growing values σ_{max} and θ_{max}. It is typical that the first limit σ_{adm} is reached at the stage that corresponds to time $\varphi = \varphi_\sigma$. At that point, the maximum temperature is still lower than θ_{adm}. Then, in the interval $(\varphi_\sigma,\varphi_p)$, the tensile thermal stress is held at admissible level σ_{adm} by the control input $u^\sigma(\varphi)$, which stepwise decreases at $\varphi = \varphi_\sigma$. Heat power $u^\sigma(\varphi)$ is rapidly reduced in comparison with U_{max} to provide the increase rate of σ_{max} equal to zero. At the same time, during interval $(\varphi_\sigma,\varphi_\theta)$, the maximum temperature θ_{max} continues to grow.

To provide equality $\sigma_{max}(\varphi) \equiv \sigma_{adm}$, it is necessary to further increase $u^\sigma(\varphi)$ from the initial value (less than U_{max}) up to the value U_{max} that would be achieved at the moment $\varphi = \varphi_p$. Further holding of σ_{max} at the level σ_{adm} is not possible. Then, the next stage of control process would be performed under maximum heating power U_{max} when θ_{max} keeps growing. At the same time, the value of σ_{max} decreases from σ_{adm}. This stage is over at the certain moment $\varphi = \varphi_\theta$ when temperature maximum becomes equal to admissible value (i.e., when the equality $\theta_{max}(\varphi_\theta) = \theta_{adm}$ is true). During the interval $(\varphi_\theta,\Delta_1^0)$, temperature maximum is kept at the maximum admissible level by applying control input $u^\theta(\varphi)$.

Similarly to $u^\sigma(\varphi)$, the heat power $u^\theta(\varphi)$ is rapidly reduced in comparison with U_{max}, providing no increase of temperature rate θ_{max}. To satisfy equality $\theta_{max}(\varphi) \equiv \theta_{adm}$, it would be necessary to reduce the heat power $u^\theta(\varphi)$ continuously

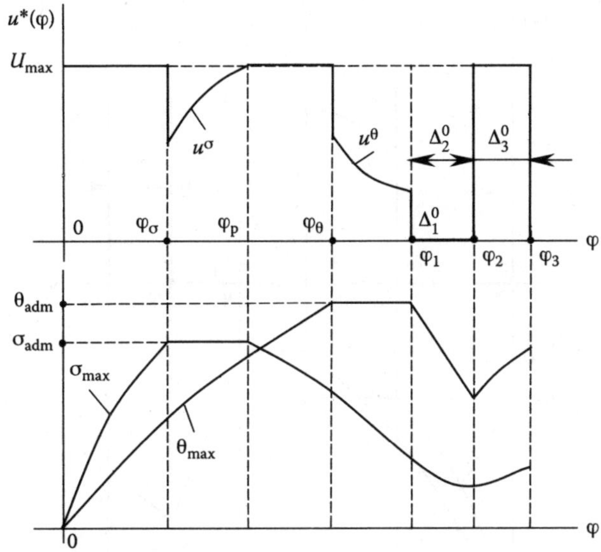

FIGURE 4.2 Optimal control $u^*(\varphi)$, maximum temperature θ_{max}, and thermal stress σ_{max} during the time-optimal process with technological constraints.

(and therefore $\sigma_{max}(\varphi)$). For $\varphi > \Delta_1^0$, the following stages are performed according to the general control algorithm (Equation 2.27). The conditions $\theta_{max}(\varphi) < \theta_{adm}$ and $\sigma_{max}(\varphi) < \sigma_{adm}$ usually are not violated over these stages because the values σ_{max} and θ_{max} decrease during even intervals under $u(\varphi) \equiv 0$. During odd intervals, these values do not reach their admissible levels.

As a result, the typical algorithm of the time-optimal heat power control with consideration of the "real-life" technological constraints can be represented by the following expression (Figure 4.2):

$$u^*(\varphi) = \begin{cases} U_{max}, \varphi \in (0, \varphi_\sigma); \\ u^\sigma(\varphi), \varphi \in (\varphi_\sigma, \varphi_p); \\ U_{max}, \varphi \in (\varphi_p, \varphi_\theta); \\ u^\theta(\varphi), \varphi \in (\varphi_\theta, \Delta_1^0); \\ \dfrac{U_{max}}{2}\left[1 + (-1)^{j+1}\right], \varphi_{j-1} < \varphi < \varphi_j, j = 2, 3, \ldots, N. \end{cases} \tag{4.1}$$

Certain optimal control functions (Equation 4.1) represent certain situations or applications. If, during the stage when thermal stress is held at its maximum

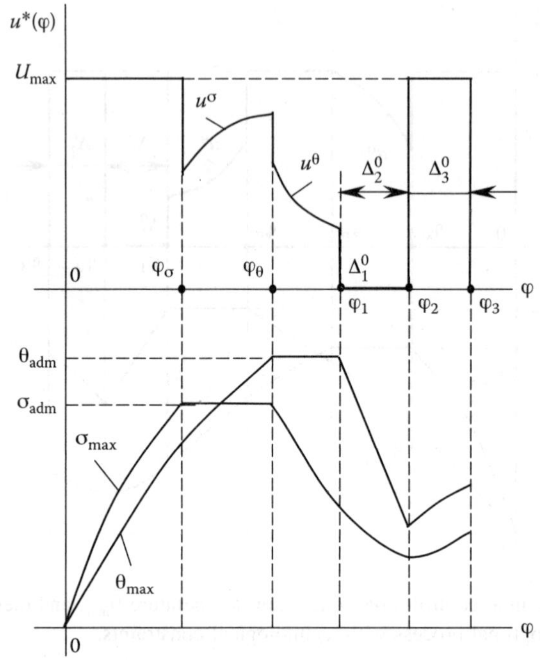

FIGURE 4.3 Optimal control $u^*(\varphi)$ with a single stage of heating under maximum power, maximum temperature θ_{max}, and thermal stress σ_{max} during the time-optimal process with technological constraints.

permissible level σ_{max}, the equality $u^{\sigma}(\varphi) = U_{max}$ is not reached until the time point $\varphi = \varphi_{\theta}$, then the interval $(\varphi_p, \varphi_{\theta})$ is excluded. In this case, instead of Equation (4.1), one can rewrite an optimal control function (Figure 4.3) as:

$$u^*(\varphi) = \begin{cases} U_{max}, \varphi \in (0, \varphi_{\sigma}); \\ u^{\sigma}(\varphi), \varphi \in (\varphi_{\sigma}, \varphi_{\theta}); \\ u^{\theta}(\varphi), \varphi \in (\varphi_{\theta}, \Delta_1^0); \\ \frac{U_{max}}{2}\left[1 + (-1)^{j+1}\right], \varphi_{j-1} < \varphi < \varphi_j, j = 2, 3, \ldots, N. \end{cases} \tag{4.2}$$

If, during the first heating stage under applying the maximum power, the temperature maximum reaches admissible value θ_{adm}, but at the same time the value σ_{max} still remains less than σ_{adm}, the interval of σ_{max} holding should be excluded (Figure 4.4):

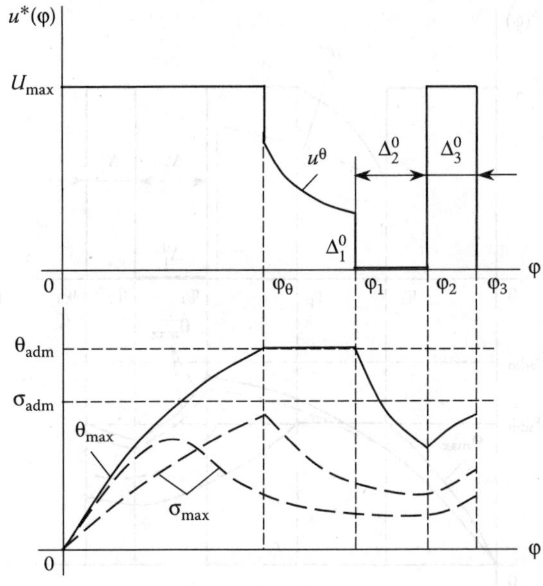

FIGURE 4.4 Time-optimal process without interval of σ_{max} holding.

$$u^*(\varphi)=\begin{cases} U_{max},\varphi\in(0,\varphi_\theta); \\ u^\theta(\varphi),\varphi\in(\varphi_\theta,\Delta_1^0); \\ \dfrac{U_{max}}{2}\left[1+(-1)^{j+1}\right],\varphi_{j-1}<\varphi<\varphi_j, j=2,3,\dots,N. \end{cases} \tag{4.3}$$

If the first heating stage is completed before θ_{max} reaches θ_{adm} (i.e., at the moment $\varphi = \Delta_1^0$), then the interval of θ_{max} holding should be excluded from Equation (4.1) or Equation (4.2) (Figure 4.5) and the optimal heat power algorithm can be written as:

$$u^*(\varphi)=\begin{cases} U_{max},\varphi\in(0,\varphi_\sigma); \\ u^\sigma(\varphi),\varphi\in(\varphi_\sigma,\varphi_p); \\ U_{max},\varphi\in(\varphi_p,\Delta_1^0); \\ \dfrac{U_{max}}{2}\left[1+(-1)^{j+1}\right],\varphi_{j-1}<\varphi<\varphi_j, j=2,3,\dots,N. \end{cases} \tag{4.4}$$

or

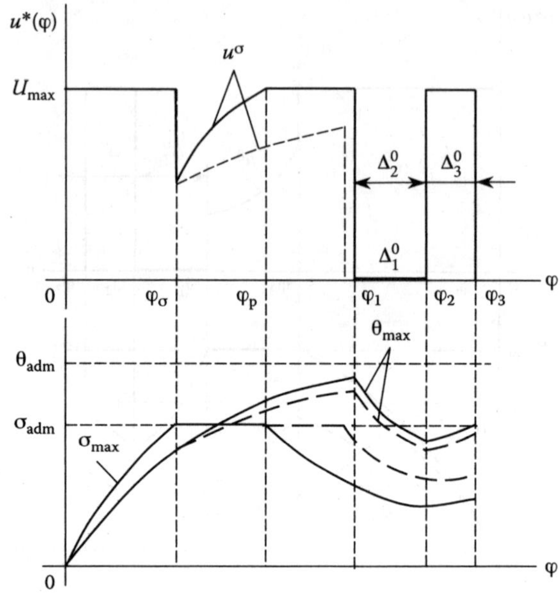

FIGURE 4.5 Time-optimal process without interval of θ_{max} holding.

$$u^*(\varphi) = \begin{cases} U_{\max}, \varphi \in (0, \varphi_\sigma); \\ u^\sigma(\varphi), \varphi \in (\varphi_\sigma, \Delta_1^0); \\ \dfrac{U_{\max}}{2}\left[1 + (-1)^{j+1}\right], \varphi_{j-1} < \varphi < \varphi_j, j = 2, 3, \ldots, N. \end{cases} \tag{4.5}$$

If temperature reaches plastic deformation range, then a limitation of σ_{max} at an appropriate control stage can be excluded. It is easy to extend an optimal control technique to this specific case, but further discussion will not be provided here.

In each particular case, the single variant of time-optimal heating power control can be chosen according to arrangement of curves $\theta_{max}(\varphi)$ and $\sigma_{max}(\varphi)$ over the optimal process computed without technological restraints (Figure 4.1). If the computations reveal that at least one of conditions $\sigma_{max}(\varphi) \leq \sigma_{adm}$ and $\theta_{max}(\varphi) \leq \theta_{adm}$ is not violated (assuming that control input is represented by Equation 2.27), then the appropriate holding interval in the algorithm in Equation (4.1) should be excluded. This allows one to obtain heat power control in one of the simplified forms of Equation (4.3), Equation (4.4), or Equation (4.5).

If restraint on σ_{max} should be taken into account, then trial OCP solution allows one to ascertain the existence of the second heating interval under $u^*(\varphi) = U_{max}$. During the first step of the computational procedure, it is assumed that

$u^{\sigma}(\varphi)$ would not reach the value U_{max}. This means that, instead of Equation (4.1) or Equation (4.4), the control inputs in Equation (4.2) or Equation (4.5) should be applied. If it turns out that the condition $u^{\sigma}(\varphi_{\theta}) > U_{max}$ is true under the control function in the form of Equation (4.2) or an expression $u^{\sigma}(\Delta_1^0) > U_{max}$ is valid in the case of Equation (4.5), then the control inputs (Equation 4.1 or Equation 4.4) should be used. Time points φ_{σ} and φ_{θ} can be found as solutions of the following equations:

$$\sigma_{max}(\varphi_{\sigma}) = \sigma_{adm};\ \theta_{max}(\varphi_{\theta}) = \theta_{adm}, \tag{4.6}$$

where the functions $\sigma_{max}(\varphi)$ and $\theta_{max}(\varphi)$ are known. Control functions $u^{\sigma}(\varphi)$ and $u^{\theta}(\varphi)$ that provide holding in Equation (4.1) through Equation (4.5) can also be determined using the following conditions:

$$\sigma_{max}(\varphi) \equiv \sigma_{adm};\ \theta_{max}(\varphi) \equiv \theta_{adm}. \tag{4.7}$$

Let us assume that φ_{σ}, φ_{θ}, $u^{\sigma}(\varphi)$, and $u^{\theta}(\varphi)$ in Equation (4.1) through Equation (4.5) are found; then, the shape of the optimal control algorithm is known. At the same time, the values of N control parameters $\Delta_i, i = \overline{1,N}$ remain unknown. Similarly to the solution described in Section 2.8, the values $\Delta_i, i = \overline{1,N}$ represent durations of control stages. However, in contrast to the time-optimal control in Section 2.8, in this case the heating mode in the first stage proves to be more complicated than heating under constant maximum power.

As a result, the considered time-optimal control problem is reduced to the special problem of mathematical programming in the form of Equation (2.34). To obtain the final temperature distribution $\theta(l,\Delta)$ in an explicit form, it is necessary to solve Equation (1.27) through Equation (1.35) with respect to power control algorithms (Equation 4.1 through Equation 4.5). The values of φ_{σ} and φ_{θ} and holding controls $u^{\sigma}(\varphi)$ and $u^{\theta}(\varphi)$ can be found in the course of the computational procedure.

4.1.2 Power Control during the Holding Stage

The holding stage in induction heating applications represents a special process when it is necessary to hold maximum temperature in any workpiece point at a certain permissible level. In other cases, in addition to temperature, the holding stage incorporates holding thermal stresses.

Variation of temperature distribution during the heating cycle can be computed from Expression (1.36) after substitution of control functions in the form of Equation (4.1) through Equation (4.5) and using the relationship in Equation (2.15) between maximum stress σ_{max} and temperature $\theta(l,\varphi)$. According to Equation (1.38) and Equation (1.40) under $\theta_o(l) \equiv \theta_{av}(\varphi) \equiv \theta_o \equiv \text{const}$, it is possible to obtain the following results.[1–5]

1. The first equality in the set in Equation (4.6) can be rewritten with respect to unknown time point φ_{σ} as the following equation:

$$U_{\max}\gamma\sum_{n-1}^{\infty}\frac{A_n^{\sigma}W_n(\xi)}{\mu_n^2}\left(1-e^{-\mu_n^2\varphi_{\sigma}}\right)=\sigma_{adm}, \tag{4.8}$$

where

$$A_n^{\sigma}=\frac{2\mu_n\left[(\Gamma+1)K_1(\mu_n)-\mu_n\right]}{\left(\mu_n^2+Bi^2+(1-\Gamma)Bi\right)K^2(\mu_n)}.$$

Equation (4.8) can be solved with required accuracy using well-known numerical methods.

2. The first equality in the set in Equation (4.7) is correct if thermal stress is held at the desired level utilizing control input $u^{\sigma}(\varphi)$, which evolves in time according to exponential law as follows:

$$u^{\sigma}(\varphi)\cong a+be^{-\beta(\varphi-\varphi_{\sigma})},\ \varphi\in(\varphi_{\sigma},\varphi_*), \tag{4.9}$$

where $\varphi_* = \varphi_p$ in Expression (4.1) or Expression (4.4), and $\varphi_* = \varphi_\theta$ in Equation (4.2).

The coefficients a, b, and β in Expression (4.9) can be calculated using the following relations:

$$a=\frac{L_1\mu_2^2+L_2\mu_1^2}{E_1\mu_2^2+E_2\mu_1^2},\ b=\frac{\left(\mu_2^2-\mu_1^2\right)\left(E_1L_2-E_2L_1\right)}{\left(E_1+E_2\right)\left(E_1\mu_2^2+E_2\mu_1^2\right)};\ \beta=\frac{E_1\mu_2^2+E_2\mu_1^2}{E_1+E_2}, \tag{4.10}$$

where

$$L_n=A_n^{\sigma}W_n(\xi)\left(1-e^{-\mu_n^2\varphi_{\sigma}}\right);\ E_n=A_n^{\sigma}W_n(\xi)\ for\ n=1\ and\ n=2. \tag{4.11}$$

These formulas take into account only the first and the second proper numbers μ_n^2 in Equation (2.29).

3. The shape of temperature profile (Figure 3.4) is unchanged under control input (Equation 4.1 through Equation 4.3) until the time $\varphi = \varphi_\theta$. Therefore, the second equality in the set in Equation (4.6) tends to be handled with the appropriate condition of obtaining maximum temperature in the workpiece's internal point $l = l_{\max}$ at the time $\varphi = \varphi_\theta$. Time φ_θ and coordinate $l_{e2}(\varphi_\theta) = l_{\max}$ can be found by considering the equalities:

$$\theta(l_{\max}, \varphi_\theta) = \theta_{adm}; \frac{\partial\theta(l_{\max}, \varphi_\theta)}{\partial l} = 0 . \quad (4.12)$$

Here, temperature distribution $\theta(l_{\max},\varphi_\theta)$ and temperature gradient $\partial\theta(l_{\max},\varphi_\theta)/\partial l$ should be calculated according to the general expression (Equation 1.36) after substitution of heating power in the form of Equation (4.1) through Equation (4.3) for time interval $(0,\varphi_\theta)$.
The expressions for $\theta(l_{\max},\varphi_\theta)$ and $\partial\theta(l_{\max},\varphi_\theta)/\partial l$ should be obtained in the explicit form as functions of $l_{\max}$ and φ_θ. This allows transforming a set of equalities in Equation (4.12) into the set of two equations that ought to be solved with respect to unknowns $l_{\max}$ and φ_θ using known numerical methods.

4. Assuming that the time of holding the maximum temperature $\theta_{\max}$ in any workpiece point at the level θ_{adm} is relatively small, it is possible to neglect a displacement of maximum temperature point during this interval and to obtain the control function $u^\theta(\varphi)$. This control function can be obtained using a condition of holding temperature in point $l_{\max}$ at its maximum permissible level. According to the second equation in the set in Equation (4.7) and similarly to Expression (4.9), the holding control function $u^\theta(\varphi)$ can be obtained using the following exponential form[1–5]:

$$u^\theta(\varphi) \cong a + be^{-\beta(\varphi-\varphi_\theta)}, \ \varphi \in (\varphi_\theta, \Delta^0{}_1). \quad (4.13)$$

The coefficients a, b, and β can be calculated using Equation (4.10).
Here, instead of Equation (4.11), it is possible to assume:

$$L_n = A_n^{(\theta)}\left(\theta_n\mu_n^2 - \frac{\mu_n\theta_0 K_1(\mu_n)}{U_{\max}}\right); E_n = A_n^{(\theta)} W_n(\xi) \text{ for } n = 1 \text{ and } n = 2 , \quad (4.14)$$

where

$$A_n^{(\theta)} = \frac{2\mu_n^2 K(\mu_n l_{\max})}{\left(\mu_n^2 + Bi^2 + (1-\Gamma)Bi\right)K^2(\mu_n)}; \theta_n = \int_0^1 \theta(l, \varphi_\theta) l^\Gamma K(\mu_n l)\, dl .$$

5. If time required for heating is appreciably long, then the holding temperature at a point $l_{\max}$ under calculated control function $u^\theta(\varphi)$ of the form in Equation (4.13) does not provide holding the maximum

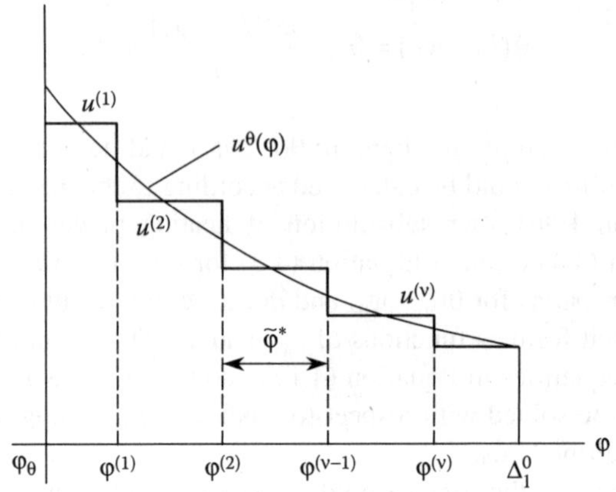

FIGURE 4.6 Approximation of holding control $u^{\theta}(\varphi)$ by piecewise constant function.

temperature in any workpiece point at the level θ_{adm}. As a rule, for such heating modes, the following condition, $\theta_{\max}(\varphi) > \theta_{adm}$, is true for $\varphi > \varphi_\theta$. In this case, the simple computational procedure is suggested to determine optimal power control function $u^{\theta}(\varphi)$, using its approximation by piecewise constant function (Figure 4.6). The accuracy of the approximation increases with the number, S, of discrete steps of duration $\tilde{\varphi}^*$ and is held at acceptable level when S is big enough.

A sequence of optimization problems that could be reduced to solution of the system of equations (Equation 4.12) for fixed values $\varphi^{(\nu)} = \varphi_\theta + \nu\tilde{\varphi}^*$, $\nu = \overline{1, S}$ should be solved with respect to the control function $u^{(\nu)}$ and the point $l_{\max}^{(\nu)}$. For each $\nu \in \{\overline{1, S}\}$, a temperature distribution at the appropriate time point $\varphi = \varphi^{(\nu)}$ can be calculated from Expression (1.36) under known control functions $u^{(1)}$, $u^{(2)}$, …, $u^{(\nu-1)}$ within the range $(\varphi_\theta, \varphi^{(\nu-1)})$. As a result, at time points $\varphi^{(1)}$, $\varphi^{(2)}$, …, $\varphi^{(S)}$, the obtained control functions $u^{(\nu)}$, $\nu = \overline{1, S}$ allow one to hold the maximum temperature at points $l_{\max}^{(1)}$, $l_{\max}^{(2)}$, …, $l_{\max}^{(S)}$ at the level of θ_{adm}. This means that, for each $\varphi \in (\varphi_\theta, \Delta_1^0)$, the condition $\theta_{\max}(\varphi) = \theta_{adm}$ takes place and its accuracy depends upon S. The values $u^{(\nu)}$ can be defined more precisely in the course of subsequent computation of optimal heating processes using the iteration method under a priori unknown value Δ_1^0.

To simplify implementation of optimal heating mode, it is possible to choose constant values of u_c^{σ} and u_c^{θ}, which only approximately hold $\sigma_{\max}$ and $\theta_{\max}$ at the extreme admissible levels. In this case, the values u_c^{σ} and u_c^{θ} should be defined using the standard method — for example, by minimizing mean square deviation of $\sigma_{\max}(\varphi)$ and/or $\theta_{\max}(\varphi)$ from σ_{adm} and θ_{adm}.

4.1.3 Computational Technique for Optimal Heating Modes, Taking into Consideration Technological Constraints

If the values φ_σ and φ_θ and functions $u^\sigma(\varphi)$ and $u^\theta(\varphi)$ are known, then the final temperature distribution $\theta(l,\Delta)$ can be calculated from the basic formula in Equation (1.36) by substituting control inputs in the form of Equation (4.1) through Equation (4.5).

Functions $u^\sigma(\varphi)$ and $u^\theta(\varphi)$ can be obtained in an exponential form as Equation (4.9) and Equation (4.13). Then, it is possible to write the following expression that describes $\theta(l,\Delta)$ under optimal control (Equation 4.1):

$$\theta(l,\Delta) = \theta_0 + U_{\max} \sum_{n=1}^{\infty} \frac{2W_n(\xi)K(\mu_n l)}{\left(\mu_n^2 + Bi^2 + (1-\Gamma)Bi\right)K^2(\mu_n)} \cdot C_n\,, \tag{4.15}$$

where

$$\begin{aligned} C_n = \Bigg[& e^{\mu_n^2\varphi_\sigma} - 1 + a_\sigma\left(e^{\mu_n^2\varphi_p} - e^{\mu_n^2\varphi_\sigma}\right) + \frac{b_\sigma\mu_n^2}{\mu_n^2 - \beta_\sigma}\left(e^{\mu_n^2\varphi_p - \beta_\sigma(\varphi_p - \varphi_\sigma)} - e^{\mu_n^2\varphi_\sigma}\right) + \\ & + e^{\mu_n^2\varphi_\theta} - e^{\mu_n^2\varphi_p} + a_\theta\left(e^{\mu_n^2\Delta_1} - e^{\mu_n^2\varphi_\theta}\right) + \frac{b_\theta\mu_n^2}{\mu_n^2 - \beta_\theta}\left(e^{\mu_n^2\Delta_1 - \beta_\theta(\Delta_1 - \varphi_\theta)} - e^{\mu_n^2\varphi_\theta}\right)\Bigg] e^{-\mu_n^2\varphi^0} + \\ & + \sum_{j=3}^{N}(-1)^{j+1}\left(1 - \exp\left(-\mu_n^2\sum_{m=j}^{N}\Delta_m\right)\right) \end{aligned} \tag{4.16}$$

Here, $\varphi^0 = \sum_{i=1}^{N}\Delta_i$, a_σ, b_σ, β_σ and a_θ, b_θ, β_θ are constant coefficients for $u^\sigma(\varphi)$ and $u^\theta(\varphi)$ in Equation (4.9) and Equation (4.13), respectively.

Therefore, we come to the problem in Equation (2.34); however, Expression (4.15) and Expression (4.16) for computation of $\theta(l,\Delta)$ become more complicated in comparison with Equation (2.28) and Equation (2.29). Expression for optimal time-dependent heating power that is considered as a control function will also be more complex when technological constraints are taken into consideration. Therefore, previously discussed conclusions (Chapter 3) regarding an optimal spatial temperature distribution cannot be easily transferred to these cases.

Computational procedures described in Section 3.7 should be used to determine the vector of optimal parameters $\Delta^0 = (\Delta_1^0, \Delta_2^0, \ldots, \Delta_i^0)$. Our experience shows[1–5] that the established (in Chapter 3) shape of final temperature distribution

$\theta(l, \Delta^0)$ and its properties (including the condition in Equation 3.3) remained unchanged when these properties were not in contradiction with constraints on σ_{max} and θ_{max}. The latter condition is always true in practice when the inequality (Equation 4.17) is valid for the required maximum admissible deviation ε:

$$\varepsilon \le \theta_{adm} - \theta^*. \tag{4.17}$$

According to the property in Equation (3.9), the maximum admissible deviations, $\theta(l_j^0,\Delta^0) - \theta^*$, of the final temperature distribution from the required one arise with different signs for each pair of sequentially located points l_j^0. This means that these deviations are alternating at the points l_j^0, where $0 \le l_1^0 < l_2^0 \ldots < l_R^0 \le 1$ (Figure 3.11, Figure 3.13, etc.). Therefore, if the condition in Equation (4.17) is not met, then equality

$$\theta(l_k^0,\Delta^0) - \theta^* = \varepsilon \tag{4.18}$$

violates the maximum temperature constraint (Equation 2.11) in a certain point $l_k^0 \in [0,1]$.

If, during the heating process, temperature drops within the heated workpiece are substantially small, then the value of σ_{max} proves to be less than maximum admissible level σ_{adm}. This allows one to consider the inequality in Equation (4.17) as the single condition under which the restraints (Equation 2.11 and Equation 2.13) do not violate alternance properties of the final temperature distribution $\theta(l,\Delta^0)$.

Let us consider the optimal heating process that satisfies the condition in Equation (4.17). The following computational procedure is suggested for determining an optimal control algorithm for induction heating processes.

1. At the beginning, constraints should be neglected and the time-optimal control problem should be solved assuming that there are no constraints. This leads to definition of dependencies $\sigma_{max}(\varphi)$ and $\theta_{max}(\varphi)$. As was mentioned in Section 4.1.1, the particular type of the optimal algorithm of power control function during the first process stage can be chosen from the set in Equation (4.1) through Equation (4.5), according to the shapes of curves $\theta_{max}(\varphi)$ and $\sigma_{max}(\varphi)$.
2. In this case, the optimal control functions $u^\sigma(\varphi)$ and/or $u^\theta(\varphi)$ and times φ_σ and φ_θ should be computed from expressions provided in Section 4.1.2. In the general case of Equation (4.1), the values φ_σ, φ_θ, and l_{max} represent roots of Equation (4.8) and the set in Equation (4.12). At the same time, instant φ_p that satisfies inequality $\varphi_\sigma < \varphi_p < \varphi_\theta$ can be determined as a solution to the following equation:

$$a_\sigma + b_\sigma e^{-\beta_\sigma(\varphi_p - \varphi_\sigma)} = U_{max}. \tag{4.19}$$

If constraint on σ_{max} is taken into account, then the optimal control should be considered in the form of Equation (4.2) or Equation (4.5), assuming that, under maximum power, the reheating stage is absent (Figure 4.2 and Figure 4.5). If OCP solution leads to such control function that $u^{\sigma}(\varphi_*) > U_{max}$ for $\varphi_* = \varphi_\theta$ or $\varphi_* = \Delta_1{}^0$ in Equation (4.2) or Equation (4.5), respectively, then the problem should be solved applying algorithms in Equation (4.1) or Equation (4.4). The value of φ_p is defined as a root of Equation (4.19).

Furthermore, the computational procedure described in Chapter 3 could be used for determining the vector of optimal control parameters Δ^0. The function $\theta(l,\Delta^0)$ in the form of Equation (4.15) should be substituted into the sets of equations. Obviously, the resulting minimax values in Equation (3.1) will differ from minimax values obtained in the problem without taking into account technological restraints.

In the case of stepwise constant approximation of $u^\theta(\varphi)$, the expression for $\theta(l,\Delta^0)$ is modified and becomes more complicated. However, it can be derived easily, using the general expression (Equation 1.36). Each iterative update of the values $u^{(v)}$ under a priori unknown Δ_1^0 adds excessive complexity to the computational procedure.

3. A unique feature emerges when the problem is solved for $\varepsilon = \varepsilon_{min}^{(1)}$ and $N = 1$ under condition $\varepsilon_{min}^{(1)} > \varepsilon_{inf}$ (just as in the general time-optimal control problem; see Section 3.3). According to Equation (4.1) through Equation (4.3), a single-stage optimal-control process comes to the end during the interval of temperature holding at the level of θ_{adm} (Figure 4.2 through Figure 4.4). In any case, the deviation of the maximum final temperature from the desired one is equal to the difference $\theta_{adm} - \theta^*$. This means that the value $\varepsilon_{min}^{(1)}$ (which is equal to $\theta_{adm} - \theta^*$) is known a priori (Figure 4.7). In addition, in compliance with properties of temperature profile $\theta(l,\Delta_1{}^0)$ for $\varepsilon = \varepsilon_{min}^{(1)}$ (see Expression 3.11), optimal heating time could be obtained from the following equation:

$$\theta(0,\Delta_1{}^0) - \theta^* = \varepsilon_{min}^{(1)} = -(\theta_{adm} - \theta^*). \tag{4.20}$$

This equation corresponds to the workpiece's center $l = 0$ and has the same form as the first equation in the set in Equation (3.11). Essentially, the system of equations for $\varepsilon = \varepsilon_{min}^{(1)}$ remains similar to the set in Equation (3.11). The shape of the temperature profile $\theta(l,\Delta_1^0)$ replicates the curve shown in Figure 3.6. It is important to mention that there are certain differences. For example, the second equation in Equation (3.11) always holds true under control function $u^\theta(\varphi)$. In addition, the value $\varepsilon_{min}^{(1)}$ proves to be known a priori. At last, in spite of the fact that the form of temperature profile remains unchanged, the expression for $\theta(l,\Delta_1^0)$ is noticeably different. It is clear that the value $\varepsilon_{min}^{(1)}$ varies when the technological constraints are taken into account.

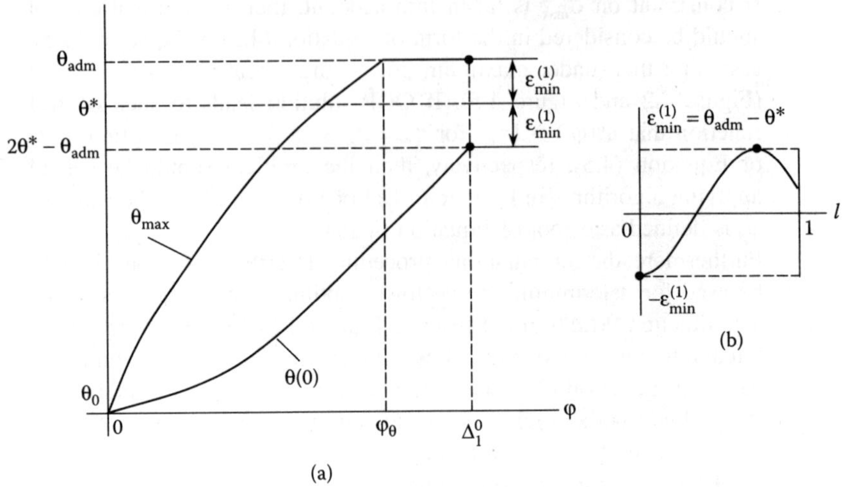

FIGURE 4.7 (a) Time–temperature history and (b) final temperature profile $\theta(l,\Delta_1^0) - \theta^*$ for the optimal control process under constraint on maximum temperature for $\varepsilon = \varepsilon_{\min}^{(1)} = \theta_{adm} - \theta^* > 0$.

4. For all values of $\varepsilon < \varepsilon_{\min}^{(1)} = \theta_{adm} - \theta^*$, satisfying the condition in Equation (4.17), it is possible to apply the previously described computational technique (Chapter 3) for the time-optimal control problem with the single constraint on maximum heating power. As a rule, all established properties of final temperature distribution and corresponding sets of equations remain unchanged. The condition in Equation (3.3) for determining the number, N, of control intervals remains valid as well. At the same time, the expressions for temperature distribution are changed. Due to this reason, a particular computation of optimal process parameters can be noticeably different in comparison with the optimal control problem without additional technological restraints.
The violation of the condition in Equation (4.17) is widely met in practice. The "limiting" case occurs when

$$\theta_{adm} = \theta^*. \tag{4.21}$$

Expression (4.21) means that no overheating is allowed anywhere within the heated workpiece during the heating process. Such a mode is realized by widespread technology of accelerated isothermal induction heating.[6,7] This mode usually corresponds to the control function of the type in Equation (4.3) under the condition in Equation (4.21) for $N = 1$ (Figure 4.8). According to that approach, during the first heating stage, the maximum temperature (which is very close to the surface temperature) is steered forcibly under maximum power to the target value θ^*. During the following stage, that temperature is held at

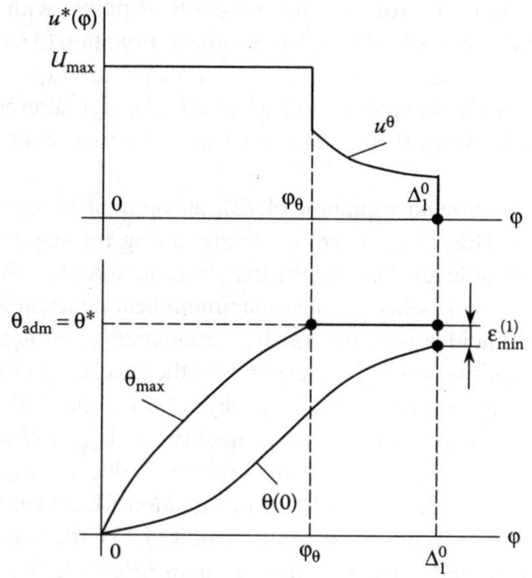

FIGURE 4.8 Optimal control $u^*(\varphi)$ and time–temperature history during accelerated isothermal induction heating.

this level until center temperature reaches the same value θ^* with required accuracy.

It is not difficult to show that, even if the inverse sign of inequality (Equation 4.17) takes place, i.e.,

$$\varepsilon > \theta_{adm} - \theta^*, \tag{4.22}$$

and, therefore, the correlations of the type in Equation (4.18) do not hold true, then even in this case all established properties of final temperature distribution $\theta(l,\Delta^0)$ remain the same. This allows extending the proposed computational technique to the special case in Equation (4.22).[1] Keep in mind, though, that essential features arise in applying the alternance method.

One of these features deals with the necessity to replace equalities in Equation (4.18), which are not allowed due to the condition in Equation (4.22), by the following equalities:

$$\theta(l_k^0, \Delta^0) - \theta^* = \theta_{adm} - \theta^*, \tag{4.23}$$

where l_k^0 are coordinates of points where the maximum admissible temperature is reached.

It is not necessary to incorporate these points into a set of equations because conditions in Equation (4.23) are satisfied by control input $u^\theta(\varphi)$. As a result, the set of equations should be solved with respect to R points with minimum final temperature. Number R can be defined according to Equation (3.9). The deviations of $\theta(l_j^0,\Delta^0)$ from θ^*, because they are the opposite sign (see Equation 3.9), remain equal to value $-\varepsilon$. Thus, in contrast to Equation (3.9), sign-alternating maximum deviations of $\theta(l,\Delta^0)$ from θ^* turn out to be asymmetrical with respect to their absolute values.

Under the condition in Equation (4.22), an optimal heating represents the one-stage process. This process will complete during the stage of holding temperature at maximum level. This means that, in Equation (4.1) through Equation (4.3), the equality $N = 1$ holds true and maximum heating accuracy (temperature uniformity) $\varepsilon_{min}^{(1)}$ coincides with the smallest attainable value ε_{inf}. Therefore, the simplest form of optimal control corresponds to the accelerated isothermal induction heating mode if constraint on σ_{max} is absent or can be neglected.

If attainable accuracy satisfies the inequality $\varepsilon < \varepsilon_{min}^{(1)}$ at $N = 2$, then on the second control stage the value θ_{max} reduces from θ_{adm} due to temperature smoothing. Therefore, for $N = 2$, the condition in Equation (4.23) cannot be provided during two-stage control. In contrast to Equation (4.17), the value $\varepsilon_{min}^{(1)} = \varepsilon_{inf}$ was not defined a priori, and it exceeds the required difference $\theta_{adm} - \theta^*$ under the condition in Equation (4.22). The temperature profile $\theta(l,\Delta_1^0)$ in Figure 4.9 represents this case (similar to Figure 3.6).

Only for a temperature profile of that shape there will be two points within the heated workpiece where minimum admissible final temperature would be met. One point would be located at the workpiece surface ($l = 1$) and another in the core of the workpiece ($l = 0$). The maximum temperature will be reached in the internal point $l = l_{e2}$ according to the equality in Equation (4.23). As a result, the following system of two equations,

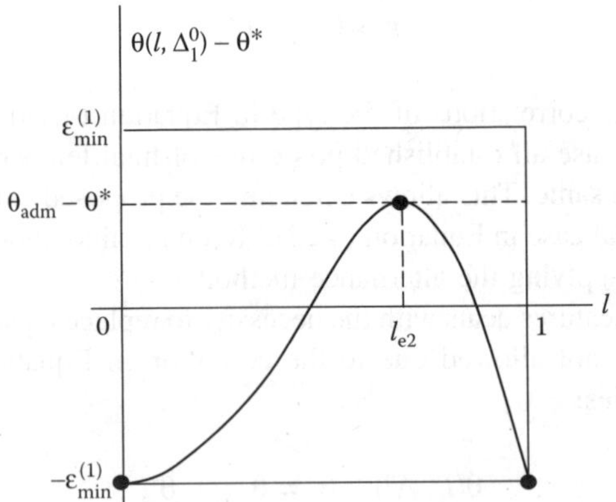

FIGURE 4.9 Temperature profile at the end of time-optimal heating under constraint on maximum temperature for $\varepsilon = \varepsilon_{min}^{(1)} = \varepsilon_{inf} > \theta_{adm} - \theta^*$.

$$\begin{cases} \theta\left(1,\Delta_1^0\right)-\theta^* = -\varepsilon_{\min}^{(1)}; \\ \theta\left(0,\Delta_1^0\right)-\theta^* = -\varepsilon_{\min}^{(1)}; \end{cases} \quad (4.24)$$

should be solved with respect to two unknowns: optimal process time $\Delta_1{}^0$ and value $\varepsilon_{\min}^{(1)}$.

Based on preliminary analysis, the following computational procedure can be put into practice to solve the time-optimal control problem for the control algorithm of the type in Equation (4.1) through Equation (4.3) when at least one constraint on $\theta_{\max}$ is reached (Figure 4.10a and Figure 4.10b).

1. At the beginning, the optimal control problem should be solved for prescribed value $\varepsilon = \varepsilon_0$ neglecting technological constraints. If this problem is solvable,* then this step should reveal that conditions in Equation (2.11) and Equation (2.13) are not violated over the computed heating cycle and, therefore, an optimal solution is obtained. Otherwise, the special type of optimal power control should be chosen from the set in Equation (4.1) through Equation (4.5). This selection will depend upon the shapes of profiles $\theta_{\max}(\varphi)$ and $\sigma_{\max}(\varphi)$ (Section 4.1). Furthermore, the next step should be taken.
2. Assuming that $\varepsilon_{\min}^{(1)} = \theta_{adm} - \theta^* > \varepsilon_{\inf}$, the time-optimal control problem should be solved under the additional technological constraints in Equation (2.11) and Equation (2.13) for providing temperature uniformity (heating accuracy) $\varepsilon = \varepsilon_{\min}^{(1)}$ (Figure 4.7). This problem can be reduced to solution of Equation (4.20).
3. The final surface temperature $\theta(l,\Delta_1^0)$ can be calculated using the results of the previous computation step and Expression (4.15) under $l = 1$, $N = 1$ (Figure 4.11). If the condition $\theta(1,\Delta_1^0) > \theta(0,\Delta_1^0) = -(\theta_{adm} - \theta^*)$ is true (curve 1, Figure 4.11), then the problem is solved as a result of the previous step, and step 5 should be performed next. Otherwise, it is necessary to perform step 4.
4. If results of the previous computation step satisfy the condition $\theta(1,\Delta_1^0) < \theta(0,\Delta_1^0) = -(\theta_{adm} - \theta^*)$ (curve 2, Figure 4.11), then this means that the assumption regarding the value of $\varepsilon_{\min}^{(1)}$ is not valid (step 2), i.e., $\varepsilon_{\min}^{(1)} = \varepsilon_{\inf} > \theta_{adm} - \theta^*$. Therefore, the case (Equation 4.22) takes place for $\varepsilon = \varepsilon_{\min}^{(1)}$ (Figure 4.9) and the set of equations in Equation (4.24) should be solved with respect to Δ_1^0 and $\varepsilon_{\min}^{(1)}$.
5. It is necessary to compare the prescribed numerical value of $\varepsilon = \varepsilon_0$ with values of $\varepsilon_{\min}^{(1)}$ obtained in step 3 or step 4. If the prescribed value of ε_0 coincides with the obtained numerical value of minimax $\varepsilon_{\min}^{(1)}$, then the optimal control problem proves to be solved during one of the previous steps. If this value of ε satisfies the condition $\varepsilon_0 < \varepsilon_{\min}^{(1)}$

* Otherwise, the initial problem is not solvable.

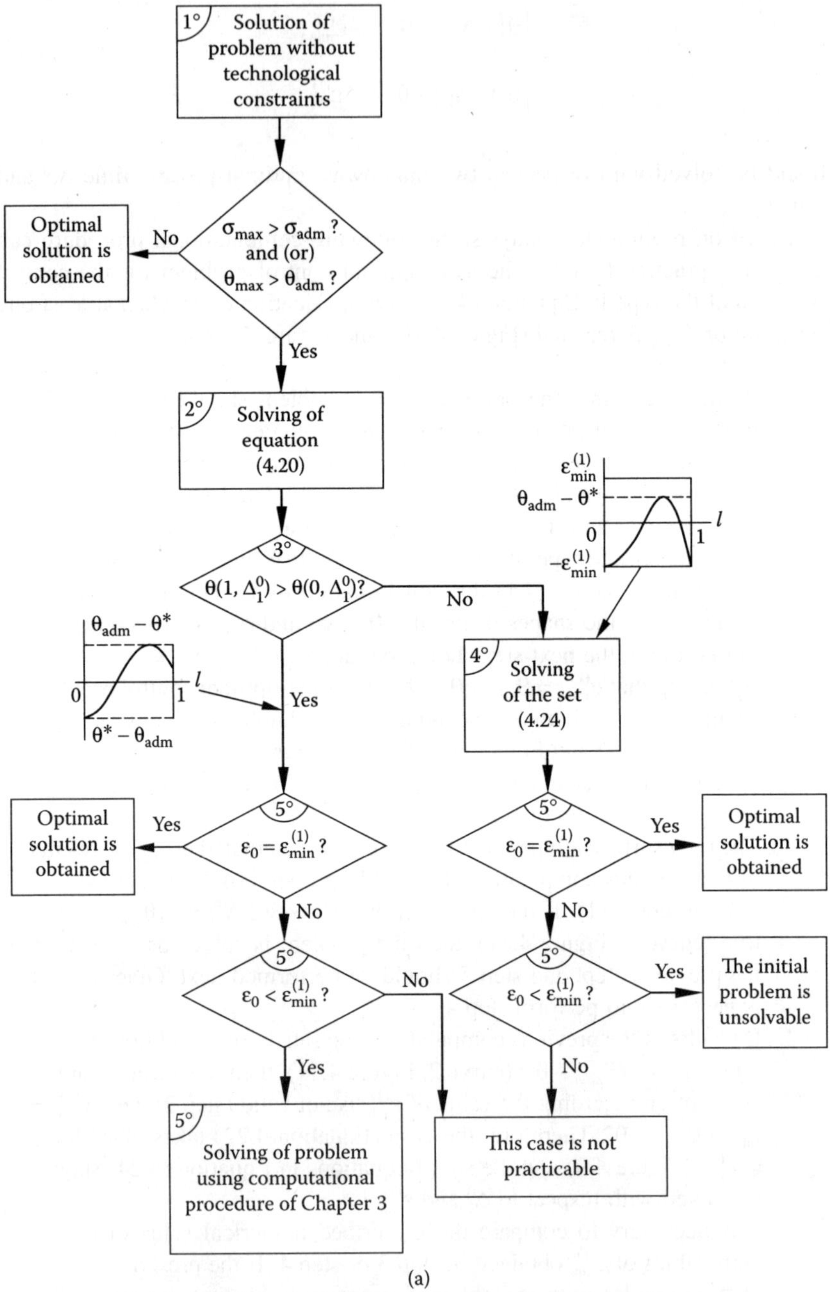

FIGURE 4.10 (a) Block diagram of the computational procedure for solution of the optimal control problem with technological constraints.

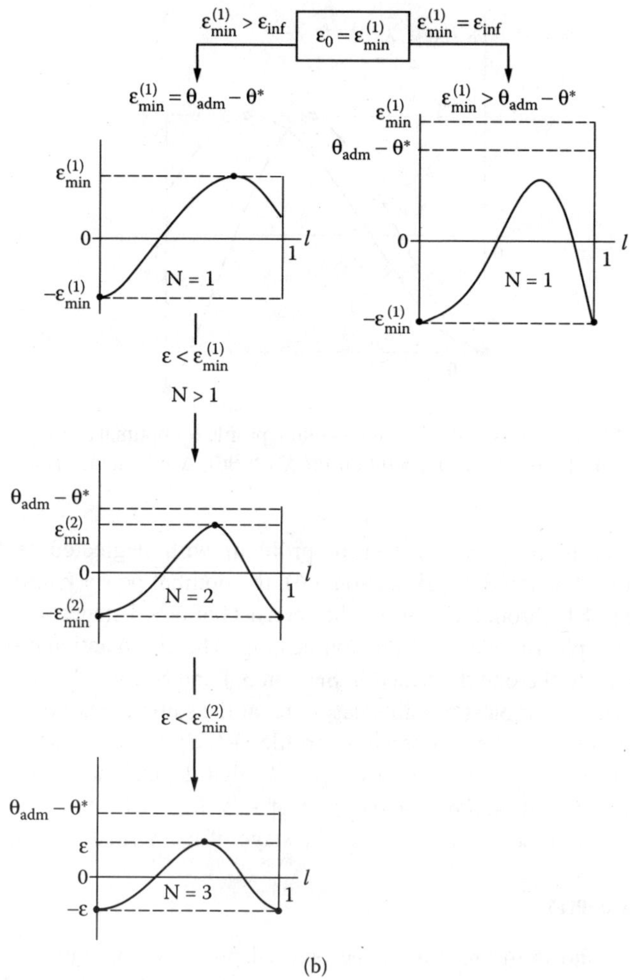

(b)

FIGURE 4.10 (Continued) (b) Variations of the final temperature curves $\theta(l,\Delta^0) - \theta^*$ for different relationships between values of ε_0 and difference $\theta_{adm} - \theta^*$.

(in the case when $\varepsilon_{min}^{(1)} = \varepsilon_{inf}$), then the initial problem is unsolvable because ε_{inf} is the best attainable heating accuracy. If $\varepsilon_0 < \varepsilon_{min}^{(1)}$ (in the case of $\varepsilon_{min}^{(1)} > \varepsilon_{inf}$), then solution of the problem should be found among optimal control functions in the form of Equation (4.1) through Equation (4.3) with $N > 1$ intervals). The previously described computational procedure (Chapter 3) can be used here. Theoretically, a possible case of $\varepsilon_0 > \varepsilon_{min}^{(1)}$ is not practicable, so it will not be considered.

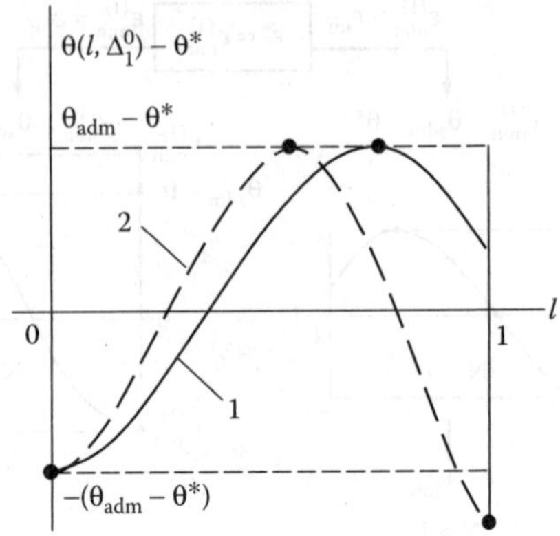

FIGURE 4.11 Variations of the final temperature profile for optimal process time obtained from Equation (4.20). 1: under condition $\theta(l,\Delta_1^0) \geq \theta(0, \Delta_1^0)$; 2: under condition $\theta(l,\Delta_1^0) < \theta(0,\Delta_1{}^0)$.

Similarly to the optimal control problem with neglected technological restraints (see Section 3.3 and Section 3.6), the optimal power control algorithm in Equation (4.1) through Equation (4.5) consists of $N \leq 3$ stages that correspond to the most typical modes of induction heating. The computational procedure is kept identical to the one described in Section 3.4 and Section 3.6. It is important to notice that the simplest possible cases are, at the same time, the most practical cases. Variations of the temperature profile $\theta(l,\Delta^0)$ for different relationships between values of ε_0 and difference $\theta_{adm} - \theta^*$ that depends upon the value of ε_0 are shown in Figure 4.10b. For the case of $\varepsilon_0 > \theta_{adm} - \theta^*$, the optimal control process comes to the end after the single stage of isothermal soaking.

4.1.4 Examples

Figure 4.12 shows the results of the optimal processes of static heating under constraints on σ_{max} and θ_{max} when the condition in Equation (4.17) is met. Computations have been performed according to the previously described technique. Expressions for final temperature distribution $\theta(l,\Delta^0)$ in the form of Equation (4.15) and Equation (4.16) were used in all sets of equations. At least 20 terms of infinite series in Equation (4.15) were included in governing equation sets. This example shows an optimal control mode when it is necessary to provide the maximum heating accuracy $\varepsilon_0 = \varepsilon_{min}^{(2)}$ using two-stage optimal control: heating with maximum power and subsequent soaking.

Figure 4.12a shows how power should be changed during the heating cycle according to the optimal control algorithm in Equation (4.1) for $N = 2$. It is obvious that the heating power $u^{\sigma}(\varphi)$ reaches maximum admissible value U_{max} before reaching maximum temperature. Figure 4.12b shows the temperature profile $\theta(l,\Delta^0)$ that takes place at the end of the optimal heating cycle. One can conclude that it is identical to the temperature profile for the similar case of

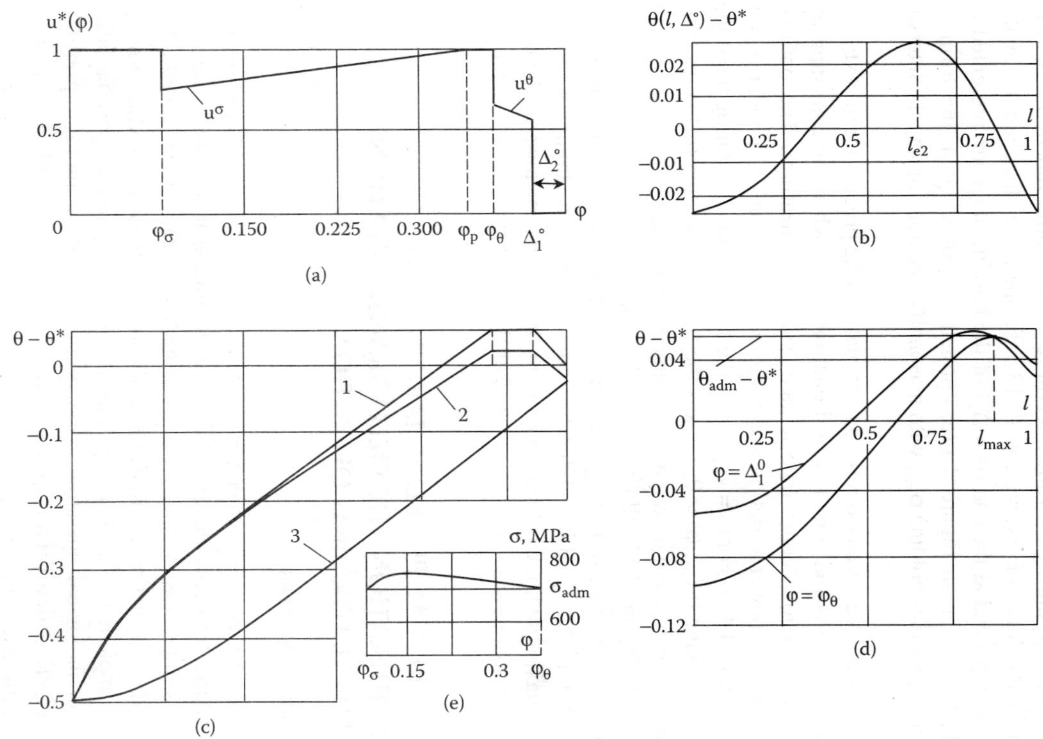

FIGURE 4.12 (a) Optimal control, (b) final temperature profile, (c, d) temperature distributions, and (e) tensile thermal stress for optimal process of heating under constraints on σ_{max} and θ_{max} for $\Gamma = 1$; $\xi = 4$; $U_{max} = 1$; $\theta_0 = -0.5$; $\theta^* = 0$; $Bi = 0.7$; $\theta_{adm} = 0.055$; $\sigma_{adm} = 700$ MPa; $\varepsilon_0 = \varepsilon_{min}^{(2)}$; $u^{\sigma}(\varphi) = 7.1 + 7.9 \exp[0{,}1(\varphi - \varphi_{\sigma})]$; $u^{\theta}(\varphi) = 0.38 + 0.28 \exp[-9, 3(\varphi - \varphi_{\theta})]$. 1: $\theta(l_{max},\varphi)$; 2: $\theta(1,\varphi)$; 3: $\theta(0,\varphi)$.

$\varepsilon_0 = \varepsilon_{\min}^{(2)}$ when the optimal control problem is solved neglecting technological constraints (see Figure 3.8). It is clear that the inequality $\varepsilon_{\min}^{(2)} < \theta_{adm} - \theta^*$ holds true under the condition in Equation (4.17).

Figure 4.12c shows time–temperature profiles for the two-stage heating cycle. Workpiece temperature reaches the maximum admissible value at some point $l_{\max}$ at time $\varphi = \varphi_\theta$. Maximum of temperature at the point $l_{\max}$ should be held at the admissible level θ_{adm} during the time interval $(\varphi_\theta, \Delta_1^0)$. Control function $u^\theta(\varphi)$ can be written in the form of Expression (4.13). Temperature distributions at the beginning $(\varphi = \varphi_\theta)$ and at the end $(\varphi = \Delta_1^0)$ of the holding stage are presented in Figure 4.12d. As seen in Figure 4.12e, optimal control $u^\sigma(\varphi)$ in the form of Equation (4.9) provides holding $\sigma_{\max}$ with reasonable accuracy during the interval $(\varphi_\sigma, \varphi_p)$.

Let us now consider as an example an optimal heating process for the extreme case $\theta_{adm} = \theta^*$ meeting the condition in Equation (4.22) when the constraint on $\sigma_{\max}$ is not taken into account (accelerated isothermal heating). The optimal program of heating power variation according to Expression (4.3) under $N = 1$ is shown in Figure 4.13a. For $\varepsilon_0 = \varepsilon_{\min}^{(1)} = \varepsilon_{\inf}$, the control function during the stage of isothermal soaking $\theta(l_{\max}, \varphi) \equiv \theta_{adm}$ takes the form of Equation (4.13). Dependence of this stage duration on required heating accuracy $(\varepsilon_0 > \varepsilon_{\min}^{(1)})$ is presented in Figure 4.13b. Figure 4.13c illustrates time–temperature history within workpiece volume for the case of $\varepsilon_0 = \varepsilon_{\min}^{(1)}$. Figure 4.13d shows the final temperature distribution for $\varepsilon_0 \geq \varepsilon_{\min}^{(1)}$. Little fallibility could be seen in Figure 4.13d for temperature maximum holding. This maximum shifts over time towards the billet center when temperature is held at the level of $\theta_{adm} = \theta^*$ at the single point $l_{\max}$.

4.2 TIME-OPTIMAL PROBLEM, TAKING INTO CONSIDERATION THE BILLET TRANSPORTATION TO METAL FORMING OPERATION

4.2.1 Problem Statement

Let us consider the time-optimal control problem for the static induction heating processes, taking into consideration billet transportation to the metal forming operation. It is imperative to have a proper temperature state of the heated billet not at the end of induction heating but at the beginning of the metal forming stage. After completion of the heating stage, it is necessary to transport the billet to the next technological operation (i.e., forging, rolling, stamping, upsetting, extrusion, etc.). Time of billet transportation depends upon the application, including but not limited to billet sizes, weight, layout of the machinery, and necessity of intermediate operation (for example, deposition of lubricant or scale removal).

Long transportation time leads to increase of the heat losses from the surface of the heated workpiece due to heat radiation and thermal convection. Therefore, transportation time should be minimized. However, duc to previously mentioned technological restrictions, a transportation time is usually kept within 5 to 20 s.

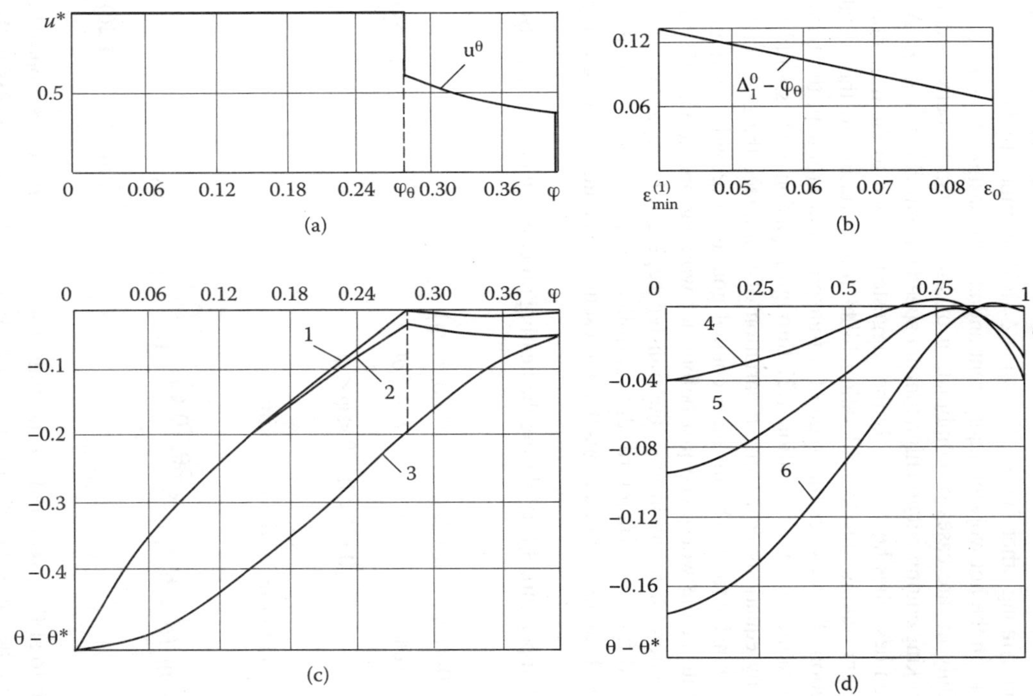

FIGURE 4.13 Results of computation for the optimal process under constraint on θ_{max} at $\theta_{adm} = \theta^* = 0$ for $\Gamma = 1$; $\xi = 4$; $U_{max} = 1$; $Bi = 0.7$; $\theta_0 = -0.5$; $u^\theta(\varphi) = 0.32 + 0.28 \exp[-9.1(\varphi - \varphi_\theta)]$. a: Optimal control ($\varepsilon_0 = \varepsilon_{min}^{(1)}$); b: optimal process time ($\varepsilon_0 \geq \varepsilon_{min}^{(1)}$); c: time–temperature history during optimal heating process ($\varepsilon_0 = \varepsilon_{min}^{(1)}$) d: final temperature distribution ($\varepsilon_0 \geq \varepsilon_{min}^{(1)}$). 1: $\theta(l_{max},\varphi)$; 2: $\theta(1,\varphi)$; 3: $\theta(0,\varphi)$; 4: $\theta(l,\Delta_1{}^0)$ for $\varepsilon_0 = \varepsilon_{min}^{(1)}$; 5: $\theta(l,\Delta_1{}^0)$ for $\varepsilon_0 = 0.095 > \varepsilon_{min}^{(1)}$; 6: $\theta(l,\varphi_\theta)$.

Therefore, when optimizing the whole cycle "induction heating–metal forming," all thermal stages should be taken into consideration. This includes heating cycle, cooling/soaking stage during billet transportation, and metal forming stage.

As was already mentioned in the Section 2.5.3, the transfer stage results in appreciable distortion of the temperature distribution obtained at the end of the heating cycle. The total workpiece process cycle time, φ_1^0, is a sum of time φ^0 required for heating and fixed time Δ_T needed for workpiece delivery to the metal forming station. Among other factors, the value Δ_T is determined by how far the IHI is located from the hot working equipment and also by design of the feeding mechanism. Only in rare cases of combination of small distances between the IHI and hot working equipment and high transfer speed it is possible to neglect Δ_T.

This section describes how to apply the suggested method to the time-optimization problem, taking into consideration billet transportation to the metal forming operation. To consider this type of a transportation problem, the basic model of the heating process (Equation 1.27 through Equation 1.35) should be supplemented by equations describing a variation of workpiece thermal conditions during a transfer time. The ultimate control goal is to provide a required temperature state of the workpiece just before hot working operations. That is why the temperature distribution after the transfer stage can be regarded as an output controlled function. Therefore, the controlled heating cycle is described by the system of heat transfer equations that represent the heating and the cooling (transfer) stages.

The billet temperature $\theta_h(l,\varphi)$ during the heating process can be described by:

$$\frac{\partial\theta_h(l,\varphi)}{\partial\varphi} = \frac{\partial^2\theta_h(l,\varphi)}{\partial l^2} + \frac{\Gamma}{l}\frac{\partial\theta_h(l,\varphi)}{\partial l} + W(\xi,l)u(\varphi);$$
$$0 < l < 1; \quad 0 < \varphi < \varphi^0; \tag{4.25}$$

with boundary conditions of the third kind:

$$\frac{\partial\theta_h(1,\varphi)}{\partial l} = Bi_h\left(\theta_a - \theta_h(1,\varphi)\right); \frac{\partial\theta_h(0,\varphi)}{\partial l} = 0; \quad \theta_h(l,0) \equiv \theta_0 \equiv \theta_a = const. \tag{4.26}$$

Here, φ^0 is the total time required for heating; Biot number Bi_h corresponds to heat losses from the surface of the billet during the whole heating process in the inductor.

During the further transfer stage, the temperature field $\theta_T(l,\varphi)$ satisfies the same equation (Equation 4.25) without heat sources:

$$\frac{\partial\theta_T(l,\varphi)}{\partial\varphi} = \frac{\partial^2\theta_T(l,\varphi)}{\partial l^2} + \frac{\Gamma}{l}\frac{\partial\theta_T(l,\varphi)}{\partial l} + 0 < l < 1; \quad \varphi^0 < \varphi < \varphi^0 + \Delta_T; \tag{4.27}$$

with boundary conditions

$$\frac{\partial\theta_T(1,\varphi)}{\partial l} = Bi_T\left(\theta_a - \theta_T(1,\varphi)\right); \frac{\partial\theta_T(0,\varphi)}{\partial l} = 0; \quad \theta_T\left(l,\varphi^0\right) \equiv \theta_h\left(l,\varphi^0\right). \tag{4.28}$$

Time Δ_T should be considered here as a total transportation time. Heat losses during the transfer stage can be estimated by the value Bi_T in Equation (4.28) that vastly increases in comparison with heat losses inside the induction coil ($Bi_T > Bi_h$). This leads to intensive billet cooling during the transfer stage.

Initial temperature distribution $\theta_T(l,\varphi^0)$ at the beginning of the transfer stage complies with final temperature distribution $\theta_h(l,\varphi^0)$ at the end of the heating stage. This fact, written in the form of equality in Equation (4.28), is imperative to bind the workpiece temperature field during the heating stage to that during the transfer stage.

At the time $\varphi = \varphi^0 + \Delta_T$, instead of Equation (2.25), the condition for the required absolute accuracy of heating can be formulated as:

$$\max_{l\in[0,1]}\left|\theta\left(l,\varphi^0 + \Delta_T\right) - \theta^*\right| \le \varepsilon . \tag{4.29}$$

The time-optimal control problem will be considered based upon the following assumptions:

1. Equation (4.25) through Equation (4.28) represent the heating process.
2. Initial billet temperature distribution prior to the induction heating stage is assumed to be uniform, and required temperature distribution is described by Equation (4.29).
3. Heating power $u(\varphi)$ is chosen as a control input function and the control constraint set is given in the form of Equation (2.7).
4. The cost function can be estimated by the time-optimal criterion (Equation 2.1).

The statement of the time-optimal control problem can be formulated as the following. It is necessary to select this control function $u(\varphi) = u^*(\varphi)$ bounded by the preassigned set in Equation (2.7) that provides steering the workpiece initial temperature distribution to the desired temperature θ^* with prescribed accuracy ε (according to Equation 4.29) for the moment $\varphi = \varphi^0 + \Delta_T$. The control function $u^*(\varphi)$ should hold cost criterion at extremum value.

Obviously, in the case of fixed value of Δ_T, the process of billet heating to the certain a priori unknown temperature distribution $\theta_h(l,\varphi^0)$ (complying with $\theta_T(l,\varphi^0)$) must be time optimal. In this case, the total process time proves to be minimal.

Similarly to Section 2.8, it is possible to prove mathematically[1] that, with regard to the considered models, the time-optimal control algorithm consists of alternating stages of heating with maximum power $u \equiv U_{max}$ (heat ON) and subsequent soaking/cooling under power $u \equiv 0$ (heat OFF) cycles. Consequently, optimal control $u^*(\varphi)$ can be written in the form of Equation (2.27) (Figure 2.8).

Let us consider the transfer stage of known fixed duration Δ_T as an additional last stage of the heating cycle under $u(\varphi) \equiv 0$. Optimal control $u^{**}(\varphi)$ of the whole heating cycle described by the system of Equation (4.25) through Equation (4.28) can be written in the same form (Equation 2.27). However, duration of the transfer stage and billet cooling rate are beyond the control — i.e., control stages of the heating cycle are supplemented by the uncontrollable transfer stage of fixed duration Δ_T. This leads to a number of the essential features in contrast to the control function $u^*(\varphi)$ when the billet was inside the induction coil.

Total number N_1 of $u^{**}(\varphi)$ stages always proves to be even because, in the last interval, the control $u^{**}(\varphi)$ is equal to zero (during billet transfer), and on the first stage it is always equal to U_{max}. The last control stage could be longer than the transfer stage. In this case, the remaining time of this stage takes place when the billet is inside the inductor section that has no power (so-called passive section). Therefore, the number N of $u^*(\varphi)$ stages proves to be equal to N_1. If the duration of the last control stage coincides with the transfer stage, then it is possible to assume the following expression: $N = N_1 - 1$.

Let us consider as an example a single-stage heating with $u^*(\varphi) \equiv U_{max}$ ($N = 1$). The control $u^{**}(\varphi)$ consists of two stages ($N_1 = N + 1$) with the no-power second interval realized under $u(\varphi) = 0$ during workpiece transfer (Figure 4.14). The two-stage heating cycle ($N = 2$) means that, after heating under maximum power, the next interval of temperature soaking takes place inside the induction coil. The following mandatory transfer stage only prolongs duration of the second (heat OFF) interval of the control function $u^{**}(\varphi)$ and N_1 remains equal to N (Figure 4.15).

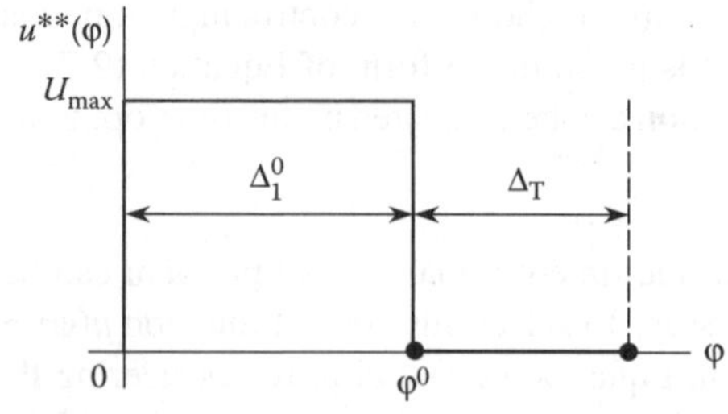

FIGURE 4.14 Time-optimal control with consideration of the transfer stage in the case of $N = 1$, $N_1 = 2$, $\alpha_T = 1$.

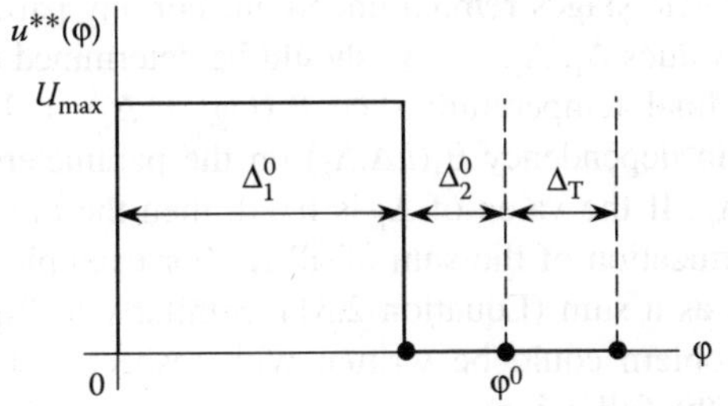

FIGURE 4.15 Time-optimal control with consideration of the transfer stage in the case of $N = 2$, $N_1 = 2$, $\alpha_T < 1$.

It is important to underline at this point that, for the same control function $u(\varphi) = 0$, the temperature field variations that take place during the soaking inside the induction coil and during billet transfer in the open air are described by different equations (Equation 4.25 through Equation 4.26 and Equation 4.27 through Equation 4.28, respectively). The temperature soaking through the billet that is in the inductor occurs usually under relatively small heat losses (i.e., under small value of Bi_h in Equation 4.26). However, during the billet's transfer stage, it is cooled intensively under the much greater value of $Bi_T > Bi_h$ in Equation (4.28).

If $N > 2$, then the optimal control function $u^{**}(\varphi)$ is not changed by nature during the last two control intervals. During these stages, the optimal control represents one of two variants shown in Figure 4.14 and Figure 4.15.

It is imperative at this point to introduce a factor, α_T, that is defined by a formula:

$$\alpha_T = \frac{\Delta_T}{\Delta_T + \Delta_{N_1}^0}. \tag{4.30}$$

Here, $\Delta_{N_1}^0$ is a duration of the last N_1-th stage of the optimal power control $u^*(\varphi)$. It is clear that, if $N_1 = N$, then we deal with duration of the last stage under $u(\varphi) = 0$ and therefore the conditions $\Delta_{N_1}^0 > 0$ and $\alpha_T < 1$ are met (Figure 4.15). However, if $N = N_1 - 1$, then the N_1-th stage is merely absent, and the equalities $\Delta_{N_1}^0 = 0$ and $\alpha_T = 1$ take place (Figure 4.14). Consequently, depending on the value of factor α_T and according to the correlation between N and N_1, two possible scenarios could exist:

$$N_1 = \begin{cases} N, if\ 0 < \alpha_T < 1; \\ N+1, if\ \alpha_T = 1. \end{cases} \tag{4.31}$$

Similarly to the general time-optimal problem (Section 2.8), the shape of the optimal control algorithm is known, but the number, N, and time durations, Δ_1,

$\Delta_2, \ldots, \Delta_N$, of the cycle stages remain uncertain. For any particular process, the number N and the values $\Delta_1, \Delta_2, \ldots, \Delta_N$ should be determined during subsequent computation. The final temperature state $\theta_T(l,\varphi^0 + \Delta_T)$ in Equation (4.29) is described by certain dependency $\theta_T(l,\Delta,\Delta_T)$ on the parameters $\Delta = (\Delta_1, \Delta_2, \ldots, \Delta_N)$ and value of Δ_T. If the value of Δ_T is fixed, then the time-optimal problem is reduced to minimization of the sum of all Δ_i. For example, the cost criterion can be determined as a sum (Equation 2.31). Similarly to Equation (2.34), the optimal control problem could be written with respect to the requirement in Equation (4.29) as the following:

$$\begin{cases} I(\Delta) = \sum_{i=1}^{N} \Delta_i \to \min\limits_{\Delta}; \\ \Phi(\Delta) = \max\limits_{l\in[0,1]} \left|\theta_T(l,\Delta,\Delta_T) - \theta^*\right| \leq \varepsilon. \end{cases} \quad (4.32, 4.33)$$

In contrast to Equation (2.34), important unique features occur when solving this problem. One of these features deals with a necessity to solve the problem under the condition of a priori unknown value of the factor α_T in Equation (4.31) — i.e., when correlation between the duration of the last $u^{**}(\varphi)$ interval and transfer time is unknown. In addition, the optimal final temperature distribution can be significantly distorted due to the workpiece cooling during the transfer stage of fixed duration Δ_T. This circumstance greatly complicates computational techniques for the "transportation" problem because it is impossible simply to apply conclusions obtained earlier for the general time-optimal problem.

The explicit form of expression for $\theta_T(l,\Delta,\Delta_T)$ in Equation (4.33) can be found as follows. At first, initial temperature distribution $\theta_T(l,\varphi^0)$ for the transfer stage should be obtained. According to Equation (4.28), this distribution would be identical to the final temperature distribution $\theta_h(l,\varphi^0)$ that corresponds to the temperature profile at the end of the heating stage. Similarly to the general time-optimal problem (Section 2.8), the temperature field $\theta_h(l,\varphi^0)$ can be described using expressions in Equation (2.28) and Equation (2.29) by substituting control input in the form of Equation (2.27). It is possible to calculate $\theta_T(l,\Delta,\Delta_T)$ by known methods as a solution of Equation (4.27) with boundary conditions (Equation 4.28) under previously found initial conditions $\theta_T(l,\varphi^0) = \theta_h(l,\varphi^0) = \theta_h(l,\Delta)$. Finally, for $Bi_T \neq Bi_h$, the temperature distribution $\theta_T(l,\Delta,\Delta_T)$ could be written in the form of the following infinite series expansion[1,2,8]:

$$\theta_T(l,\Delta,\Delta_T) = \theta_0 + 4(Bi_T - Bi_h)U_{\max} \sum_{s=1}^{\infty}\sum_{n=1}^{\infty} N_{ns} M_n K(\mu_s l) \cdot e^{-\mu_s^2 \Delta_T}, \quad (4.34)$$

where

$$N_{ns} = W_n(\xi)\mu_s^2\Big[\left(\mu_s^2 + Bi_T^2 + (1-\Gamma)Bi_T\right)\cdot\left(\mu_n^2 + Bi_h^2 + (1-\Gamma)Bi_h\right)\cdot \left(\mu_s^2 - \mu_n^2\right)\cdot K(\mu_s)\cdot K(\mu_n)\Big]^{-1}, \tag{4.35}$$

$$M_n = \sum_{r=1}^{N}(-1)^{r+1}\left(1 - \exp\left(-\mu_n^2\sum_{m=r}^{N}\Delta_m\right)\right), \tag{4.36}$$

and μ_n and μ_s represent corresponding roots of the equations:

$$Bi_h K(\mu) - \mu K_1(\mu) = 0;\ Bi_T K(\mu) - \mu K_1(\mu) = 0. \tag{4.37}$$

Thus, this optimal control problem is reduced to solution of the mathematical programming problem (Equation 4.32 through Equation 4.37), which should be considered as a particular case of the general problem (Equation 3.38).

4.2.2 Computational Technique for the "Transportation" Problem of Time-Optimal Heating

The basic properties (Equation 3.39 and Equation 3.40) remain valid for the final temperature distribution $\theta_T(l,\Delta,\Delta_T)$ in the transportation problem of time-optimal heating (see conclusions made in Section 3.7). The proposed engineering technique is based on these universal features as they are used in the course of computational procedure.

It is important to notice that the fixed value Δ_T is not included in the set of sought parameters. Because N_1 is always an even value, the appropriate inequalities sequence (Equation 3.41) contains only minimax temperature deviations $\tilde{\varepsilon}_{\min}^{(k)},\ k = \overline{1, N^*}$ with even indices:

$$\tilde{\varepsilon}_{\min}^{(2)} > \tilde{\varepsilon}_{\min}^{(4)} > \ldots > \tilde{\varepsilon}_{\min}^{(N^*)} = \tilde{\varepsilon}_{\inf} > 0\,. \tag{4.38}$$

The sign "~" is used to indicate a minimax temperature deviation considered in the class of "general" control inputs $u^{**}(\varphi)$ with N_1 intervals. It is possible to prove[1] that in this case the basic relation in Equation (3.3) holds true after substituting N_1 and $\tilde{\varepsilon}_{\min}^{(s)}$ instead of N and $\varepsilon_{\min}^{(s)}$, respectively:

$$N_1 = S \ \textit{for all}\ \varepsilon : \tilde{\varepsilon}_{\min}^{(S)} \le \varepsilon < \tilde{\varepsilon}_{\min}^{(S-1)}\,. \tag{4.39}$$

Number N_1 of $u^{**}(\varphi)$ constancy intervals depends on the place of ε in the sequence in Equation (4.38) and can be defined according to Equation (4.39).

On the basis of the universal properties in Equation (3.39) and Equation (3.40), the particular temperature profile of final temperature distribution should be revealed depending on the given time of transfer stage that represents the specifics of the transportation time-optimal control problem.

It is important to note that the previously mentioned deal with the case of $\alpha_T = 1$. It is easy to understand that, under $\alpha_T < 1$, all basic rules and properties of the temperature profile stated in Chapter 3 for time-optimal problem (with $\Delta_T = 0$) remain the same for the considered case of $\Delta_T > 0$. Obviously, if $\alpha_T = 1$, then the sum $\Delta_{N_1}^0 + \Delta_T$ represents total duration of the last interval of control function $u^*(\varphi)$ (Figure 4.15). Therefore, it turns out that duration of the last interval of $u^*(\varphi)$ can be changed by varying the value $\Delta_{N_1}^0 > 0$. This means that the last interval time is under control as well as all intervals $\Delta_i^0, i = \overline{1, N}$. All constancy intervals of control function $u^{**}(\varphi)$ can be chosen freely as well as in the general time-optimal problem.

Conclusions regarding dependency of number N on ε prove to be the same. The shape of the curve that represents temperature distribution at the end of the optimal process will also be the same as in the general OCP. At the same time, only the case of $\alpha_T = 1$ (under $\Delta_{N_1}^0 = 0$ [Figure 4.14]) should be considered as an exception because duration of the last N_1-th interval of control $u^{**}(\varphi)$ becomes fixed and equal to Δ_T. Essentially, the number of sought parameters decreases by one. This is the main reason that the properties of final temperature distribution are changed under $\alpha_T = 1$.

If the value of the factor α_T in Equation (4.30) is unknown a priori, then the following solution algorithm is suggested for the transportation time-optimal control problem (Figure 4.16):

1. The first term, $\tilde{\varepsilon}_{\min}^{(2)}$, of the inequalities series (Equation 4.38) represents the best heating accuracy for the class of two-stage control inputs $u^{**}(\varphi)$. Obviously, $\tilde{\varepsilon}_{\min}^{(2)}$ depends on the value Δ_T. The accuracy $\tilde{\varepsilon}_{\min}^{(2)}$ can be reached using control inputs with different values Δ_T for $\alpha_T = 1$ or $\alpha_T < 1$ (Figure 4.14 and Figure 4.15). It is clear that condition $\alpha_T = 1$ is met if the value Δ_T is appreciably large and billet soaking inside the induction coil becomes unnecessary because sufficient temperature equalization takes place during the transfer stage. On the other hand, the condition $\alpha_T < 1$ is true if the value Δ_T is substantially small, leading to too short duration of this stage for providing required temperature equalization. Taking this fact into consideration, it is possible to choose certain minimum (unknown a priori) value $\Delta_T = \Delta_{T2}^*$ from all Δ_T values for which $\alpha_T = 1$ (Figure 4.17). This value will correspond to a certain value of $\tilde{\varepsilon}_{\min}^{(2)}(\Delta_{T2}^*)$.
   The problem in Equation (4.32) through Equation (4.37) can be solved assuming that required heating accuracy complies with $\tilde{\varepsilon}_{\min}^{(2)}(\Delta_{T2}^*)$ and given transportation time is equal to Δ_{T2}^*. Let us consider two values of Δ_T: $\Delta_T = \Delta_{T2}^*$ and $\Delta_T = \Delta_{T2}^* - \delta\Delta_T$, where $\delta\Delta_T > 0$ is a small enough value.

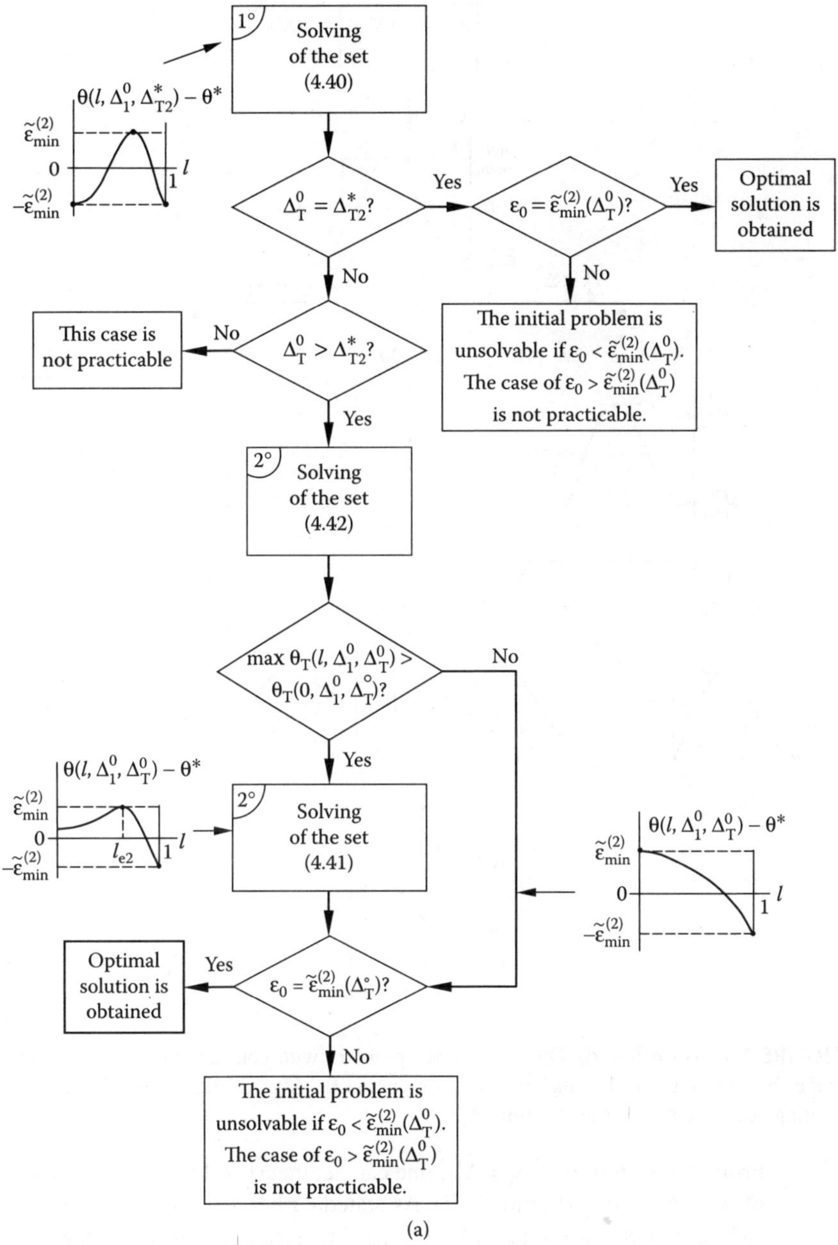

FIGURE 4.16 Optimal control problem with consideration of the transfer stage. a: Block diagram of the computational procedure for problem solution.

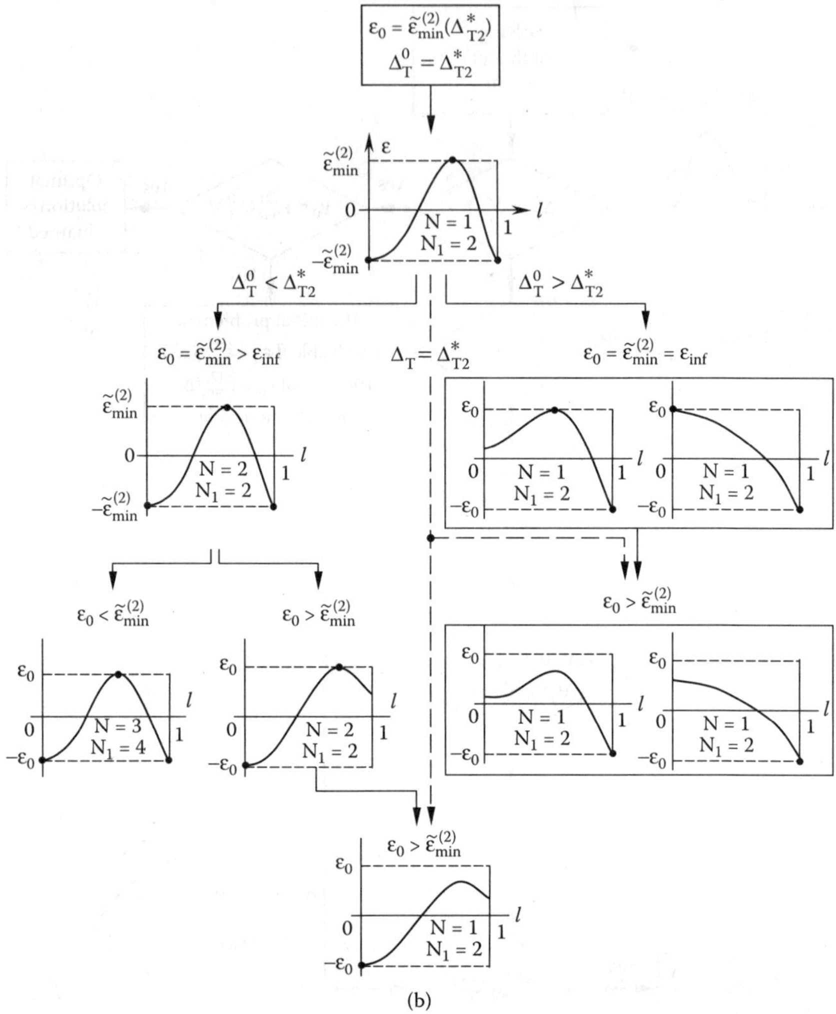

FIGURE 4.16 (Continued) Optimal control problem with consideration of the transfer stage. b: Variations of the final temperature curves $\theta(l,\Delta^0) - \theta^*$ depending on required heating accuracy ε_0 and transfer time Δ_T^0.

From the definition of $\Delta_T = \Delta_{T2}^*$, one can get that $\alpha_T < 1$ under condition of $\Delta_T = \Delta_{T2}^* - \delta\Delta_T$ (Figure 4.17). As stated earlier, the condition $N = 2$ can be written for $\varepsilon = \tilde{\varepsilon}_{\min}^{(2)}(\Delta_{T2}^* - \delta\Delta_T)$. The properties of the temperature distribution curve $\theta_T(l,\Delta^0,\Delta_{T2}^* - \delta\Delta_T)$ remain similar to properties of $\theta_T(l,\Delta^0)$ for $\varepsilon = \varepsilon_{\min}^{(2)}$ under $\Delta_T = 0$ (this represents the general time-optimal problem). Due to the continuous nature of dependency of temperature distribution $\theta_T(l,\Delta^0,\Delta_T)$ on Δ_T, the temperature distribution $\theta_T(l,\Delta^0,\Delta_{T2}^*)$ also exhibits these properties for $N = 1$ and $\varepsilon =$

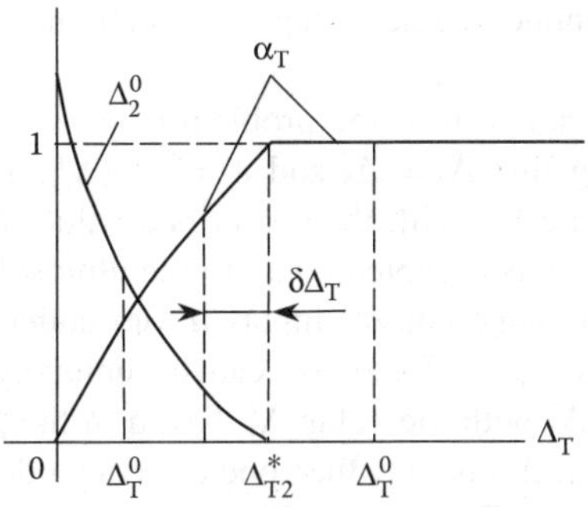

FIGURE 4.17 Optimal control parameters as functions of transfer time for $\tilde{\varepsilon}^{(2)}_{\min}(\Delta_T)$.

$\tilde{\varepsilon}^{(2)}_{\min}(\Delta^*_{T2})$. Thus, it is possible to use here a computational procedure described in Chapter 3.

Figure 4.14 shows the sought optimal control input $u^{**}(\varphi)$. The shape of the temperature curve $\theta_T(l,\Delta^0,\Delta^*_{T2})$ corresponds to the one shown in Figure 3.8 for $\varepsilon = \varepsilon^{(2)}_{\min}$. Based on these conclusions, the computational set of equations can be written in the form similar to Equation (3.16):

$$\begin{cases} \theta_T\left(0,\Delta_1^0,\Delta_{T2}^*\right)-\theta^* = -\tilde{\varepsilon}^{(2)}_{\min}; \\ \theta_T\left(l_{e2},\Delta_1^0,\Delta_{T2}^*\right)-\theta^* = +\tilde{\varepsilon}^{(2)}_{\min}; \\ \theta_T\left(1,\Delta_1^0,\Delta_{T2}^*\right)-\theta^* = -\tilde{\varepsilon}^{(2)}_{\min}; \\ \dfrac{\partial\theta_T\left(l_{e2},\Delta_1^0,\Delta_{T2}^*\right)}{\partial l} = 0. \end{cases} \tag{4.40}$$

This system of four equations can be solved with respect to four unknowns: duration Δ_1^0 of the billet heating inside the induction coil; time of transfer stage Δ^*_{T2} being optimal duration of the second control interval for $\alpha_T = 1$; the minimax value $\tilde{\varepsilon}^{(2)}_{\min}(\Delta^*_{T2})$; and the coordinate l_{e2} of the location of maximum temperature at the end of heating. In contrast to the general time-optimal problem, here it is necessary to substitute expressions for $\theta_T(l,\Delta_1^0,\Delta^*_{T2})$ under $l = 0$, $l = l_{e2}$, and $l = 1$ in the form of Equation (4.34) through Equation (4.37) for $N = 1$, instead of Equation (2.28).

If the values $\varepsilon = \varepsilon_0$ and $\Delta_T = \Delta_T^0$ are preset in the initial problem (Equation (4.32) and Equation (4.33)) and, at the same time, conditions $\Delta_T^0 = \Delta^*_{T2}$ and $\varepsilon_0 = \varepsilon^{(2)}_{\min}(\Delta_T) = \tilde{\varepsilon}^{(2)}_{\min}(\Delta^*_{T2})$ remain valid, then the solution of this problem would be obtained at this stage. Otherwise, it is necessary to conduct the next step of the computation procedure (Figure 4.16).

2. Let us find the solution of the problem (Equation 4.32 and Equation 4.33), assuming that $\Delta_T = \Delta_T^0$ and $\varepsilon = \tilde{\varepsilon}_{\min}^{(2)}(\Delta_T^0)$. This means that the value of Δ_T coincides with the prescribed value of the transfer time, and the value of ε is accepted equal to the utmost heating accuracy in the class of two-stage control inputs $u^{**}(\varphi)$ under the given value of Δ_T. Here, the value of factor α_T can be uniquely defined based on comparison of Δ_T^0 with the value Δ_{T2}^* found in the previous step.
If condition $\Delta_T{}^0 \geq \Delta_{T2}^*$ is true, then one can conclude that $\alpha_T = 1$ (Figure 4.17). According to Equation (4.39), control input $u^{**}(\varphi)$ has a single control interval ($N = 1$) of $u^*(\varphi)$ and takes the form shown in Figure 4.14. As shown in Rapoport,[1] in this case the minimax $\tilde{\varepsilon}_{\min}^{(2)}(\Delta_T^0)$ under $\alpha_T = 1$ already complies with $\varepsilon_{\inf}$, and heating with accuracy better than $\tilde{\varepsilon}_{\min}^{(2)}(\Delta_T^0)$ becomes impossible.
In the case of $\alpha_T = 1$, the heating accuracy $\tilde{\varepsilon}_{\min}^{(2)}$, attainable under control inputs $u^{**}(\varphi)$ with fixed Δ_T, can be considered as a minimax $\varepsilon_{\min}^{(1)}$ for the class of power control inputs $u^*(\varphi)$ applied to billets located inside the induction coil (Figure 4.14). In this case, the required heating time, Δ_1^0 ($N = 1$), represents a single sought parameter of optimal process. Therefore, at the end of the optimal process, according to the basic properties in Equation (3.39) and Equation (3.40), there are two points, l_j^0, $j = 1, 2$ ($R = N + 1 = 2$), where the maximum deviations of final temperature from the desired one are reached.
To find out the exact location of these points, it is necessary to find the shape of curve $\theta_T(l,\Delta^0,\Delta_T)$, which is now different from those shown in Figure 3.8. In contrast to the final temperature distribution for $\Delta_T{}^0 = \Delta_{T2}^*$, the longer transfer stage results in lower billet surface temperature and higher temperature in its center (core). Only points $l_1^0 = l_{e2}$, $l_2^0 = 1$ or $l_1^0 = 0$, $l_2^0 = 1$ can be considered as extremum points $l = l_j^0$. Finally, in the case of $\Delta_T^0 > \Delta_{T2}^*$ and depending on the particular value of Δ_T^0, the radial temperature profile at the end of the optimal process will be as shown in Figure 4.18a or Figure 4.18b.

The appropriate computational equation sets can be written for the case of Figure 4.18a as:

$$\begin{cases} \theta_T\left(l_{e2}, \Delta_1^0, \Delta_T^0\right) - \theta^* = +\tilde{\varepsilon}_{\min}^{(2)}; \\ \theta_T\left(1, \Delta_1^0, \Delta_T^0\right) - \theta^* = -\tilde{\varepsilon}_{\min}^{(2)}; \\ \dfrac{\partial \theta_T\left(l_{e2}, \Delta_1^0, \Delta_T^0\right)}{\partial l} = 0; \end{cases} \tag{4.41}$$

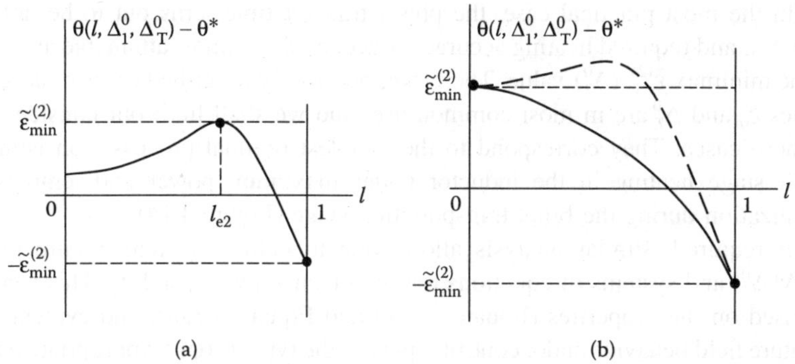

FIGURE 4.18 Variations of the temperature curve $\theta(l, \Delta_1^0, \Delta_T^0) - \theta^*$ at the end of the optimal process for $\Delta_T^0 > \Delta_{T2}^*$: a $-\theta(l_{e2}, \Delta_1^0, \Delta_T^0) - \theta^* = \tilde{\varepsilon}_{min}^{(2)}$, $l_{e2} > 0$; b $-\theta(0, \Delta_1^0, \Delta_T^0) - \theta^* = \tilde{\varepsilon}_{min}^{(2)}$ (dashed curve is not optimal temperature distribution).

and for the case of Figure 4.18b as:

$$\begin{cases} \theta_T\left(0, \Delta_1^0, \Delta_T^0\right) - \theta^* = \tilde{\varepsilon}_{min}^{(2)}; \\ \theta_T\left(1, \Delta_1^0, \Delta_T^0\right) - \theta^* = -\tilde{\varepsilon}_{min}^{(2)}. \end{cases} \tag{4.42}$$

Thus, if the transfer stage is sufficiently long, there is a surface area ($l = 1$) with the minimum admissible temperature that deviates from the required one by value of $\tilde{\varepsilon}_{min}^{(2)}$ (Δ_T^0) (maximum underheating). At the same time, there is an internal area with the maximum admissible temperature that is higher by the same value. With increase of Δ_T^0, the maximum temperature could be shifted further away from the billet surface towards its core ($l = 0$). At the same time, the core temperature always remains higher than minimum admissible level. In contrast to the discussed case, at the end of the time-optimal process, without consideration for transfer stage, the region with the minimum temperature always occurs in the billet center ($l = 0$) as well as on its surface ($l = 1$) (Figure 3.8).

Unambiguous choice between temperature profiles (Figure 4.18) should be made using the tested solution of the system in Equation (4.42) with respect to unknown variables Δ_1^0 and $\tilde{\varepsilon}_{min}^{(2)} = \tilde{\varepsilon}_{inf}$. If it leads to such value, Δ_1^0, that an expression $\max\limits_{l\in[0,1]} \theta_T(l, \Delta_1^0, \Delta_T) > \theta_T(0, \Delta_1^0, \Delta_T)$ is valid (Figure 4.18b, dashed line), then this means that the shape of curve $\theta_T(l, \Delta_1^0, \Delta_T^0)$ would correspond to Figure 4.18a. Therefore, it would be necessary to proceed solving the set in Equation (4.41) with respect to unknown variables Δ_1^0, $\tilde{\varepsilon}_{min}^{(2)} = \tilde{\varepsilon}_{inf}$, and l_{e2}. Otherwise, the system of equations (Equation 4.42) has been chosen properly. The expression $\theta_T(l,\Delta_1^0,\Delta_T^0)$ in the form of Equation (4.34) through Equation (4.37) should be substituted in both systems (Equation 4.41 and Equation 4.42) for $N = 1$.

In the most practical case, the preset transfer time turns out to be not less than Δ^*_{T2}, and required heating accuracy is accepted as utmost attainable and equal to the minimax $\tilde{\varepsilon}^{(2)}_{\min}(\Delta^0_T)$ value. Therefore, previously described cases of accepted values ε_0 and Δ^0_T are in most common use, and we shall limit our consideration to these cases. They correspond to the simplest optimal process, consisting of single-stage heating in the inductor under maximum power and temperature equalization during the billet transportation stage (Figure 4.14).

If required, similar analysis allows one to define the temperature profile $\theta(l,\Delta^0,\Delta^0_T)$ and systems of equations for any given values ε_0 and Δ^0_T. This analysis is based on the properties (Equation 3.39 and Equation 3.40) and evident temperature field behavior under control inputs of the type $u^{**}(\varphi)$.[1] Appropriate results are shown in Figure 4.16b, which illustrates changes in the shape of the temperature profile depending on values ε_0 and Δ^0_T. It is important to notice that the value $\tilde{\varepsilon}^{(2)}_{\min}(\Delta^0_T)$ exceeds $\varepsilon_{\inf}$ only for $\Delta^0_T < \Delta^*_{T2}$. Therefore, there is a possibility to further improve the heating accuracy relative to $\tilde{\varepsilon}^{(2)}_{\min}$. The condition $\alpha_T < 1$ would be valid, and, as was stated earlier, all properties of temperature profile $\theta(l,\Delta^0,\Delta^0_T)$ as well as computational sets of equations remain unchanged in comparison to the general optimal control problem with $\Delta_T = 0$.

4.2.3 Technological Constraints in "Transportation" Problem

Previously defined optimal heating modes could result in the necessity for appreciable overheating of the billet surface to compensate core soaking and surface thermal losses that take place during its transportation stage. Therefore, admissible temperature maximums can be exceeded during the heating stage, and appropriate technological constraints should be taken into account. It is typically required that maximum workpiece temperature $\theta_{\max}(\varphi)$ should be below a certain admissible value θ_{adm}. If this value will be exceeded, then undesirable and sometimes irreversible adverse changes in material structural properties and even metal melting could take place. This constraint can be written in the form of Equation (2.11).

On the other hand, the temperature differences within the heated workpiece should be restricted during heating in such a way that the maximum value $\sigma_{\max}$ of tensile thermal stresses $\sigma(l,\varphi)$ (due to thermal gradients) would not exceed prescribed admissible value σ_{adm} that corresponds to the ultimate stress limit of the heated material. Therefore, the constraint on the thermal stresses field in the form of Equation (2.13) should be satisfied. Violation of these conditions could result in irreversible product damage, i.e., crack development.

Additional technological constraints add complexity to finding a solution for the optimal control problem resulting in modification of appropriate computational procedures (Chapter 2).

The technological constraints (Equation 2.11 and Equation 2.13) now will be taken into account. Therefore, the optimal power control algorithm prior to the transfer stage can be described in the form of Equation (4.1). The final temperature distribution $\theta_T(l,\Delta^0,\Delta_T^0)$ at the end of this stage is described by Expression (4.34), in which it is necessary to replace M_n to C_n in the form of Equation (4.16).

If condition $\Delta_T^0 < \Delta_{T2}^*$ holds true and Expression (4.17) is valid for given heating accuracy, then the optimal computational technique described in Section 4.1 remains unchanged for the transportation problem. The foregoing conclusion follows from the fact that lengths of all $u^{**}(\varphi)$ intervals remain under control when α_T becomes less than 1. This conclusion can be extended to the case of $\Delta_T^0 = \Delta_{T2}^*$. Here, Δ_{T2}^* makes the same sense as in Section 4.2.2. The difference lies in the fact that the earlier obtained expression for final temperature distribution $\theta_T(l,\Delta^0,\Delta_T^0)$ should be substituted into computational sets of equations. Noticeable distinctions emerge only when these conditions are not met.

1. Let us first consider the case when condition $\varepsilon > \theta_{adm} - \theta^*$ holds true (Section 4.1.3). Because temperature maximum, which is limited during the whole heating process by value θ_{adm}, can only decrease during the transfer stage, then only negative admissible deviations of final temperature from a required one could be reached. These deviations are equal to $-\varepsilon$.
   Let us assume that $\varepsilon = \tilde{\varepsilon}_{\min}^{(2)}(\Delta_T^0)$. If technological constraints will be taken into account, the shape of temperature distribution $\theta_T(l,\Delta^0,\Delta_T^0)$ typically remains the same. Then the area with the minimum temperatures can be reached only at two points: $l = 0$ and $l = 1$ for $N = 1$ (if $\alpha_T = 1$) or $N = 2$ (if $\alpha_T < 1$) (Figure 4.16b). At the same time, if $\alpha_T < 1$ and $N = 2$, then the value $\varepsilon = \tilde{\varepsilon}_{\min}^{(2)}$ in the class of controls $u^{**}(\varphi)$ can be considered as $\varepsilon_{\min}^{(2)}$ in the class of $u^*(\varphi)$ for fixed value Δ_T. In that case, according to Equation (3.39) and Equation (3.40), three points with minimum temperature $\theta^* - \tilde{\varepsilon}_{\min}^{(2)}$ should exist.
   Therefore the condition $\alpha_T < 1$ is impossible and $\alpha_T = 1$, $N = 1$ (Figure 4.14). Similarly to other cases for $\alpha_T = 1$ (Section 4.2.2), the condition $\tilde{\varepsilon}_{\min}^{(2)} = \tilde{\varepsilon}_{\inf}$ would hold true. Minimax $\tilde{\varepsilon}_{\min}^{(2)}$ can be treated as a value $\varepsilon_{\min}^{(1)}$ in the class of controls $u^*(\varphi)$ for the same fixed value Δ_T. As a

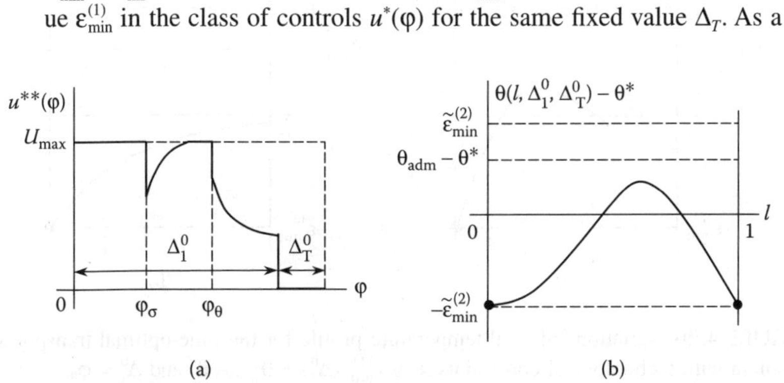

FIGURE 4.19 Time-optimal transportation problem with technological restraints for $\varepsilon = \tilde{\varepsilon}_{\min}^{(2)} > \theta_{adm} - \theta^*$ and $\Delta_1^0 > \varphi_\theta$. a: optimal control; b: final temperature profile.

result, in practice an optimal process with ultimately allowable deviation $\tilde{\varepsilon}_{\inf} > \theta_{adm} - \theta^*$ can be applied relatively easily. It consists of two stages. The first one is heating with maximum power under constrained control, thermal stresses, and maximum temperature according to algorithm (Equation 4.1). The second stage is workpiece transfer during fixed time Δ_T^0 (Figure 4.19a).

According to Equation (3.39) and Equation (3.40), at the end of optimal process the minimum temperatures $\theta^* - \tilde{\varepsilon}_{\min}^{(2)}$ should be reached at two points: in the center, $l = 0$, and on the surface, $l = 1$, of the billet (Figure 4.19b; compare to Figure 4.9). An appropriate set of equations takes the following form:

$$\begin{cases} \theta_T\left(0, \Delta_1^0, \Delta_T^0\right) - \theta^* = -\tilde{\varepsilon}_{\min}^{(2)}; \\ \theta_T\left(1, \Delta_1^0, \Delta_T^0\right) - \theta^* = -\tilde{\varepsilon}_{\min}^{(2)}. \end{cases} \tag{4.43}$$

This set can be solved with respect to Δ_1^0 and $\tilde{\varepsilon}_{\min}^{(2)}$.

If transfer stage duration Δ_T^0 is sufficiently long, only the second equality in Equation (4.43) can be met. This is so in our case because only a single point of minimum temperature exists on the billet surface (Figure 4.20). Therefore, one can obtain that $R = 1$ for $\varepsilon = \varepsilon_{\min}^{(N)}$ in Equation (3.39) and Equation (3.40). This means that optimal control proves to be completely determined a priori and there are no unknown parameters (i.e., $N = 0$!). This conclusion holds true for $\varphi_\theta = \Delta_1^0$ in Equation (4.1) — i.e., when the transfer stage begins at the moment of obtaining maximum admissible temperature θ_{adm}.

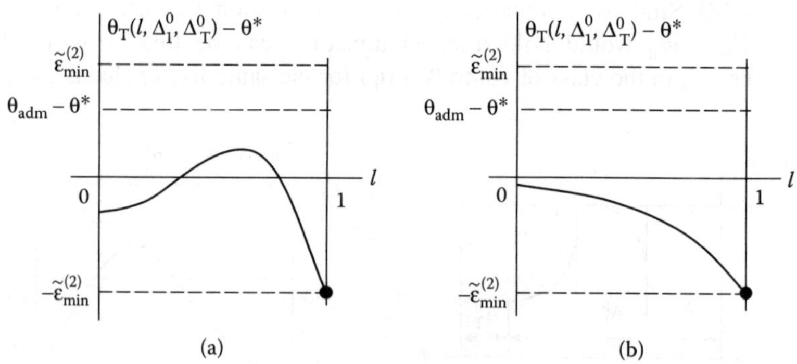

FIGURE 4.20 Variations of final temperature profile for the time-optimal transportation problem with technological constraints: $\varepsilon = \tilde{\varepsilon}_{\min}^{(2)}(\Delta_T^0) > \theta_{adm} - \theta^*$ and $\Delta_1^0 = \varphi_\theta$.

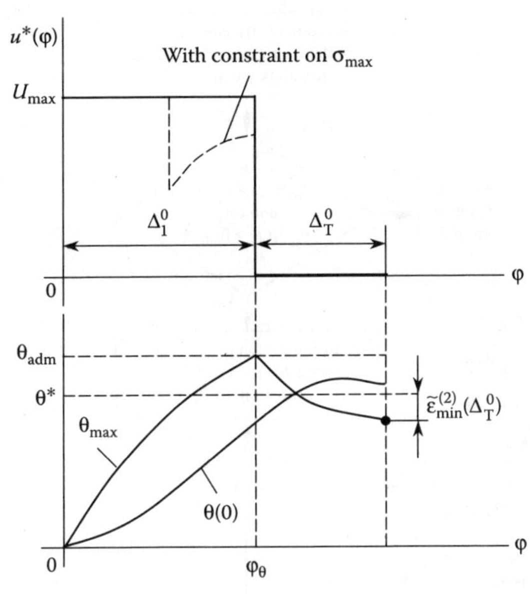

FIGURE 4.21 Optimal control and time–temperature profiles for the time-optimal transportation problem with constraint on maximum temperature. $\varepsilon = \tilde{\varepsilon}^{(2)}_{min}(\Delta_T^0)$ and $\Delta_1^0 = \varphi_\theta$.

In this case, duration of the isothermal soaking stage under control input $u^\theta(\varphi)$ (Figure 4.2 through Figure 4.4) proves to be equal to zero. Then the optimal process is reduced to the billet heating stage with maximum power up to the moment φ_θ, determined by relations in Equation (4.12), and further transfer stage of fixed duration. All parameters of control function $u^{**}(\varphi)$ turn out to be known a priori, and the value $\tilde{\varepsilon}^{(2)}_{min}$ can be calculated as $|\theta_T(l, \Delta_1^0, \Delta_T^0) - \theta^*|$ under $l = 1$, $\Delta_1^0 = \varphi_\theta$ (Figure 4.21).

2. Due to the continuous nature of dependency of final temperature distribution on Δ_T and θ_{adm}, the described properties of temperature distribution $\theta_T(l, \Delta^0, \Delta_T)$ can be extended to the case of $\varepsilon_0 < \theta_{adm} - \theta^*$ under condition $\Delta_T > \Delta_{T2}^*$. In this case, the problem is reduced to searching for only a single unknown parameter Δ_1^0 of optimal control.

As a result, the following general computational procedure can be put into practice for determining parameters of optimal heating processes that involve the transfer stage and intervals of movement along the constraints (Figure 4.22). We shall limit our consideration to the cases of $\Delta_T{}^0 \geq \Delta_{T2}^*$ and $\varepsilon = \tilde{\varepsilon}^{(2)}_{min}(\Delta_T^0)$ (Section 4.2.2).

1. In the beginning, the transportation time-optimal control problem ought to be solved for preset value $\varepsilon = \varepsilon_0$ without additional technological constraints on temperature distribution during a process of heating.

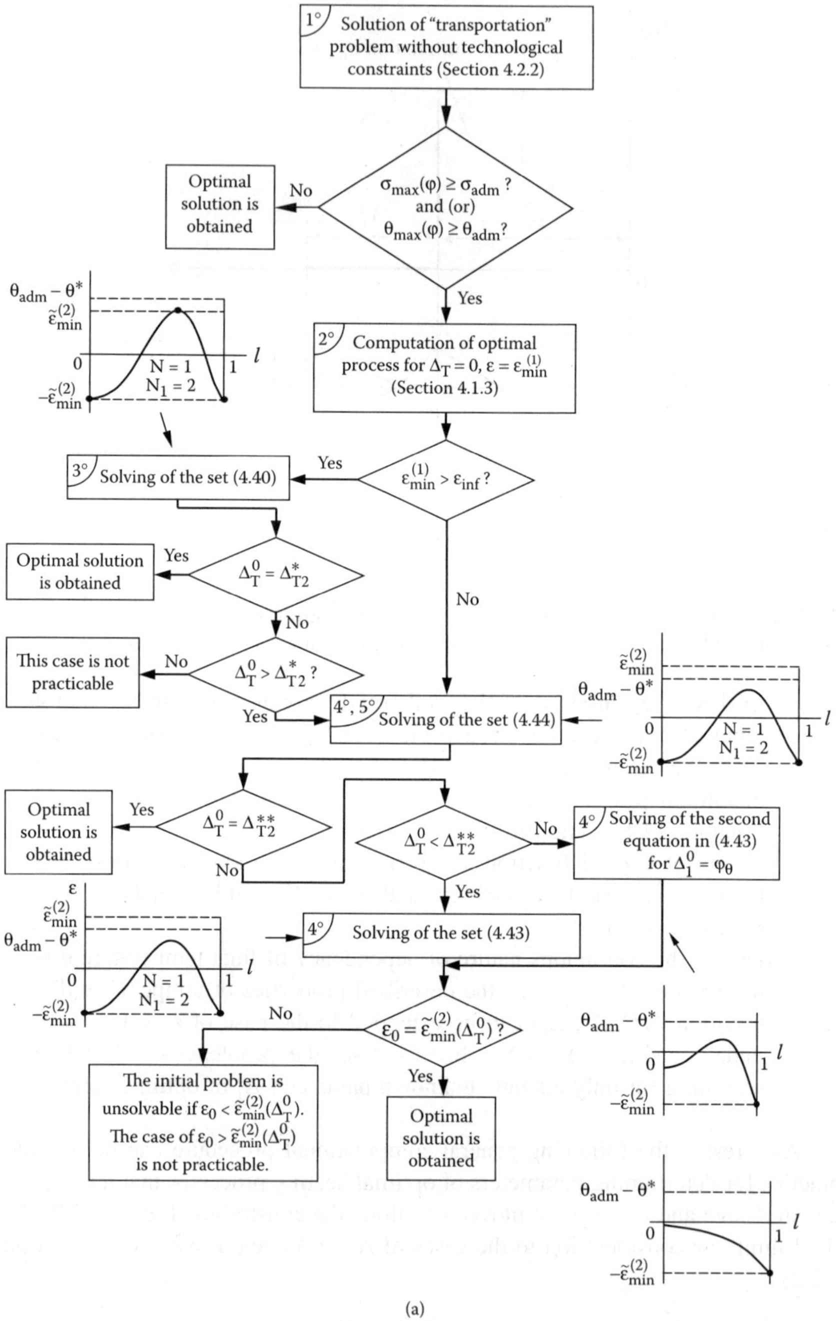

FIGURE 4.22 Time-optimal transportation problem with constraints on θ_{max} and σ_{max}. a: Block diagram of the computational procedure for problem solution.

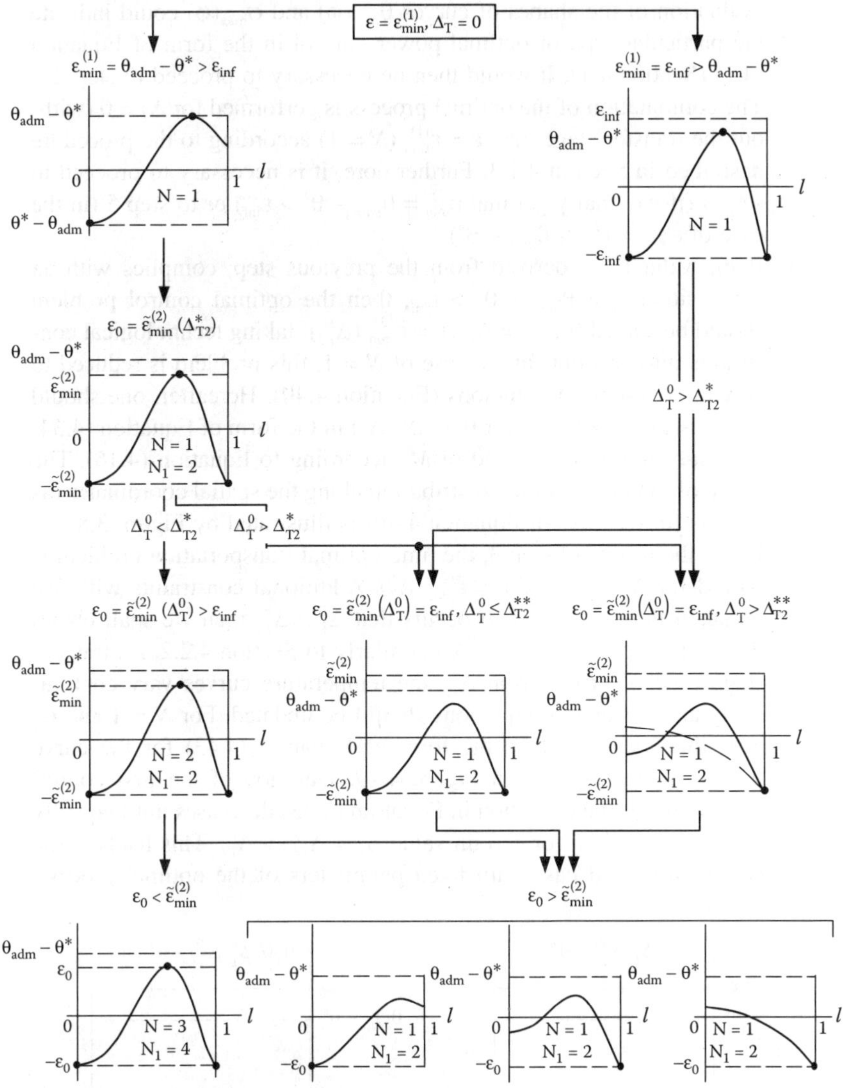

(b)

FIGURE 4.22 (Continued) Time-optimal transportation problem with constraints on θ_{max} and σ_{max}. b: Variations of the final temperature curves $\theta(l,\Delta^0,\Delta_T^0) - \theta^*$ depending upon the values of ε_0 and Δ_T^0.

The technique described in Section 4.2.2 can be extended to treat this problem. If careful consideration at this point reveals that conditions in Equation (2.11) and Equation (2.13) are not violated during the whole heating process, then only in this case it may be concluded that the obtained OCP solution should be considered as optimal. Otherwise, evaluation of the shapes of curves $\theta_{max}(\varphi)$ and $\sigma_{max}(\varphi)$ could indicate the particular type of optimal power control in the form of Equation (4.1) (Section 4.1). It would then be necessary to proceed to step 2.

2. The computation of the optimal process is performed for $\Delta_T = 0$ (without the transfer stage) and $\varepsilon = \varepsilon_{min}^{(1)}$ ($N = 1$) according to the procedure described in Section 4.1.3. Furthermore, it is necessary to proceed to step 3 (if we finally get that $\varepsilon_{min}^{(1)} = \theta_{adm} - \theta^* > \varepsilon_{inf}$) or to step 5 (in the case of $\varepsilon_{min}^{(1)} = \varepsilon_{inf} > \theta_{adm} - \theta^*$).
3. If the value $\varepsilon_{min}^{(1)}$, derived from the previous step, complies with an expression $\varepsilon_{min}^{(1)} = \theta_{adm} - \theta^* > \varepsilon_{inf}$, then the optimal control problem should be solved for $\Delta_T = \Delta_{T2}^*$, $\varepsilon = \tilde{\varepsilon}_{min}^{(2)}(\Delta_{T2}^*)$, taking technological constraints into account. In the case of $N = 1$, this problem is reduced to solving a system of equations (Equation 4.40). Hereafter, one should use only the expression for $\theta_T(l, \Delta^0, \Delta_T)$ in the form of Equation (4.34) with substitution C_n instead of M_n according to Equation (4.16). The shape of final temperature distribution along the spatial coordinates (as well as for the case in Equation 4.40) is illustrated by Figure 3.8.
4. Using the results of step 3, the time-optimal transportation problem is solved for $\Delta_T = \Delta_T^0$ and $\varepsilon = \tilde{\varepsilon}_{min}^{(2)}(\Delta_T^0)$. Additional constraints will also be taken into account. If it occurs that $\Delta_T^0 > \Delta_{T2}^*$ then we shall obtain the equality $\tilde{\varepsilon}_{min}^{(2)}(\Delta_T^0) = \tilde{\varepsilon}_{inf}(\Delta_T^0)$ similarly to Section 4.2.2. As transfer stage time Δ_T^0 grows from Δ_{T2}^*, the temperature curves vary continuously and systems of equations should be updated. For $N = 1$ and $\alpha_T = 1$, these systems take the form of Equation (4.43) for the curve $\theta_T(l, \Delta_1^0, \Delta_T)$ represented in Figure 4.19. Duration of the first control interval under the condition in Equation (4.43) decreases until equality $\Delta_1^0 = \varphi_\theta$ is reached for certain value $\Delta_T^0 = \Delta_{T2}^{**} > \Delta_{T2}^*$. This leads to the earlier discussed case with fixed parameters of the optimal process.

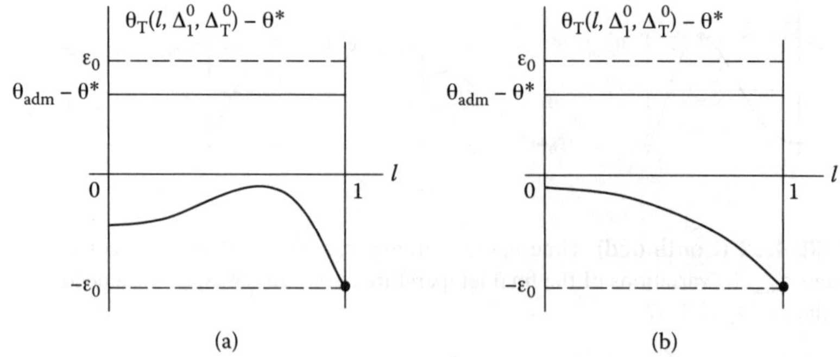

FIGURE 4.23 Variations of final temperature profile for the time-optimal transportation problem with constraints on θ_{max} and σ_{max}: $\varepsilon = \varepsilon^0 = \tilde{\varepsilon}_{min}^{(2)}$, $\varepsilon_0 > \theta_{adm} - \theta^*$; $\Delta_T^0 > \Delta_{T2}^{**}$.

Then, the system of equations (Equation 4.43) for $\Delta_1^0 = \varphi_\theta$ takes the following form:

$$\begin{cases} \theta_T\left(0, \varphi_\theta, \Delta_{T2}^{**}\right) - \theta^* = -\tilde{\varepsilon}_{\min}^{(2)}; \\ \theta_T\left(1, \varphi_\theta, \Delta_{T2}^{**}\right) - \theta^* = -\tilde{\varepsilon}_{\min}^{(2)}. \end{cases} \tag{4.44}$$

This set can be solved with respect to unknown values Δ_{T2}^{**} and $\tilde{\varepsilon}_{\min}^{(2)}(\Delta_{T2}^{**}) = \tilde{\varepsilon}_{\inf}(\Delta_{T2}^{**})$.

For all $\Delta_T^0 > \Delta_{T2}^{**}$, the equalities $N = 1$, $\alpha_T = 1$, and $\varphi_\theta = \Delta_1^0$ hold true as before and optimal control proves to be known a priori. In this case, the maximum deviation of the final temperature from the required one is equal to $-\tilde{\varepsilon}_{\min}^{(2)}(\Delta_T^0)$ and it is reached at the unique point $l = 1$ for the shape of curve $\theta_T(l, \Delta_1^0, \Delta_T^0)$ shown in Figure 4.23a or Figure 4.23b. The value of minimax $\tilde{\varepsilon}_{\min}^{(2)}(\Delta_T^0) = \tilde{\varepsilon}_{\inf}(\Delta_T^0)$ can be computed directly from the second equation in Equation (4.43).

As Δ_T^0 increases from Δ_{T2}^{*}, first, $\tilde{\varepsilon}_{\min}^{(2)}(\Delta_T^0) < \theta_{adm} - \theta^*$ and, second, the sign of this inequality is changed to the inverse one. Therefore, depending on the relationship between Δ_T^0 and Δ_{T2}^{**}, the problem for $\Delta_T = \Delta_T^0$ and $\varepsilon = \tilde{\varepsilon}_{\min}^{(2)}(\Delta_T^0)$ is reduced now to solution of the equations system (Equation 4.43) under $\Delta_T^0 < \Delta_{T2}^{**}$. If $\Delta_T^0 > \Delta_{T2}^{**}$, then the second equation in Equation (4.43) should be solved for $\Delta_1^0 = \varphi_\theta$.

5. If the value $\varepsilon_{\min}^{(1)}$ derived from step 2 satisfies the expression $\varepsilon_{\min}^{(1)} = \varepsilon_{\inf} > \theta_{adm} - \theta^*$ at $\Delta_T = 0$, then for any $\Delta_T > 0$ the following expression $\tilde{\varepsilon}_{\min}^{(2)}(\Delta_T) = \tilde{\varepsilon}_{\inf}(\Delta_T) > \theta_{adm} - \theta^*$ would be valid. In the case of values $\varepsilon = \tilde{\varepsilon}_{\min}^{(2)}(\Delta_T^0)$ the optimal heating process is completed within one stage ($N = 1$) under $u^*(\varphi) \equiv U_{\max}$ at $\alpha_T = 1$. In this case, the "transportation" problem is solved again for $\Delta_T = \Delta_{T2}^{**}$ and $\varepsilon = \tilde{\varepsilon}_{\min}^{(2)}(\Delta_{T2}^{**})$ under technological constraints. Solution of the equations set (Equation 4.44) can be found similarly to one described at step 4. If $\Delta_T^0 = \Delta_{T2}^{**}$, $\varepsilon_0 = \tilde{\varepsilon}_{\min}^{(2)}(\Delta_{T2}^{**})$, then the initial problem proves to be solved at this stage. Under $\Delta_T^0 < \Delta_{T2}^{**}$ or $\Delta_T^0 > \Delta_{T2}^{**}$, the computational procedure is continued for $\varepsilon = \tilde{\varepsilon}_{\min}^{(2)}(\Delta_T^0)$ according to the technique described in step 4 (the case of $\Delta_T^0 > \Delta_{T2}^{**}$).

   In a similar manner, the alternance method can be generalized to all other possible cases. Computational sets are updated according to a required heating accuracy ε_0, transfer time Δ_T^0, and admissible temperature θ_{adm}. Figure 4.22b shows the variations of temperature curves $\theta_T(l, \Delta^0, \Delta_T^0)$ depending upon the values of ε_0 and Δ_T^0.

4.2.4 Examples

Figure 4.24 and Figure 4.25 illustrate some computational results of time-optimal heating processes, taking into consideration the transfer stage without technological

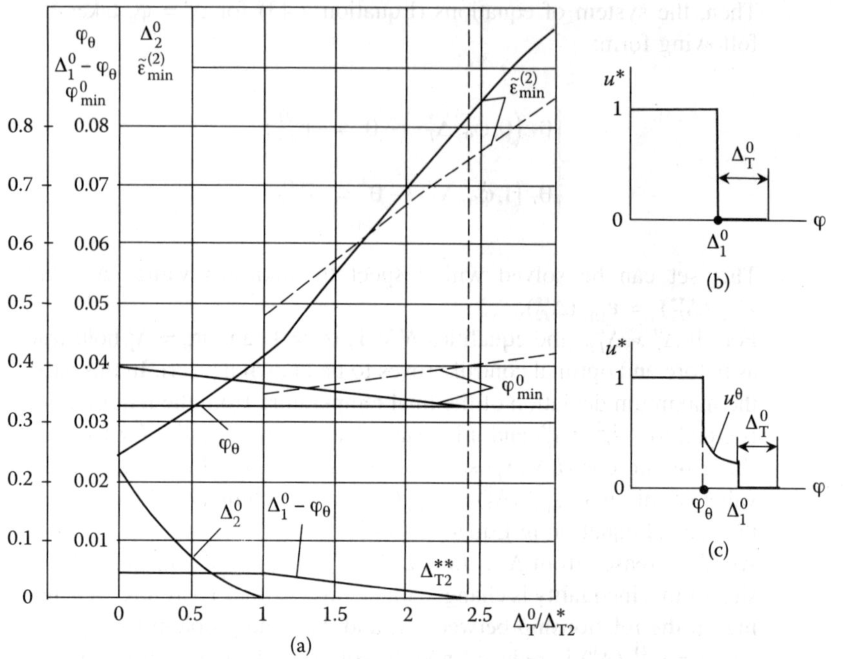

FIGURE 4.24 (a) Optimal control parameters and (b, c) optimal control inputs under $\Delta_T^0 > \Delta_{T2}^*$ for the time-optimal transportation problem without constraints (a, dashed line and b) and with constraint on θ_{max} (a, solid lines and c): $\Gamma = 1$; $\xi = 4$; $Bi_h = 0.7$; $Bi_T = 1.8$; $\theta_0 = -0.5$; $\theta^* = 0$; $U_{max} = 1$; $\theta_{adm} = 0.055$; $\Delta_{T2}{}^* = 0.0073$; $\varepsilon_0 = \tilde{\varepsilon}_{min}^{(2)}$; $u^\theta(\varphi) = 0.36 + 0.27 \exp[-9.27(\varphi - \varphi_\theta)]$.

constraints (Figure 4.24a, dashed line; Figure 4.24b) and under these constraints (Figure 4.24a, solid line; Figure 4.24c and Figure 4.25). Computations have been performed for $\varepsilon_0 = \tilde{\varepsilon}_{min}^{(2)} (\Delta_T^0)$ using the technique described in Sections 4.2.2 and 4.2.3.

Figure 4.24a shows optimal process parameters as functions of Δ_T^0. There is an obvious difference comparing it with the case of $\Delta_T^0 = 0$. In particular, heating accuracy dramatically decreases with increase of Δ_T^0, leading to increase of the minimax value $\tilde{\varepsilon}_{min}^{(2)} (\Delta_T^0)$.

According to previously discussed conclusions, time Δ_2^0 of temperature equalization inside the induction coil tends to zero as Δ_T^0 increases up to value Δ_{T2}^*. Duration $\Delta_1^0 - \varphi_\theta$ of the isothermal soaking interval under control $u^\theta(\varphi)$ becomes equal to zero when Δ_T^0 grows up to value Δ_{T2}^{**}. Figure 4.24 clearly shows that minimax value $\tilde{\varepsilon}_{min}^{(2)}$ varies when technological constraint on θ_{max} is taken into account.

Figure 4.25 illustrates the optimal heating process for $\tilde{\varepsilon}_{min}^{(2)} (\Delta_T^0)$ under $\theta_{adm} = \theta^* = 0$ and $\Delta_T{}^0 > \Delta_{T2}^{**}$. Here, according to the aforesaid analysis, the first heating interval (with maximum power) is completed at the moment φ_θ, when equality $\theta_{max} = \theta_{adm} = \theta^* = 0$ becomes true. The following stage entirely coincides with

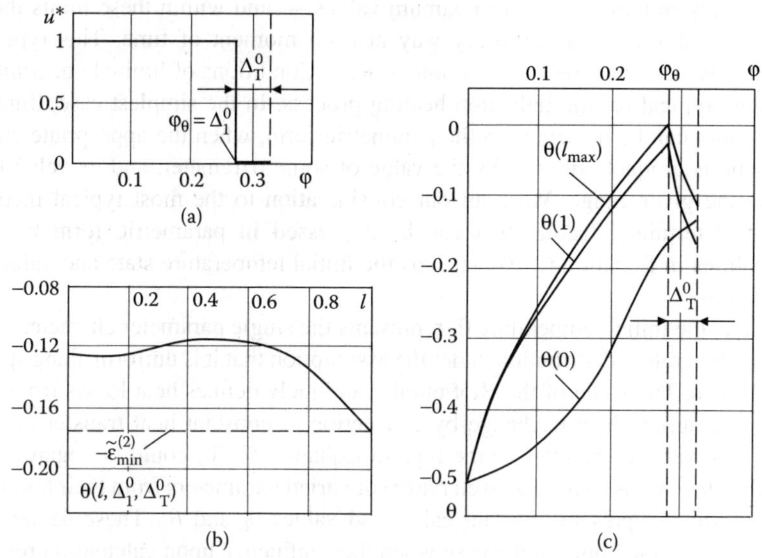

FIGURE 4.25 (a) Optimal control, (b) final temperature curve $\theta(l,\Delta_1^0,\Delta_T^0) - \theta^*$, and (c) time–temperature history for the time-optimal transportation problem with constraint on θ_{max}: $\theta_{adm} - \theta^* = 0$ and $\Delta_T{}^0 > \Delta_{T2}^{**}$ ($\Gamma = 1$; $\xi = 4$; $U_{max} = 1$; $\theta_0 = -0.5$; $\theta^* = 0$; $Bi_h = 0.7$; $Bi_T = 1.8$; $\Delta_T^0 = 5\Delta_{T2}^*$; $\varepsilon_0 = \tilde{\varepsilon}_{min}^{(2)}$).

the stage of workpiece delivery (Figure 4.25a). Final temperature distribution is shown in Figure 4.25b. Figure 4.25c shows the time-temperature profiles during an optimal heating.

4.3 TIME-OPTIMAL HEATING UNDER INCOMPLETE INFORMATION WITH RESPECT TO CONTROLLED SYSTEMS

In real production situations, we often face incompleteness of source information with respect to the induction heating system. Incompleteness of information is caused at first by an imperfection of our knowledge about certain subtle features of heating systems. This includes but is not limited to uncertainties of electromagnetic and thermal–physical properties of heated billets, power supply limitations, presence of harmonics, etc.

On the other hand, there are uncertainties with respect to particular conditions of the technological process. Initial temperature distribution of the heated workpiece and/or value of heat losses from its surface to the ambient area could be regarded as other examples of incompletely defined operating characteristics of the heating process.

It was mentioned earlier (see Section 2.6) that a priori information regarding process characteristics includes only a limited range of their possible variations

— i.e., only minimum and/or maximum values — and within these limits these values could vary in an arbitrary way at each moment of time. This type of uncertainty is often called *interval uncertainty.* Conditions of limited uncertainty are quite typical for the induction heating process. In the simplest case, limited uncertainty can be presented in the parametric form, when the appropriate characteristic is uniquely defined by the value of some parameter, and its value lies within the given range. We limit our consideration to the most typical incompletely determined factors that can be expressed in parametric form by the inequalities in Equation (2.20), such as the initial temperature state and value of heat losses.

Here, the initial temperature θ_0 represents the single parameter characterizing initial temperature distribution under the assumption that it is uniform along space coordinates. The value of the Biot number uniquely defines heat losses from the surface assuming heat exchange by convection at constant heat transfer coefficient. Interval uncertainties of the type in Equation (2.20) could be regarded as possible deviations (within a given range) of varied parameters from their nominal values, which represent constant calculated values θ_0 and Bi. These deviations could be neglected only in the case when their influence upon calculation results proves to be insignificant. Optimization problems considered previously have been solved under such assumption. However, the range of uncertainty cannot be estimated in advance. Therefore, the solution of the optimal control problem under condition of incomplete information with respect to the heating system characteristics is of vital importance.

This section shows that the proposed approach of optimal mode calculation could be used for incompletely determined induction heating systems. To facilitate understanding of the proposed method, the explanations will be given with respect to the simplest IHP model and time-optimal control function.

4.3.1 Problem Statement

Similarly to Section 2.8, let us consider the time-optimal control problem of billet heating with given accuracy up to required temperature (the additional technological constraints and transportation time will not be taken into account). We shall assume that complete information with respect to initial temperature and heat losses is not available and that their interval uncertainty can be written in the form of Equation (2.20). Let us suppose that each separate billet is heated under constant heat losses from its surface estimated by Bi number, which can take on any invariable in time value within the range $[Bi_{min}, Bi_{max}]$. Obviously, the values θ_0 and Bi can differ from one billet to another for different heating cycles. Keep in mind that results obtained next can be extended to other types of uncertainties. The technique discussed in Section 4.1 and Section 4.2 can be effectively applied to the problems subject to additional constraints and the billet transfer stage.

Under conditions in Equation (2.20), the problem description should be revised. Because specific values of θ_0 and Bi were not defined a priori for the heating cycle of a particular billet, all possible values within given intervals would be equally probable. Therefore, a set of controlled heating processes could be considered instead of one controlled system under uncertainty (Equation 2.20). Each process of this set provides the full information regarding values of θ_0 and Bi chosen from the range of all possible values of these uncertain factors.

In other words, each combination of fixed admissible values of θ_0 and Bi (the number of such combinations is infinite) should be in accordance with the appropriate heating process described by Equation (1.27) through Equation (1.31) and Equation (1.34). This means that the infinite set of such "auxiliary" heating systems with full information about their properties will replace real IHP under the condition in Equation (2.20).

When governing the whole set of these objects by means of a single power control, the requirement (Equation 2.25) to final temperature must be provided under any admissible values θ_0 and Bi. If we denote the final temperature distribution for each combination of fixed values θ_0 and Bi as $\theta(l,\theta_0,Bi,\varphi^0)$, then, instead of Equation (2.25), the requirement to temperature distribution with given accuracy at the end of the heating cycle can be obtained in the following form:

$$\max_{Bi,\theta_0}\left[\max_{l\in[0,1]}\left|\theta\left(l,\theta_0,Bi,\varphi^0\right)-\theta^*\right|\right]\le\varepsilon;$$
$$Bi\in\left[Bi_{\min},Bi_{\max}\right],\theta_0\in\left[\theta_{0_{\min}},\theta_{0_{\max}}\right]. \tag{4.45}$$

In contrast to Section 2.8, the time-optimal control problem for the heating process described by Equation (1.27) through Equation (1.31) and Equation (1.34) could be formulated as follows. It is necessary to choose time-dependent power control restricted by Equation (2.7) that provides steering workpiece initial temperature distribution to the target state (Equation 4.45) in minimal possible time $\varphi^0 = \varphi^0_{\min}$.

The optimal control algorithm, which is the solution of this problem, guarantees that, under any possible combination of values θ_0 and Bi, the required heating accuracy is provided at the end of the heating process. Obviously, heating time $\varphi^0_{\min}$, obtained by this algorithm, will comply with minimal possible time for the most unfavorable combination of values θ_0 and Bi. However, this time certainly exceeds the minimal possible time for all other admissible values of uncertain factors. Therefore, the value of the time-optimal cost function exceeds the appropriate value attainable for the system with complete initial information with respect to θ_0 and Bi. This cost loss is a "compensation" for guaranteed heating quality under conditions of interval uncertainty. This control algorithm implies the guaranteed cost control approach, well known in control theory.[9]

It can be shown[1,10] that the optimal control algorithm has the same form of stepwise function (Equation 2.27) (Figure 2.8) as in the time-optimal problem for the IHP model with full information about the heating system. Therefore, the time-optimal control algorithm consists of alternating stages of heating with maximum power $u \equiv U_{max}$ (heat ON) and subsequent soaking (heat OFF) cycles, $u \equiv 0$. The number of stages ($N \geq 1$) is defined uniquely by given heating accuracy ε. Smaller ε requires a larger number of stages N. Therefore, the profile of the optimal control algorithm is known, but the number, N, and time durations, Δ_1, Δ_2, …, Δ_N, of the described stages remain unknown. With regard to the inequality in Equation (4.45), the problem is reduced to the following form (similar to Equation 2.34):

$$\begin{cases} I(\Delta) = \sum_{i=1}^{N} \Delta_i \to \min_{\Delta}; \\ \Phi(\Delta) = \max_{Bi,\theta_0} \left[\max_{l \in [0,1]} \left| \theta(l, \theta_0, Bi, \Delta) - \theta^* \right| \right] \leq \varepsilon. \end{cases} \tag{4.46}$$

Here, $\theta(l,\theta_0,Bi,\Delta)$ is defined as $\theta(l,\Delta)$ in Equation (2.34); however, in this case θ_0 and Bi represent unknown parameters meeting conditions in Equation (2.20).

4.3.2 Technique for Time-Optimal Problem Solution under Interval Uncertainties

Let us introduce the vector parameter $v = (l, Bi, \theta_0)$ that denotes a set of any fixed values:

$$l \in [0;1],\ Bi \in [Bi_{min};\ Bi_{max}],\ \theta_0 \in [\theta_{0min};\ \theta_{0max}].$$

Then, similarly to Equation (3.38), the problem (Equation 4.46) can be written in the following form:

$$\begin{cases} I(\Delta) = \sum_{i=1}^{N} \Delta_i \to \min_{\Delta}; \\ \Phi(\Delta) = \max_{v} \left| \theta(v, \Delta) - \theta^* \right| \leq \varepsilon. \end{cases} \tag{4.47}$$

Resemblance between the problems in Equation (3.38) and Equation (4.47) consists of the fact that, in the general statement (Equation 3.38), it is also possible to explore the maximum of temperature deviation inside multidimensional area Ω — i.e., within the set of spatial coordinate vectors x. At the same time, the problem in Equation (4.47) differs from Equation (3.38) in a way that the vector parameter v contains the parameters θ_0 and Bi in addition to spatial coordinate l.

Finally, this problem becomes sufficiently complicated in comparison with Equation (2.34) due to consideration of the maximum of temperature deviation within the set of vector v values (instead of the set of spatial coordinate l values).

Analysis shows that, according to conclusions of Section 3.7, the basic properties (Equation 3.39 and Equation 3.40) of the final temperature distribution remain valid for the problem in Equation (4.47).

It is important to underline that $\theta(v,\Delta)$ in Equation (4.47) for each Δ represents the whole set of final temperature distributions along spatial coordinate l for all possible values of uncertain factors θ_0 and Bi. Therefore, properties in Equation (3.39) and Equation (3.40) pertain exactly to this whole set, but corresponding points v_j (instead of x_j^0 in Equation 3.39) can contain combinations of values $(l_j^0, Bi_j^0, \theta_{0j}^0)$ with different values of θ_0 and Bi for different j. In other words, temperature maximums and minimums can be reached on solutions of the problem (Equation 4.47) at different points of coordinate axis for different values of uncertain factors.

The specific problem should be solved to find values Bi_j^0 and θ_{0j}^0 for each point l_j^0 and to reveal the particular temperature profile $\theta(v,\Delta^0)$ within the set of values v. The proper choice of particular temperature distribution $\theta(v,\Delta^0)$ allows transforming Expression (3.39) and Expression (3.40) into appropriate equations set according to the alternance method.

This solution can be easily found for the most typical cases based on common physical sense and the shape of temperature profile $\theta(l,\Delta^0)$, which is known from the problem with full information.

In the sequence of inequalities (Equation 3.41) for the problem (Equation 3.39 and Equation 3.40), one can define each value $\varepsilon_{\min}^{(k)}, k=\overline{1,N^*}$ as follows:

$$\varepsilon_{\min}^{(k)} = \min_{\Delta}\left[\max_{x\in\Omega}\left|t(x,\Delta)-t^*\right|\right], \Delta=(\Delta_1,\Delta_2,\ldots,\Delta_k)\,. \tag{4.48}$$

With regard to the one-dimensional statement of the problem (Equation 2.34), it is necessary to search for the maximum of temperature along one coordinate l:

$$\varepsilon_{\min}^{(k)} = \min_{\Delta}\left[\max_{l\in[0,1]}\left|\theta(l,\Delta)-\theta^*\right|\right], \Delta=(\Delta_1,\Delta_2,\ldots,\Delta_k)\,. \tag{4.49}$$

Similarly to Equation (4.49), the definition of minimax $\bar{\varepsilon}_{\min}^{(k)}$ in the problem (Equation 4.47) can be written as

$$\bar{\varepsilon}_{\min}^{(k)} = \min_{\Delta}\left[\max_{v}\left|\theta(v,\Delta)-\theta^*\right|\right], \Delta=(\Delta_1,\Delta_2,\ldots,\Delta_k)\,. \tag{4.50}$$

This means that, for the whole set of possible values of initial temperature distribution and heat losses, the minimax value $\bar{\varepsilon}_{\min}^{(k)}$ represents the best heating accuracy or minimum temperature deviation within the billet for the considered class of control inputs with k constancy intervals.

In contrast to Equation (4.49), where the values θ_0 and Bi are considered as precisely known, the search for maximum in Equation (4.50) should be implemented on extended argument set $v = (l, Bi, \theta_0)$ rather than along the spatial coordinate l only.

Similarly to Equation (3.1) and Equation (3.41), the following inequality sequence holds true for values $\bar{\varepsilon}_{\min}^{(k)}$:

$$\bar{\varepsilon}_{\min}^{(1)} > \bar{\varepsilon}_{\min}^{(2)} > \ldots > \bar{\varepsilon}_{\min}^{(N)} > \bar{\varepsilon}_{\min}^{(N+1)} > \ldots > \bar{\varepsilon}_{\min}^{(N^*)} = \varepsilon_{\inf} \geq 0 \,. \tag{4.51}$$

Based on previously discussed conclusions, the following computational procedure is suggested to determine optimal control algorithms for the problem (Equation 4.46).

1. At the beginning, the problem should be solved assuming that required heating accuracy ε in Equation (4.46) is equal to $\bar{\varepsilon}_{\min}^{(1)}$ in Equation (4.51). Obviously, in this case the equality $N = 1$ is true, and the time-optimal heating process with accuracy $\bar{\varepsilon}_{\min}^{(1)}$ consists of a single-stage heating under constant maximum power.
When the surface heat losses are small enough, the temperature distribution $\theta(l, \Delta^0)$ has the shape of solid curves shown in Figure 3.4 and Figure 3.6 for all possible combinations of values θ_0 and Bi. In this case, the inequality $\bar{\varepsilon}_{\min}^{(1)} > \varepsilon_{\inf}$ is met, and heating accuracy could be improved comparatively to $\bar{\varepsilon}_{\min}^{(1)}$ by supplementing the process with the further stage of temperature soaking.
According to basic properties (Equation 3.39 and Equation 3.40), for $\varepsilon = \bar{\varepsilon}_{\min}^{(1)}$ under $N = 1$, two points, $v_1^0 = (l_1^0, Bi_1^0, \theta_{01}^0)$ and $v_2^0 = (l_2^0, Bi_2^0, \theta_{02}^0)$, should be found that correspond to maximum admissible absolute deviation of final temperature from required temperature θ^*. If the temperature profile $\theta(l, \Delta_1^0)$ has the shape shown in Figure 3.6, then equalities $l_1^0 = 0$ and $l_2^0 = l_{e2}$ follow for any fixed pair of values θ_0 and Bi. The appropriate values Bi_1^0 and θ_{01}^0 and Bi_2^0 and θ_{02}^0 should be found to reveal particular temperature profiles for which the temperature minimum at the point $l_1^0 = 0$ and maximum at the point $l_2^0 = l_{e2}$ (Figure 4.26) are reached under the conditions in Equation (2.20). Values Bi_1^0 and θ_{01}^0 and Bi_2^0 and θ_{02}^0 respectively specify the "lower" and "upper" temperature curves between which the possible final temperature profiles for all other combinations of variables Bi' and θ_0' satisfying Equation (2.20) are situated.
It is clear that temperature maximum at the point $l_2^0 = l_{e2}$ is reached under maximal possible initial temperature and minimal level of heat losses. The temperature minimum at the point $l_1^0 = 0$, on the contrary, is reached under minimal initial temperature and maximal losses. Therefore, it follows that

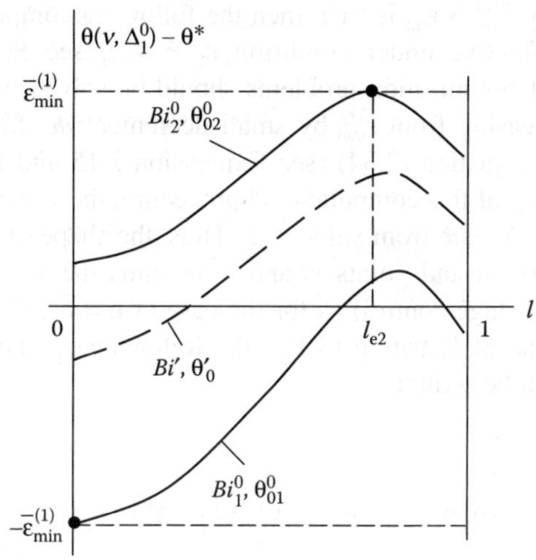

FIGURE 4.26 Variations of temperature profile for $\varepsilon = \bar{\varepsilon}^{(2)}_{\min}$ at the end of time-optimal heating under incomplete information with respect to initial temperature and heat loss level.

$$Bi_1^0 = Bi_{\max}, \theta_{01}^0 = \theta_{0_{\min}}; \quad Bi_2^0 = Bi_{\min}, \theta_{02}^0 = \theta_{0_{\max}} . \tag{4.52}$$

As a result, instead of Equation (3.12), we get the system of equations:

$$\begin{cases} \theta\left(0, \theta_{0\min}, Bi_{\max}, \Delta_1^0\right) - \theta^* = -\bar{\varepsilon}^{(1)}_{\min}; \\ \theta\left(l_{e2}, \theta_{0\max}, Bi_{\min}, \Delta_1^0\right) - \theta^* = +\bar{\varepsilon}^{(1)}_{\min}; \\ \dfrac{\partial\theta\left(l_{e2}, \theta_{0\max}, Bi_{\min}, \Delta_1^0\right)}{\partial l} = 0, \end{cases} \tag{4.53}$$

that can be solved in the same manner as Equation (3.12) with respect to Δ_1, $\bar{\varepsilon}^{(1)}_{\min}$ and l_{e2}.

If it is necessary to provide the best heating accuracy $\bar{\varepsilon}^{(1)}_{\min}$ attainable under a single-stage control, then the initial problem is already solved at this stage. The optimal control problem is also solved for case 1 if the prescribed numerical value of $\varepsilon = \varepsilon_0$ in Equation (4.46) coincides with the obtained numerical value of minimax $\bar{\varepsilon}^{(1)}_{\min}$. Even though $\varepsilon_0 < \bar{\varepsilon}^{(1)}_{\min}$, it is still necessary to perform step 2.

2. If inequality $\bar{\varepsilon}_{\min}^{(1)} > \varepsilon_{\inf}$ is true, then the following computational algorithm is effective under condition $\varepsilon_0 < \bar{\varepsilon}_{\min}^{(1)}$ (see Section 3.7). A sequence of optimization problems should be solved for fixed values $\varepsilon = \varepsilon'$ decreasing from $\bar{\varepsilon}_{\min}^{(1)}$ by small decrement $d\varepsilon$. Similarly to the problem in Equation (2.34) (see Expression 3.15 and Figure 3.7), at the beginning of this computational procedure, the value of ε' does not appreciably deviate from value $\bar{\varepsilon}_{\min}^{(1)}$. Thus, the shape of final temperature distribution and points v_1^0 and v_2^0 remains the same (even under optimal two-stage control) as for the case of $\varepsilon = \bar{\varepsilon}_{\min}^{(1)}$ (Figure 4.26). Then, instead of Equation (3.15), the following computational equations set can be written:

$$\begin{cases} \theta\left(0, \theta_{0\min}, Bi_{\max}, \Delta_1^0, \Delta_2^0\right) - \theta^* = -\varepsilon'; \\ \theta\left(l_{e2}, \theta_{0\max}, Bi_{\min}, \Delta_1^0, \Delta_2^0\right) - \theta^* = +\varepsilon'; \\ \dfrac{\partial\theta\left(l_{e2}, \theta_{0\max}, Bi_{\min}, \Delta_1^0, \Delta_2^0\right)}{\partial l} = 0. \end{cases} \tag{4.54}$$

In the case of fixed value ε', this system can be solved with respect to optimal durations Δ_1^0 and Δ_2^0 of heating/soaking intervals and coordinate l_{e2} of the temperature maximum. Analysis shows[1] that under $N = 2$, the temperature distribution profile is not distorted and the set of equations (Equation 4.54) remains valid up to the value of $\varepsilon' = \bar{\varepsilon}_{\min}^{(2)}$ in Equation (4.51). According to basic properties (Equation 3.39 and Equation 3.40), for $\varepsilon' = \bar{\varepsilon}_{\min}^{(2)}$, the third point, $v_3^0 = (l_3^0, Bi_3^0, \theta_{03}^0)$, appears, where maximum deviation of final temperature from the prescribed one is reached. Similarly to the problem in Equation (2.34), it is possible to show that the expression $l_3^0 = 1$ (see Equation 3.16 and Figure 3.8) takes place under the conditions[1]:

$$Bi_3^0 = Bi_1^0 = Bi_{\max},\ \theta_{03}^0 = \theta_{01}^0 = \theta_{0\min}.$$

Therefore, for $\varepsilon' = \bar{\varepsilon}_{\min}^{(2)}$ at the end of the time-optimal two-stage control process, the minimal admissible temperatures are reached in the center and on the surface of the billet if its initial temperature is minimal and the heat loss level is maximal. At the same time, the temperature maximum will be located at some internal point that corresponds to the values $Bi = Bi_{\min}$, $\theta_0 = \theta_{0\max}$ (Figure 4.27).

Then, instead of Equation (3.16), the computational set of equations for $\varepsilon' = \bar{\varepsilon}_{\min}^{(2)}$ can be written as:

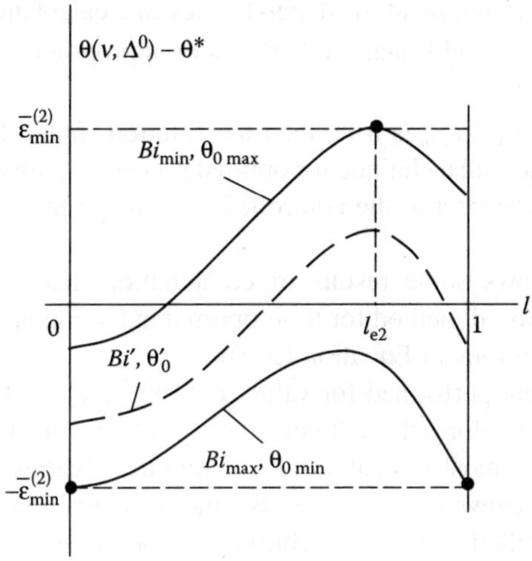

FIGURE 4.27 Variations of temperature profile for $\varepsilon = \bar{\varepsilon}^{(2)}_{min}$ at the end of time-optimal heating under incomplete information with respect to initial temperature and heat loss levels.

$$\begin{cases} \theta\left(0, \theta_{0\min}, Bi_{\max}, \Delta_1^0, \Delta_2^0\right) - \theta^* = -\bar{\varepsilon}^{(2)}_{\min}; \\ \theta\left(l_{e2}, \theta_{0\max}, Bi_{\min}, \Delta_1^0, \Delta_2^0\right) - \theta^* = +\bar{\varepsilon}^{(2)}_{\min}; \\ \theta\left(1, \theta_{0\min}, Bi_{\max}, \Delta_1^0, \Delta_2^0\right) - \theta^* = -\bar{\varepsilon}^{(2)}_{\min}; \\ \dfrac{\partial\theta\left(l_{e2}, \theta_{0\max}, Bi_{\min}, \Delta_1^0, \Delta_2^0\right)}{\partial l} = 0. \end{cases} \tag{4.55}$$

This set of four equations can be solved with respect to four unknowns: Δ_1^0, Δ_2^0, $\bar{\varepsilon}^{(2)}_{min}$, and l_{e2}.

3. Because sets of equations are defined unambiguously depending on the value of $\varepsilon' \in [\,\bar{\varepsilon}^{(2)}_{min}\,;\ \bar{\varepsilon}^{(1)}_{min}\,]$, for any particular value of required accuracy $\varepsilon = \varepsilon_0 \in [\,\bar{\varepsilon}^{(2)}_{min}\,;\ \bar{\varepsilon}^{(1)}_{min}\,]$ in Equation (4.46), the computational procedure turns out to be quite simple.

   At the beginning, two problems should be solved for required accuracy values $\varepsilon = \bar{\varepsilon}^{(1)}_{min}$ and $\varepsilon = \bar{\varepsilon}^{(2)}_{min}$. These problems are reduced to solution of the equations systems (Equation 4.53 and Equation 4.55), which allows one to find minimax values $\bar{\varepsilon}^{(1)}_{min}$ and $\bar{\varepsilon}^{(2)}_{min}$.

   The optimal control problem is solved if the value of ε_0 coincides with one of obtained minimaxes $\bar{\varepsilon}^{(1)}_{min}$ or $\bar{\varepsilon}^{(2)}_{min}$. Otherwise, if the value of ε_0 satisfies the condition $\bar{\varepsilon}^{(2)}_{min} < \varepsilon_0 < \bar{\varepsilon}^{(1)}_{min}$, then it would still be necessary to solve the set (Equation 4.54) for required heating accuracy. In all equations, the temperature dependencies are calculated by formulas (Equation 2.28 and Equation 2.29) under appropriate values of θ_0 and *Bi*.

If condition $\varepsilon_0 < \bar{\varepsilon}_{\min}^{(2)}$ is true, then the solution should be found using the computational technique for optimal processes with complete information with respect to the controlled system (Chapter 3).

Figure 4.28 shows some results of computation that are provided as an example of the proposed method for time-optimal induction heating of an infinite cylinder under conditions in Equation (2.20).

Calculations were performed for values $\varepsilon \in [\ \bar{\varepsilon}_{\min}^{(2)}\ ,\ \bar{\varepsilon}_{\min}^{(1)}\ ]$. Optimal final temperature distributions along the cylinder radius are shown in Figure 4.28a and Figure 4.28b for the most typical values of required heating accuracy $\bar{\varepsilon}_{\min}^{(1)}$ and $\bar{\varepsilon}_{\min}^{(2)}$. Figure 4.28c shows optimal process time as a function of ε compared to results for the controlled system with complete information with respect to initial data. Figure 4.28d (as well as Figure 4.28c) illustrates the total process time loss in contrast to optimal process calculated for precisely determined values of θ_0 and *Bi*.

4.4 HEATING PROCESS WITH MINIMUM PRODUCT COST

The cost of product can be regarded as one of the IHP quality indexes in certain production situations (Section 2.2). When a criterion of minimum cost is required, depending upon application, all major IHI performance characteristics can be considered, including *productivity*, *quality* of product, *power cost*, and *material losses*. The overall product cost can be estimated by a combined criterion (Equation 2.5) that includes all cost components (Equation 2.2 through Equation 2.4) with weighting coefficients C_i making sense of a relative costing and representing the importance of each of the items of the overall cost.

Some aspects of the minimum cost problem in application to static induction heating prior to hot forming of metals are discussed in this section. To make understanding easier, the one-dimensional linear model of the heating process will be used here. To simplify discussion, the technological constraints can be neglected during the first stage of the process of finding the OCP solution. The heated billet's transfer stage and incomplete information regarding the heating system will also not be considered. It is important to keep in mind that obtained results can be extended onto problems subject to technological constraints and other complicating factors by using techniques described in Section 4.1 through Section 4.3.

Heating accuracy, energy consumption, metal loss due to scale, and heating process time represent main components of product cost (Section 2.2). It was shown earlier that the alternance method has been successfully applied for optimization of the heating process with respect to maximum accuracy or minimum

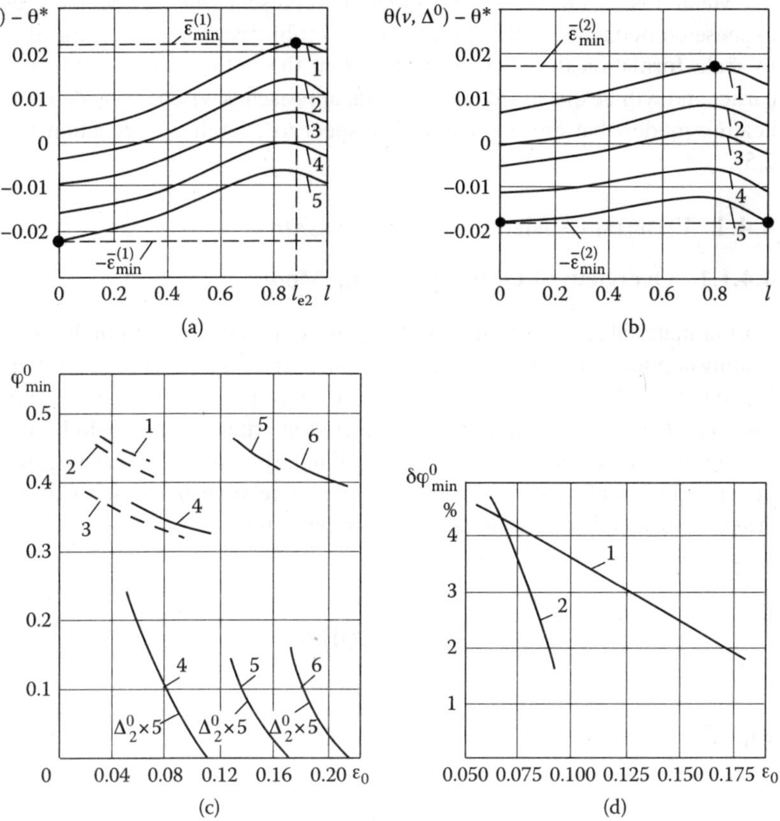

FIGURE 4.28 (a, b) Final temperature profiles and (c, d) optimal control parameters under incomplete information control. ($\Gamma = 1$; $U_{max} = 1$; $Bi \in [0.4;0.7]$; $\theta_0 \in [-0.7;-0.5]$; $\theta^* = 0$). a: $\xi = 4$; $\varepsilon_0 = \varepsilon_{min}^{(1)}$. b: $\xi = 4$; $\varepsilon_0 = \bar{\varepsilon}_{min}^{(2)}$ (1: $Bi = 0.4$; $\theta_0 = -0.5$; 2: $Bi = 0.5$; $\theta_0 = -0.55$; 3: $Bi = 0.55$; $\theta_0 = -0.6$; 4: $Bi = 0.64$; $\theta_0 = -0.65$; 5: $Bi = 0.7$); $\theta_0 = -0.7$. c: $\xi = 4$; $[\bar{\varepsilon}_{min}^{(2)}, \bar{\varepsilon}_{min}^{(1)}]$(1–3: for precisely determined values of Bi and θ_0; 1: $Bi = 0.7$; $\theta_0 = -0.6$; 2: $Bi = 0.56$; $\theta_0 = -0.6$; 3: $Bi = 0.56$; $\theta_0 = -0.5$; 4–6: control under interval uncertainty with respect to values of Bi and θ_0; 4: $Bi \in [0.4;0.7]$; $\theta_0 = -0.5$; 5: $Bi = 0.7$; $\theta_0 \in [-0.7;-0.5]$; 6: $Bi \in [0.4;0.7]$; $\theta_0 \in [-0.7;-0.5]$). d: Total process time loss under incomplete information (1: $Bi \in [0.02;0.06]$; $\theta_0 = -2$; $\xi = 10$; 2: $Bi \in [0.4;0.7]$; $\theta_0 = -0.5$; $\xi = 4$) in contrast to the optimal process calculated for precisely determined values Bi and θ_0 (1: $Bi = 0.04$; $\theta_0 = -2$; $\xi = 10$; 2: $Bi = 0.56$; $\theta_0 = -0.5$; $\xi = 4$).

energy consumption (Section 2.9 and Equation 3.5). As discussed earlier, these criteria algorithms of power variation in time have the same shape as for the time-optimal process. The problem for providing maximum heating accuracy can be reduced to the time-optimal problem. Computation procedure for determining OCP solutions according to minimum energy consumption and minimum required heating time can be noticeably different.

Metal losses due to scale formation represent one of the most essential expense components of IHP product cost. Problems of minimum metal loss due to scale formation differ significantly from the optimization problems studied earlier and will be discussed later. Several approaches will be considered to define heating modes that are optimal with respect to overall cost function (Equation 2.5).

4.4.1 Problem of Metal Scale Minimization

4.4.1.1 Overview of Optimal Heating Modes

Cost of material represents an essential part of product cost. For high-temperature heating applications, the appreciable portion of material expenses deals with metal loss due to scale formation. As shown in Butkovskij et al.,[11,12] the value of metal loss can also be considered as the optimization criterion and could be written in integral form (Equation 2.4). The integrand in Expression (2.4) can be represented in an exponential form. Using the system of relative units (see Chapter 1), a criterion of metal scale minimization can be written as:

$$I = \int_{0}^{\varphi^0} f\left(\theta_{sur}\left(\varphi\right)\right) d\varphi \rightarrow \min \tag{4.56}$$

where

$$f\left(\theta_{sur}\left(\varphi\right)\right) = \begin{cases} 0, \; if \; \theta_{sur} \le \theta_q; \\ \left(\theta_{sur} - \theta_q\right)^{S+1}, \; if \; \theta_{sur} \ge \theta_q. \end{cases} \tag{4.57}$$

Expression (4.57) is obtained under assumption that the process of metal oxidation (scale formation) occurs only when the billet's surface temperature θ_{sur} exceeds the given value θ_q.

Let us consider the heating process that can be described by Equation (1.27) through Equation (1.35). The surface temperature θ_{sur} can be easily found by substituting $l = 1$ in the function $\theta(l,\varphi)$. This process calls for total consumed heating power $u(\varphi)$ as a lumped control input. Formula (2.25) can be used for temperature requirements at the end of the heating cycle.

The statement of the problem for minimizing metal loss due to scale can be formulated as follows. It is necessary to select such power control function $u^*(\varphi)$ bounded by restraint (Equation 2.7) that provides steering workpiece initial temperature distribution to the desired temperature θ^* with prescribed accuracy ε (according to Equation 2.25) for the given time φ^0 and holds the cost criterion (Equation 4.56 and Equation 4.57) at the minimal level.

At the beginning, it is necessary to define the shape of optimal control function $u^*(\varphi)$. In this case, the optimal power control does not consist of alternating stages of heating with maximum power $u \equiv U_{\max}$ (heat ON) and subsequent soaking/cooling under $u \equiv 0$ (heat OFF) cycles. Therefore, control will look quite different compared to solutions of OCP for minimum time, maximum heating accuracy, and minimum energy consumption criteria.

If desired final temperature θ^* exceeds θ_q (usually $\theta_{sur} \cong 600$ to $650°$), then the minimal amount of scale is obtained for the fixed time φ^0, if the billet's surface temperature will vary in time, in the following way. On the one hand, the surface temperature at any moment φ^0 (see Equation 4.56 and Equation 4.57) must be as close to the value of θ_q as possible. On the other hand, to achieve the required temperature distribution that satisfies the condition $\theta^* > \theta_q$, the expression $\theta_{sur} > \theta_q$ should be valid during a certain stage of the process. It is clear that accelerated heating mode under $u(\varphi) \equiv U_{\max}$ does not meet this purpose because it results in maximum temperature gradients (under maximum values of θ_{sur}) and, therefore, leads to a greater amount of metal loss due to scale.

It is possible to show[11,12] that if the required final temperature θ^* exceeds temperature θ_q of intensive oxidation, then metal losses become minimum under exponential variation of θ_{sur} over time:

$$\theta_{sur}(\varphi) = Ae^{B\varphi} + \theta_q, \tag{4.58}$$

where $A, B > 0$ are positive coefficients.

Let us notice that the optimal surface temperature already exceeds θ_q at initial moment $\varphi = 0$. Optimal power control $u^*(\varphi)$, which provides variation of billet surface temperature according to Expression (4.58), can be found based on the general expression (Equation 1.36) for temperature distribution. If function $\theta_{sur}(\varphi)$ is given in the form of Equation (4.58) then Equation (1.36) at $l = 1$ could be considered as the *integral equation* with respect to control function $u^*(\varphi)$. Its solution can be found in the same way as in the case of the functions $u^{\sigma}(\varphi)$ and $u^{\theta}(\varphi)$ (Section 4.1).

However, it turns out that, as a rule, realization of the algorithm in Equation (4.58) by means of admissible controls over the whole duration of the heating process is impossible. This conclusion is stipulated by a number of certain practical conditions.

In reality, the initial surface temperature is less than the oxidation threshold θ_q. Obviously, it is impossible to raise the value of θ_{sur} momentarily up to the value $\theta_{sur}(0) > \theta_q$ under limited heating power $U_{\max}$ in Expression (2.7) as is required in Equation (4.58). Thus, the initial stage of the optimal process can be conducted under maximum power $u^*(\varphi) \equiv U_{\max}$ that leads to the maximal possible rise of surface temperature up to the values, which are required according to the algorithm (Equation 4.58).

Optimal heating mode depends to a greater extent upon given duration φ^0 of the heating process. Expression (4.58) remains valid if sufficient slack time exists because it provides minimal possible excess of θ_{sur} over θ_q. If there is no extra time, then it is impossible to apply this algorithm during the whole process.

As an example, let us consider the case of value φ^0 that is equal to minimum time φ^0_{min} required for heating with the same accuracy ε up to the same final temperature θ^* in Equation (2.25). As shown earlier, the optimal control algorithm $u^*(\varphi)$ in this case could be represented by only the time-dependent stepwise function in the form of Equation (2.27). Such control does not allow reducing the scale. This means that for $\varphi^0 = \varphi^0_{min}$ the value of the criterion (Equation (4.56) should be considered as optimal under control function Equation (2.27)).

If the condition $\varphi^0 > \varphi^0_{min}$ is met, then the optimal process includes separate stages of heating according to the algorithm (Equation 4.58). The optimal algorithm comprises stages of heating with maximum power $u^*(\varphi) \equiv U_{max}$ and subsequent soaking under $u^*(\varphi) \equiv 0$ that conform to time-optimal control.

Therefore, one can conclude that the sought optimal algorithm consists of stages of heating with maximum power $u \equiv U_{max}$ and subsequent soaking/cooling under $u \equiv 0$, as well as different intervals, that provide the exponential law of surface temperature growth. The latter intervals can be called singular intervals and appropriate control functions $u_s(\varphi)$ can be called singular control functions.[13] In the general case, the number of control stages and their sequence depend upon given accuracy ε in Equation (2.25) and cannot be easily defined. Calculation of singular control functions adds additional complexity to this problem compared to the conventional one. An approximate solution of such a problem is considered next for the most typical values of required heating accuracy.

4.4.1.2 Two-Parameter Power Control Algorithm of Scale Minimization

Let us consider the case of $\varphi^0 = \varphi^0_{min}$ in Equation (4.56) that allows one to transform the problem of minimum scale into the time-optimal problem (Equation 2.34). Let us assume that $\varepsilon = \varepsilon^{(2)}_{min}$. Figure 3.8 shows the shape of the time-optimal control algorithm that can be defined by *two* parameters: durations Δ^0_1 and Δ^0_2 of heating and soaking intervals. Parameters Δ^0_1 and Δ^0_2 can be found after solving the set of equations in Equation (3.16). Figure 3.8a (curve 2) shows the appropriate spatial temperature distribution at the end of the optimal process. For temperature profile of that shape, there will be *three* points within the heated workpiece ($l = 0$, $l = l_{e2}$, and $l = 1$) where maximum admissible deviations of the final temperature from the required one exist.

Let us assume that the given heating time exceeds the time $\varphi^0_{min}(\varepsilon^{(2)}_{min})$ by the relatively small value of $d\varphi^0 > 0$. We shall consider the problem of minimum scale under $\varphi^0 = \varphi^0_{min}(\varepsilon^{(2)}_{min}) + d\varphi^0$ for a certain attainable value ε in Equation (2.25) that differs from $\varepsilon^{(2)}_{min}$ by value $d\varepsilon$ of order $d\varphi^0$.

The continuous nature of all optimal process parameters as functions of φ^0 and ε allows one to conclude that small variations of $d\varphi^0$ and $d\varepsilon$ lead to small variations in the optimal control algorithm and the final temperature distribution profile. This means that there will be a maximum of three points within the heated workpiece where maximum admissible deviations of the final temperature from the required one exist. In contrast to the time-optimal problem, the minimum

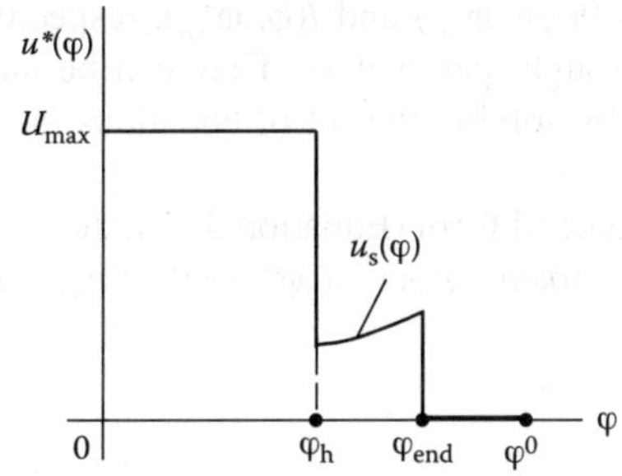

FIGURE 4.29 Two-parameter minimum scale control by heat power.

scale control algorithm is supplemented only by one singular interval of small duration $\varphi_{end} - \varphi_h$ (Figure 4.29).

"Switching" from the stage of heating where maximum power is applied to the singular interval under control $u_s(\varphi)$ occurs at the moment $\varphi = \varphi_h$. The singular control interval is finished at the time $\varphi = \varphi_{end}$. The temperature soaking stage should be performed further under $u(\varphi) \equiv 0$. If condition $\varphi_{end} = \varphi_h$ is met, then, for heating accuracy $\varepsilon = \varepsilon^{(2)}_{\min}$, it is possible to obtain typical two-stage time-optimal control without the singular interval. The value $\varphi_{end} = \varphi_h + d\varphi$ (where $d\varphi$ is a small value of order $d\varphi^0$ and $d\varepsilon$) corresponds to increase of φ^0 by $d\varphi^0 > 0$ from the level of $\varphi^0_{\min}(\varepsilon^{(2)}_{\min})$ for accuracy $\varepsilon = \varepsilon^{(2)}_{\min} + d\varepsilon$. It is important to underline at this point that singular control $u_s(\varphi)$ is uniquely determined if Expression (4.58) is known.

The optimal control algorithm (Figure 4.29) can be defined for given control $u_s(\varphi)$ and value φ^0 as a function of two sought parameters, φ_h and φ_{end}, which represent times of the beginning and the end of the singular control interval. If the optimal control algorithm includes more than one singular interval, then the number of optimal control parameters will at least double (beginnings and ends of two singular intervals). Therefore, there is a contradiction with the basic property (Equation 3.40). According to Equation (3.40), there will be no more than three points located within the heated body, where the maximum admissible absolute deviations of final temperature from the required one would meet. Obviously, only one singular interval exists during the whole optimal process with power control $u^*(\varphi)$ if the value of φ^0 differs slightly from the value $\varphi^0_{\min}(\varepsilon^{(2)}_{\min})$.

The optimal control problem can be formulated more precisely, assuming that for the given time φ^0, the required heating accuracy ε should be equal to the best attainable heating accuracy $\hat{\varepsilon}^{(2)}_{\min}$ in the class of considered control inputs $u(\varphi)$. The shape of the optimal control algorithm $u(\varphi)$ in Figure 4.29 represents this case. However, two parameters (φ_h and φ_{end}) remain unknown.

It is clear that, for $\varphi^0 = \varphi^0_{\min}(\varepsilon^{(2)}_{\min}) + d\varphi^0$, the value of $\hat{\varepsilon}^{(2)}_{\min}$ slightly differs from the minimax $\varepsilon^{(2)}_{\min}$, i.e., $\hat{\varepsilon}^{(2)}_{\min} = \varepsilon^{(2)}_{\min} + d\varepsilon$, and the shape of the optimal control algorithm is similar to the case shown in Figure 4.29. The final temperature distribution and objective function (Equation 4.56 and Equation 4.57) can be considered as functions $\theta(l,\varphi_h,\varphi_{end})$ and $I(\varphi_h,\varphi_{end})$, respectively. These functions depend only upon the sought parameters. They can be unambiguously defined for given control input by solving the set of equations that describe the heating process.

With regard to the general form (Equation 3.38), the optimal control problem formulated earlier under fixed values of $\varphi^0 > \varphi^0_{\min}(\varepsilon^{(2)}_{\min})$ and $\varepsilon = \hat{\varepsilon}^{(2)}_{\min}$, can be rewritten as:

$$\begin{cases} I(\varphi_h,\varphi_{end}) \to \min\limits_{\varphi_h,\varphi_{end}}; \\ \Phi(\varphi_h,\varphi_{end}) = \max\limits_{l\in[0,1]} \left|\theta(l,\varphi_h,\varphi_{end}) - \theta^*\right| \le \hat{\varepsilon}^{(2)}_{\min}. \end{cases} \tag{4.59, 4.60}$$

Components Δ_1 and Δ_2 of parameter vector Δ in Equation (3.38) are represented here by φ_h and φ_{end}. According to the basic property (Equation 3.40), there will be exactly three points within the heated workpiece where maximum deviations $\pm\,\hat{\varepsilon}^{(2)}_{\min}$ from θ^* are reached at the end of the optimal process. Temperature profile of that shape corresponds to optimal values of parameters φ^0_h and φ^0_{end} that should be considered instead of Δ^0_1 and Δ^0_2 in Equation (3.39). In this case, the number of extremum points is equal to maximum possible number under conditions $\varphi^0 = \varphi^0_{\min}(\varepsilon^{(2)}_{\min}) + d\varphi$ and $\varepsilon = \varepsilon^{(2)}_{\min} + d\varepsilon = \hat{\varepsilon}^{(2)}_{\min}$.

The shape of optimal final temperature distribution remains similar to the time-optimal problem (Figure 3.8). Keep in mind that the time–temperature profile for process with minimum scale differs from the profile of the time-optimal problem due to presence of the singular control interval incorporated in the power control algorithm (Figure 4.29). Such a situation remains within a certain range of variation of the value φ^0 that increases from $\varphi^0_{\min}(\varepsilon^{(2)}_{\min})$ with regard to preset accuracy $\varepsilon = \hat{\varepsilon}^{(2)}_{\min}$. This conclusion is based on the continuous nature of all optimal process parameters as the function of process time. Figure 4.29 shows optimal heat power $u^*(\varphi)$ as following the function of time:

$$u^*(\varphi) = \begin{cases} U_{\max}, 0 < \varphi < \varphi^0_h; \\ u_s(\varphi), \varphi^0_h < \varphi < \varphi^0_{end}; \\ 0, \varphi^0_{end} < \varphi \le \varphi^0. \end{cases} \tag{4.61}$$

If the shape of the optimal control function (Equation 4.61) is known, the final temperature distribution can be defined as function $\theta(l,\varphi^0_h,\varphi^0_{end})$, using the general integral dependency (Equation 1.36) for each fixed value φ^0.

If the final temperature profile replicates the curve in Figure 3.8, then, according to basic properties (Equation 3.39 and Equation 3.40) and similar to Equation (3.16), the computational set of equations should be considered in the following form with respect to φ_h^0 and φ_{end}^0:

$$\begin{cases} \theta\left(0,\varphi_h^0,\varphi_{end}^0\right)-\theta^* = -\hat{\varepsilon}_{\min}^{(2)}; \\ \theta\left(l_{e2},\varphi_h^0,\varphi_{end}^0\right)-\theta^* = +\hat{\varepsilon}_{\min}^{(2)}; \\ \theta\left(1,\varphi_h^0,\varphi_{end}^0\right)-\theta^* = -\hat{\varepsilon}_{\min}^{(2)}; \\ \dfrac{\partial\theta\left(l_{e2},\varphi_h^0,\varphi_{end}^0\right)}{\partial l} = 0. \end{cases} \tag{4.62}$$

This system differs from Equation (3.16) by expression $\theta(l,\varphi_h,\varphi_{end})$ with respect to a final temperature distribution in comparison with $\theta(l,\Delta_1^0,\Delta_2^0)$. Similarly to Equation (3.16), the set in Equation (4.62) can be solved with respect to four unknown variables: optimal values φ_h^0 and φ_{end}^0 of the beginning and the end of the singular control interval; minimax $\hat{\varepsilon}_{\min}^{(2)}$; and an intermediate unknown variable — coordinate l_{e2} of the point of temperature maximum.

The optimal power control algorithm in the form of Equation (4.61) can be found only under a priori known singular control $u_s(\varphi)$ during the interval $[\varphi_h;\varphi_{end}]$. It is possible to show that if two terms of the infinite series are used in expression for temperature field, then singular control can be found as a solution of the appropriate integral equation. In the conventional case of uniform initial temperature distribution $\theta_0(l) \equiv \theta_0 = \theta_a = const$, the singular control function can be obtained under the boundary condition (Equation 1.34) in Equation (1.27) through Equation (1.35) as follows[1,14]:

$$u_s(\varphi) = C_1 e^{\frac{A_1}{s}(\varphi-\varphi_h)} + C_2 e^{A_2(\varphi-\varphi_h)} + C_3, \quad \varphi \in \left[\varphi_h, \varphi_{end}\right]. \tag{4.63}$$

Constant coefficients C_1, C_2, C_3, and A_1, A_2 should be calculated according to the following formulas:

$$C_1 = \frac{\left(\frac{A_1}{s}+\mu_1^2\right)\left(\frac{A_1}{s}+\mu_2^2\right)\left(a_1 z_1(\varphi_h)+a_2 z_2(\varphi_h)+\theta_0-\theta_q\right)}{\frac{A_1}{s}\left(a_1 W_1(\xi)+a_2 W_2(\xi)\right)+a_1 W_1(\xi)\mu_2^2+a_2 W_2(\xi)\mu_1^2};$$

$$C_2 = \frac{a_1 a_2\left(\mu_2^2-\mu_1^2\right)\left\{\frac{A_1}{s}\left(\theta_q-\theta_0\right)W_1(\xi)W_2(\xi)\left(\mu_2^2-\mu_1^2\right)-\left[z_1(\varphi_h)\left(\frac{A_1}{s}+\mu_1^2\right)\times\right.\right.}{\left(a_1\mu_2^2 W_1(\xi)+a_2\mu_1^2 W_2(\xi)\right)\left(a_1 W_1(\xi)+a_2 W_2(\xi)\right)\times}\cdots$$

$$\cdots\frac{\left.\left.W_2(\xi)-z_2(\varphi_h)\left(\frac{A_1}{s}+\mu_2^2\right)W_1(\xi)\right]\left(a_1\mu_2^2 W_1(\xi)+a_2\mu_1^2 W_2(\xi)\right)\right\}}{\left[a_1\mu_2^2 W_1(\xi)+a_2\mu_1^2 W_2(\xi)+\frac{A_1}{s}\left(a_1 W_1(\xi)+a_2 W_2(\xi)\right)\right]};$$

$$C_3 = \frac{\mu_1^2\mu_2^2\left(\theta_q-\theta_0\right)}{a_1\mu_2^2 W_1(\xi)+a_2\mu_1^2 W_2(\xi)};$$

$$A_1 = \frac{(\Gamma+1)W_0^*(\xi)K(\mu_{*1})\mu_{*1}^2}{(\Gamma+1)W_0^*(\xi)K(\mu_{*1})+2W_1^*(\xi)};\quad A_2 = -\frac{a_1\mu_2^2 W_1(\xi)+a_2\mu_1^2 W_2(\xi)}{a_1 W_1(\xi)+a_2 W_2(\xi)},\tag{4.64}$$

where

$$a_1 = \frac{2Bi^2 K(\mu_1)}{\left(\mu_1^2+Bi^2+(1-\Gamma)Bi\right)K_1^2(\mu_1)};\quad a_2 = \frac{2Bi^2 K(\mu_2)}{\left(\mu_2^2+Bi^2+(1-\Gamma)Bi\right)K_1^2(\mu_2)};$$

$$z_1(\varphi_h) = \frac{U_{\max}W_1(\xi)}{\mu_1^2}\left(1-e^{-\mu_1^2\varphi_h}\right);\quad z_2(\varphi_h) = \frac{U_{\max}W_2(\xi)}{\mu_2^2}\left(1-e^{-\mu_2^2\varphi_h}\right).$$

μ_{*1}; W_n^*, $n = 0,1$ are defined for the boundary conditions of the second kind.

All other denotations correspond to ones discussed in Chapter 1.

Substitution of $u_s(\varphi)$ in the form of Equation (4.63) into the control algorithm (Equation 4.61) allows obtaining the expression for $\theta(l,\varphi_h^0,\varphi_{end}^0)$ in the explicit form. Temperature distribution with regard to the model (Equation 1.27 through Equation 1.35) under $\theta_0(l) \equiv \theta_0 = \theta_a < \theta_q < \theta^*$ can be written using Equation (1.36), Equation (1.38), Equation (1.40) through Equation (1.43), and Equation (1.45) as the following function of the parameters φ_h^0 and φ_{end}^0:

$$\theta\left(l,\varphi_h^0,\varphi_{end}^0\right) = \theta_0 + \sum_{n=1}^{\infty}\frac{2W_n(\xi)K(\mu_n l)e^{-\mu_n^2\varphi^0}}{\left(\mu_n^2+Bi^2+(1-\Gamma)Bi\right)K^2(\mu_n)}Q\left(\varphi_h^0,\varphi_{end}^0\right),\tag{4.65}$$

where

$$Q\left(\varphi_h^0, \varphi_{end}^0\right) = U_{\max}\left(e^{\mu_n^2\varphi_h^0} - 1\right) + \frac{C_1 e^{-\frac{A_1}{s}\varphi_h^0}}{\frac{A_1}{s\mu_n^2}+1}\left(e^{\left(\frac{A_1}{s}+\mu_n^2\right)\varphi_{end}^0} - e^{\left(\frac{A_1}{s}+\mu_n^2\right)\varphi_h^0}\right) +$$

$$+\frac{C_2 e^{-A_2\varphi_h^0}}{\frac{A_2}{\mu_n^2}+1}\left(e^{\left(A_2+\mu_n^2\right)\varphi_{end}^0} - e^{\left(A_2+\mu_n^2\right)\varphi_h^0}\right) + C_3\left(e^{\mu_n^2\varphi_{end}^0} - e^{\mu_n^2\varphi_h^0}\right). \tag{4.66}$$

Expression (4.66) should be used to solve the set of equations in Equation (4.62).

It is important to underline that the optimal algorithm (Equation 4.61) is found approximately using a singular control in the form of Equation (4.63) and taking into account only two terms of the Fourier series that describes temperature field within the billet. However, the final temperature distribution is precisely defined based on the mathematical model in the form of an infinite series in (Equation 4.65). A sufficiently large number of series terms in Equation (4.65) must be considered for calculating the temperature distribution with satisfactory accuracy. Therefore, exact OCP solution provides the desired final temperature distribution under the control input that can be considered only as approximately optimal with respect to the criterion of scale minimization.

Let us consider as an example the optimal control mode when it is necessary to provide the best heating accuracy $\varepsilon = \hat{\varepsilon}_{\min}^{(2)}$. Figure 4.30 shows the results of

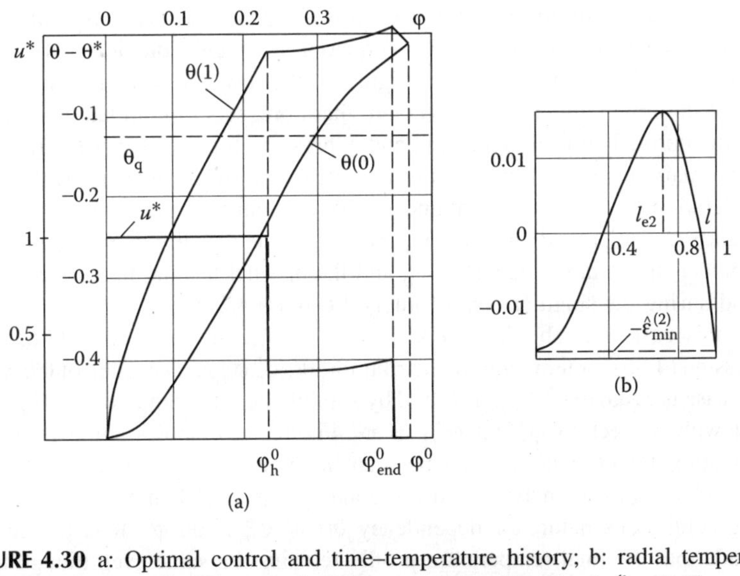

FIGURE 4.30 a: Optimal control and time–temperature history; b: radial temperature profile $\theta(l,\varphi_h^0,\varphi_{end}^0) - \theta^*$ at the end of minimum scale heating. $\Gamma = 1$; $\xi = 4$; $U_{\max} = 1$ $\theta_0 = -0.5$; $\theta^* = 0$; $\theta_q = -0.125$; $S = 3.5$; $\varepsilon_0 = \hat{\varepsilon}_{\min}^{(2)}$; $\varphi^0 = 0.44 > \varphi_{end}^0$.

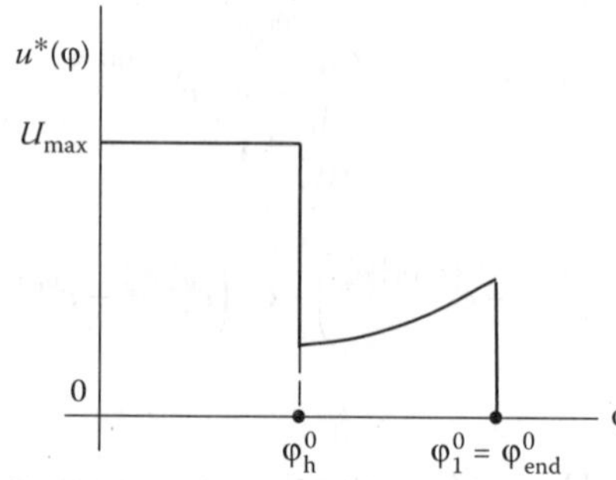

FIGURE 4.31 Two-parameter minimum scale control by heat power when the optimal heating process comes to the end during the singular control interval.

the optimal problem for minimizing metal loss due to scale formation. Computations have been performed according to the previously described technique. Expressions for final optimal control $u^*(\varphi)$ and temperature distribution $\theta(l,\varphi_h^0,\varphi_{end}^0)$ in the form of Equation (4.61), Equation (4.63), and Equation (4.65) were used in all sets of equations. Figure 4.30a shows optimal control and time–temperature profiles for the billet surface ($l = 1$) and its center during the heating process. Figure 4.30b represents radial temperature distribution that takes place at the end of the optimal heating cycle.

A sequence of optimization problems should be solved for fixed values φ^0, increasing from $\varphi_{min}^0(\varepsilon_{min}^{(2)})$ by a relatively small step. Study of obtained results could reveal a sensitivity of optimal control algorithms as a function of values of φ^0 with regard to basic properties (Equation 3.39 and Equation 3.40). Computational results for initial data in Figure 4.30 allow one to conclude that the shape of the final temperature curve and, therefore, sets of computational equations, the optimal control algorithm, and expression for final temperature distribution remain unchanged in spite of the increase of given value φ^0. At the same time, duration of soaking interval under $u^*(\varphi) \equiv 0$ reduces monotonously, and the difference $\varphi^0 - \varphi_{end}^0$ tends to zero (Figure 4.29). The equality $\varphi_1^0 = \varphi_{end}^0$ becomes valid under the certain value $\varphi^0 = \varphi_1^0$, and the optimal heating process comes to the end during the singular control interval (Figure 4.31).

In this case, one should accept $\varphi^0 = \varphi_1^0 = \varphi_{end}^0$ in Expression (4.65) and Expression (4.66) for temperature distribution $\theta(l,\varphi_h^0,\varphi_{end}^0)$. The value of φ^0, which in this case is equal to φ_1^0, can be found by solving the set of equations in Equation (4.62) with respect to φ_h^0 and $\varphi_{end}^0 = \varphi_1^0$. Figure 4.32 shows some results of computation for $\varphi^0 = \varphi_1^0$. As can be seen in Figure 4.32b, the shape of curve $\theta(l,\varphi_h^0,\varphi_{end}^0)$ remains unchanged if the condition $\varphi^0 = \varphi_1^0$ is not violated. Based on the continuous nature of dependency $\theta(l,\varphi_h^0,\varphi_{end}^0)$ on φ^0, it is possible to conclude that the final temperature profile remains the same even though there was an increase of the time φ^0 by a small value (in comparison with φ_1^0).

Similarly to the previous discussion, the optimal control algorithm for corresponding values of φ^0 should include only one singular interval. At the same

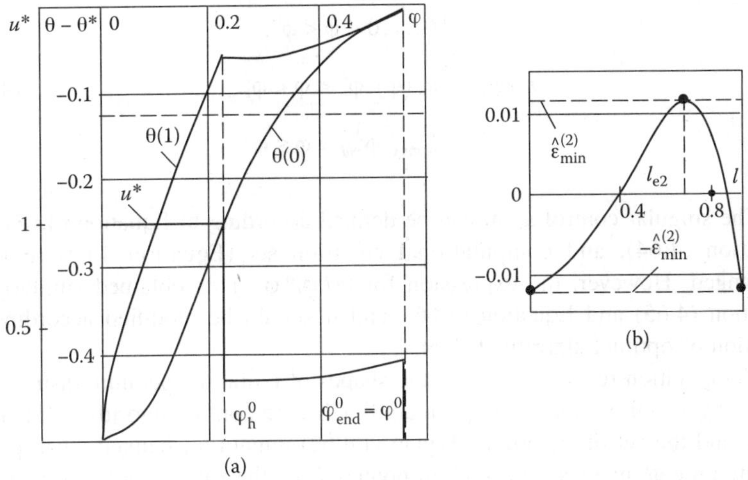

FIGURE 4.32 (a) Optimal control, time–temperature history, and (b) final temperature profile $\theta(l,\varphi_h{}^0,\varphi^0_{\text{end}}) - \theta^*$ for minimum scale heating. $\Gamma = 1$; $\xi = 4$; $U_{\max} = 1$; $\theta_0 = -0.5$; $\theta^* = 0$; $\theta_q = -0.125$; $S = 3.5$; $\varepsilon_0 = \hat{\varepsilon}^{(2)}_{\min}$; $\varphi^0 = \varphi^0_{\text{end}} = 0.55$.

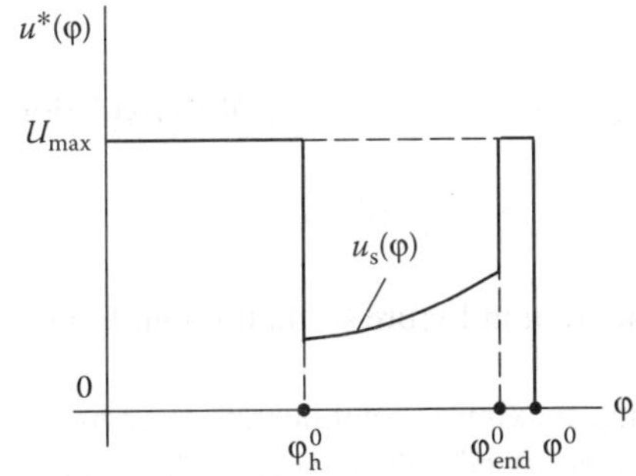

FIGURE 4.33 Two-parameter minimum scale control by heat power with interval of reheating under maximum power $u^*(\varphi) \equiv U_{\max}$.

time, control input $u^*(\varphi)$ can be defined by two parameters: φ_h and φ_{end}. Under condition of utilizing a maximum heating power during the first stage, there will be a single possible variant of the "time-power" profile of optimal algorithm $u^*(\varphi)$ (Figure 4.33). This algorithm appears different compared to the optimal solution in the form of Equation (4.61). In this case, the soaking interval $[\varphi^0_{end}, \varphi^0]$ in the algorithm in Equation (4.61) is replaced by the reheating interval that takes place under utilizing maximum power during the last stage of the optimal process. As a result, the optimal power control can be represented by the following expression:

$$u^*(\varphi) = \begin{cases} U_{\max}, 0 < \varphi < \varphi_h^0; \\ u_s(\varphi), \varphi_h^0 < \varphi < \varphi_{end}^0; \\ U_{\max}, \varphi_{end}^0 < \varphi \le \varphi^0. \end{cases} \quad (4.67)$$

The singular control $u_s(\varphi)$ can be defined according to Equation (4.63) and Equation (4.64), and computational equation set (Equation 4.62) remains unchanged. However, the expression for $\theta(l,\varphi_h^0,\varphi_{end}^0)$ is obtained similarly to Equation (4.65) and Equation (4.66), and it should be modified according to variation of optimal algorithm $u^*(\varphi)$.

Computation results show that the shapes of radial temperature distribution $\theta(l,\varphi_h^0,\varphi_{end}^0) - \theta^*$ (compare with Figure 4.32b), the control algorithm (Equation 4.67), and the set of equations (Equation 4.62) remain unchanged when preset heating time φ^0 increases beyond all bounds from the value of φ_1^0. The minimal value of cost function $I(\varphi_h^0,\varphi_{end}^0)$ in Equation (4.59) asymptotically decreases with increase of the time φ^0. The time φ_h^0 required for heating under maximum power also decreases to limiting value

$$\varphi_{h\min}^0 = \lim_{\varphi^0 \to \infty} \varphi_h^0,$$

when billet surface temperature becomes equal to oxidation threshold:

$$\theta_{sur}(\varphi_{h\min}^0) = \theta_q.$$

Some results of computation in Figure 4.34a through Figure 4.34d confirm these conclusions.

Practically speaking, maximum total heating time has certain limitations and does not exceed the value of φ_1^0. Therefore, in the majority of practical cases, the optimal control algorithm can be written in the form of Equation (4.61) (Figure 4.29). This algorithm can be calculated for prescribed heating accuracy $\varepsilon = \hat{\varepsilon}_{\min}^{(2)}$ by solving the set in Equation (4.62) after substitution of the expression for $\theta(l,\varphi_h^0,\varphi_{end}^0)$ in the form of Equation (4.65) and Equation (4.66). Because minimal scale losses $I(\varphi_h^0,\varphi_k^0)$ (as a function of φ^0) vastly diminish under $\varphi^0 > \varphi_1^0$ (Figure 4.34d), the value of φ_1^0 could be regarded to a first approximation as the optimal value of heating process time. Under condition $\varphi^0 = \varphi_1^0$, the optimal heating mode is simplified. It consists of the first stage that represents an accelerated heating interval with maximum power. The next stage represents the stage with an exponential increase of the surface temperature (see Figure 4.31).

If it would be necessary to solve the problem of minimization of scale formation for another value of heating accuracy, $\varepsilon = \varepsilon_0$, then the time φ^0 should be accordingly increased by small steps, starting from the value $\varphi_{\min}^0(\varepsilon_0)$.

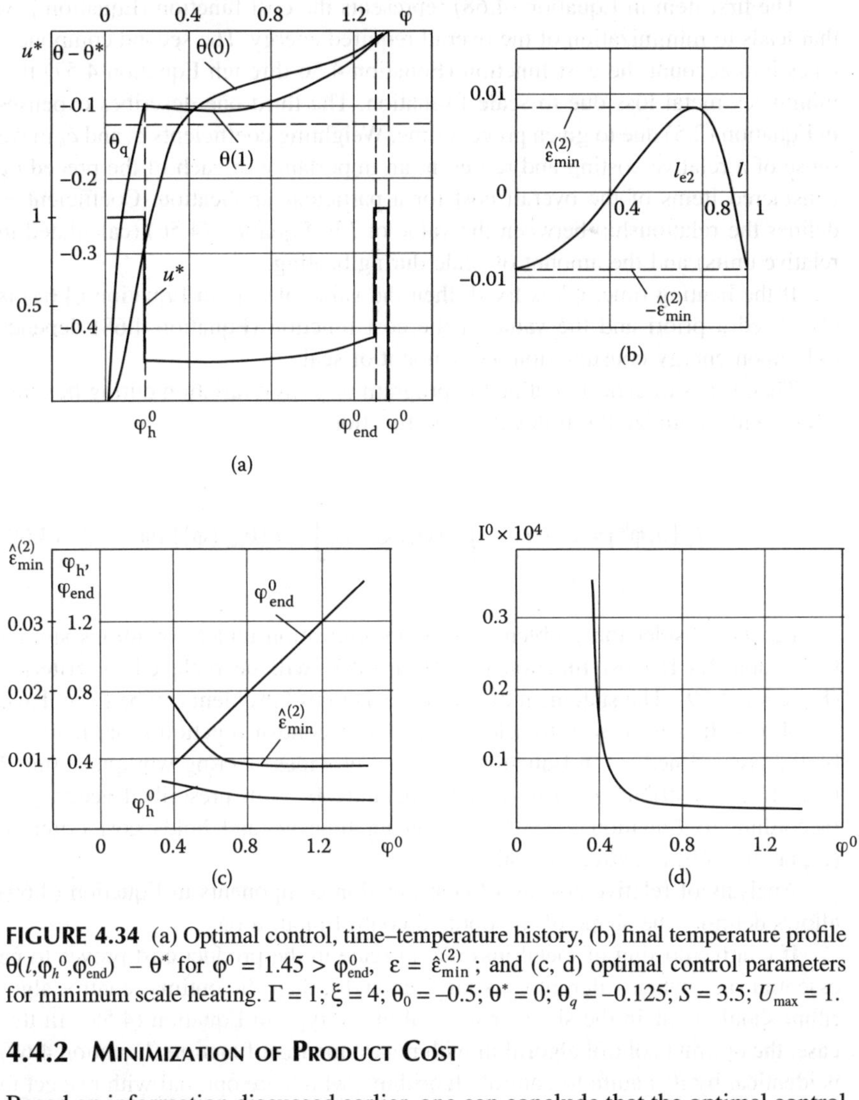

FIGURE 4.34 (a) Optimal control, time–temperature history, (b) final temperature profile $\theta(l,\varphi_h{}^0,\varphi^0_{end}) - \theta^*$ for $\varphi^0 = 1.45 > \varphi^0_{end}$, $\varepsilon = \hat{\varepsilon}^{(2)}_{min}$; and (c, d) optimal control parameters for minimum scale heating. $\Gamma = 1$; $\xi = 4$; $\theta_0 = -0.5$; $\theta^* = 0$; $\theta_q = -0.125$; $S = 3.5$; $U_{max} = 1$.

4.4.2 Minimization of Product Cost

Based on information discussed earlier, one can conclude that the optimal control algorithm in the problem of maximum absolute heating accuracy is identical to a time-optimal control algorithm. Therefore, it seems reasonable to combine cost functions (Equation 2.1, and Equation 2.3 through Equation 2.5). The following overall function can be used as a criterion for minimum product cost:

$$I_\Sigma\left(u,\varphi^0\right) = \int_0^{\varphi^0} u(\varphi)\,d\varphi + c_0 \int_0^{\varphi^0} c_1 f\left(\theta_{sur}(\varphi)\right)d\varphi + c_b \int_0^{\varphi^0} d\varphi\,. \qquad (4.68)$$

The first item in Equation (4.68) represents the cost function (Equation 2.3) that leads to minimization of the overall required energy. The second component takes into account the cost function (Equation 4.56 through Equation 4.57) that minimizes metal loss due to scale formation. The third one describes expenses in Equation (2.5) due to given process time. Weighting coefficients c_0 and c_b make sense of a relative costing and represent an importance of each of the preceding considered items of the overall cost for a particular application. Coefficient c_1 defines the relationship between the value of I in Equation (4.56) (calculated in relative units) and the amount of scale during heating.

If the heating time, φ^0, is fixed, then the value of $c_b\varphi^0$ in Equation (4.68) is also fixed a priori and the value of the cost function (Equation 4.68) depends only upon energy consumption and amount of scale.

Then it is sufficient to define the program of power variation during heating, which will minimize the following cost function:

$$I_\Sigma\left(u,\varphi^0\right)=I_3+I_4=\int_0^{\varphi^0}u(\varphi)\,d\varphi+c_0\int_0^{\varphi^0}c_1 f\left(\theta_{sur}(\varphi)\right)d\varphi\,. \tag{4.69}$$

Let us consider the problem of cost minimization under conditions similar to Section 4.4.1. Cost function (Equation 4.56) will be replaced by criterion (Equation 4.69). The statement of the optimal control problem can be formulated as follows. It is necessary to select such power control function $u^*(\varphi)$ bounded by the preassigned set in Equation (2.7) that provides steering workpiece initial temperature distribution to desired temperature θ^* with prescribed accuracy ε (according to Equation 2.25) for the given time φ^0 and holds cost criterion (Equation 4.69) at extremum value.

Analysis of relative costing of cost function components in Equation (4.69) allows defining the shape of the optimal control input $u^*(\varphi)$.

If relative costing of metal loss due to scale in the product cost proves to be essential (i.e., when value of c_0 is large enough), then the optimal control algorithm should contain the singular interval of the type in Equation (4.58). In this case, the optimal control algorithm with regard to target function (Equation 4.69) is identical by its nature to control algorithms, which are optimal with respect to minimum scale (Section 4.4.1.2). The additional calculation using the technique described in Section 4.4.1.2 leads to the same computational sets — for example, to the set in Equation (4.62). Because these sets always have a unique solution, we conclude that, under given values ε and φ^0, the solution of the minimum cost problem is identical to the solution of the minimum scale problem.

If coefficient c_0 in Equation (4.69) is small enough, then expenses due to energy consumption can be considered as the main part of overall product cost I_Σ. Therefore, optimal power control does not contain a singular control interval and becomes identical to the algorithm that is optimal with respect to minimum energy consumption. Analysis similar to that provided previously leads to the conclusion that, under given values ε and φ^0, solution of the problem for minimizing cost function I_Σ is identical to the solution of the minimum energy consumption problem.

As a result, optimal with regard to overall cost I_Σ control can be considered as optimal by minimizing the scale formation or by minimizing energy consumption for each given pair of values ε and φ^0. In the first case, the optimal algorithm contains the singular interval (Section 4.4.1), but in the second case, only intervals where $u^*(\varphi) \equiv U_{max}$ and $u^*(\varphi) \equiv 0$ (Section 2.9).

For given time φ^0, two particular optimization problems can be solved for minimum scale or minimum energy consumption by using optimal control techniques that have been described earlier. Then it is possible to choose the sought solution of the initial minimum cost problem from obtained ones, which provides smaller value of overall cost function. The value of $I_\Sigma(u^*)$ can be calculated by Expression (4.68) in order to compare obtained solutions of separate optimization problems. Keep in mind that shapes of final spatial temperature distribution can be different in these problems, as well as computational sets. Examples that have been provided in the previous section show that the continuous nature of dependency of the heating process parameters upon time φ^0 leads to the same shape of these curves as for the time-optimal control. With respect to minimum scale formation, this conclusion does not remain valid for the problem of minimum energy consumption.

Algorithms of heating power variation, which are optimal with respect to time and energy consumption criteria, have the same form (Equation 2.27) with alternating intervals of heating and temperature soaking (Section 2.9.2), but their durations Δ_i^0 and $\bar{\Delta}_i^0, i = \overline{1, N}$ in the general case do not coincide. This can lead to different profiles of final temperature distribution (Section 3.5). In particular, under $\varepsilon_{min}^{(2)} < \varepsilon \le \varepsilon_{min}^{(1)}$, the shapes of curves $\theta(l, \Delta_1^0, \Delta_2^0) - \theta^*$ and $\theta(l, \bar{\Delta}_1^0, \bar{\Delta}_2^0) - \theta^*$ (Figure 3.11 and Figure 3.17), and appropriate computational equation sets (Equation 3.12 and Equation 3.15; Equation 3.34 and Equation 3.35) differ from each other for the same value of ε.

Let us underline that the time required for heating with minimum energy consumption is not fixed a priori. Its optimal value, φ_e^0, is calculated, after computation of optimal values $\bar{\Delta}_i^0, i = \overline{1, N}$, as:

$$\varphi_e^0 = \sum_{i=1}^{N} \bar{\Delta}_i^0 . \tag{4.70}$$

Therefore, solution of this problem will comply with solution of the particular optimization problem for this component of expenses in the overall cost, I_Σ, under condition of fixed time, φ^0, in the case when φ^0 is equal to φ_e^0. Obviously, if $\varphi^0 > \varphi_e^0$, then the same result could be obtained by delaying the beginning of the heating by $\varphi^0 - \varphi_e^0$. If $\varphi^0 < \varphi_e^0$, then the new problem of minimum energy

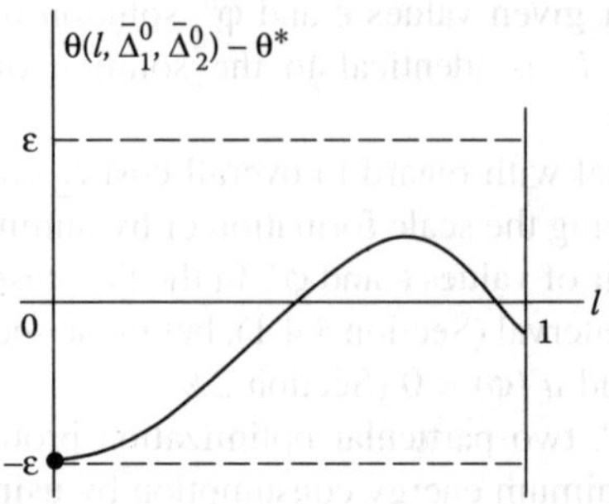

FIGURE 4.35 Temperature profile at the end of the minimum energy heating process of fixed duration $\varphi^0 < \varphi_e^{\,0}$ for $\varepsilon^{(2)}_{\min} < \varepsilon \le \varepsilon^{(1)}_{\min}$.

consumption with predefined sum φ^0 of all values $\bar{\Delta}_i^0, i = \overline{1, N}$ would appear. In this case, the number of sought optimal control parameters decreases by one.

If the value of φ^0 can be chosen freely under condition $\varepsilon^{(2)}_{\min} < \varepsilon \le \varepsilon^{(1)}_{\min}$, then durations of two power constancy intervals of $u^*(\varphi)$ can be considered as two sought parameters. Because the sum of these parameters is a known fixed value, only one of them represents the single unknown variable. According to basic properties (Equation 3.39 and Equation 3.40), the maximum deviation ε of final temperature from desired one is reached only at one point of the coordinate axis.

Similar analysis provided in Section 3.5, leads to the conclusion that this point will be located in the billet center $l = 0$. Figure 4.35 shows the shape of temperature profile $\theta(l, \bar{\Delta}_1^0, \bar{\Delta}_2^0) - \theta^*$. An appropriate computational set of equations can be reduced to the first equation of the set in Equation (3.35).

In the majority of typical cases, it is necessary to provide the best heating accuracy $\varepsilon = \varepsilon^{(2)}_{\min}$ attainable under the optimal control input that belongs to the class of two-parameter functions. Consideration of these typical cases allows simplifying the solution of the optimal control problem, which becomes optimal with respect to minimum time or minimum energy consumption criteria (Section 3.5). If given heating time φ^0 in Equation (4.69) complies with minimum possible time $\varphi^0_{\min}(\varepsilon^{(2)}_{\min})$ necessary to provide heating accuracy $\varepsilon^{(2)}_{\min}$, then this control function can be considered as optimal with respect to the cost function (Equation 4.69) as well.

The following case will take place under condition $\varphi^0 > \varphi^0_{\min}(\varepsilon^{(2)}_{\min})$. The optimal control can be defined as the two-parameter function that depends upon φ_h and φ_{end}, according to Equation (4.61) or Equation (4.67), for required accuracy $\varepsilon = \hat{\varepsilon}^{(2)}_{\min}(\varphi^0)$ when solving the particular optimization problem for minimum scale.

It is clear that, under $\varphi^0 = \varphi^0_{\min}(\varepsilon^{(2)}_{\min})$, the value $\hat{\varepsilon}^{(2)}_{\min}$ coincides with minimax $\varepsilon^{(2)}_{\min}$, and the optimal control algorithm is defined by parameters $\Delta_1^{\,0}$ and $\Delta_2^{\,0}$ instead of parameters φ_h and φ_{end}. If the condition $\varphi^0 > \varphi^0_{\min}(\varepsilon^{(2)}_{\min})$ is true, then the optimal algorithm of the type in Equation (4.61) and Equation (4.67) with singular intervals of reduced power (Figure 4.30 and Figure 4.34) leads to smaller heating nonuniformity $\hat{\varepsilon}^{(2)}_{\min}$ in comparison with $\varepsilon^{(2)}_{\min}$ (Figure 4.34c). However, in this case of $\varepsilon = \hat{\varepsilon}^{(2)}_{\min} < \varepsilon^{(2)}_{\min}$, algorithms optimal by energy consumption usually comply with time-optimal algorithms (Section 3.5) and contain no less than three power constancy intervals. In other words, under condition $\varphi^0 > \varphi^0_{\min}(\varepsilon^{(2)}_{\min})$ optimal by energy consumption heating modes that provide heating accuracy, $\hat{\varepsilon}^{(2)}_{\min}$, do not correspond to the considered class of two-parameter control inputs. Thus, one can conclude that the heat power control that is optimal by objective function (Equation 4.69) minimizes metal loss due to scale formation as well.

Therefore, the optimal control problem should be solved (according to the technique described in Section 4.4.1.2) for each given value $\varphi^0 > \varphi^0_{\min}(\varepsilon^{(2)}_{\min})$ and $\varepsilon = \hat{\varepsilon}^{(2)}_{\min}(\varphi^0)$ with respect to cost functions (Equation 4.56 and Equation 4.57). The value of overall cost I_Σ in Equation (4.68) can be calculated using control inputs that have been found as a result of OCP solution.

Dependencies $I_\Sigma(\varphi^0)$ presented in Figure 4.36 correspond to a number of values $c_0 > 0$ and $c_b = 1$ using results of optimal induction heating of steel billets with respect to minimum scale (examples in Figure 4.31, Figure 4.32, and Figure 4.34).[1] The value of $c_1 = 2.8 \cdot 10^4\, \Delta_1^0(\varepsilon^{(2)}_{\min})$ is determined for the case of metal loss that can be described by the parabolic law of oxidation in 0.5% of carbon component at $\varphi^0 = \varphi^0_{\min}(\varepsilon^{(2)}_{\min})$.

Obtained curves allow one to find optimal heating time φ^0_{opt} that corresponds to minimum value of I_Σ under the condition that any value of $\varepsilon = \hat{\varepsilon}^{(2)}_{\min}(\varphi^0)$ can be considered as satisfactory accuracy:

$$\varphi^0_{opt} = \min_{\varphi^0}\left(\min_{u} I_\Sigma\left(u,\varphi^0\right)\right), \varphi^0 \ge \varphi^0_{\min}\left(\varepsilon^{(2)}_{\min}\right), \varepsilon = \hat{\varepsilon}^{(2)}_{\min}\left(\varphi^0\right). \qquad (4.71)$$

Based on results shown in Figure 4.36, it is possible to conclude that condition $\varphi^0_{opt} = \varphi^0_{\min}(\varepsilon^{(2)}_{\min})$ is true for all values $c_0 \le c_0^* = 0.15$. Therefore, the time-optimal process proves to be optimal with respect to minimum product cost under the

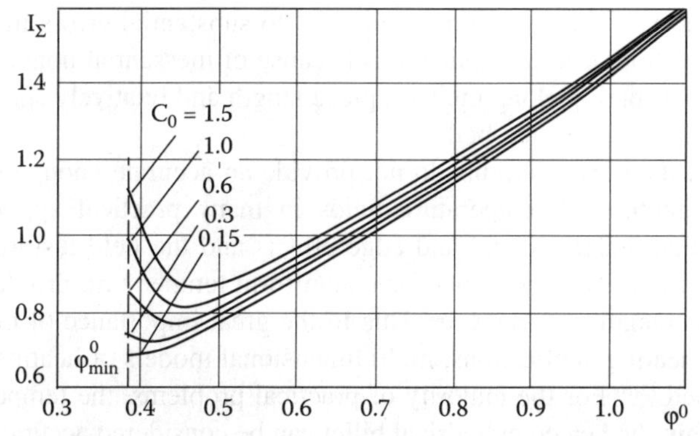

FIGURE 4.36 Minimum product cost as a function of process time. $\Gamma = 1$; $\xi = 4$; $\theta_0 = -0.5$; $\theta^* = 0$; $\theta_q = -0.125$; $S = 3.5$; $\varepsilon_0 = \hat{\varepsilon}^{(2)}_{\min}$; $U_{\max} = 1$.

appropriate values of c_0. For $c_0 > c_0^*$, the minimum product cost is reached under control, minimizing metal loss due to scale, for any value of process time.

The control algorithm, which provides minimum energy consumption, under particular conditions would result in minimum product cost. This algorithm can be found using the computational procedure described above with neglected restraint on the value of admissible temperature deviation. Control input should be considered in the class of N-parameter functions ($N > 2$).

4.5 OPTIMAL CONTROL OF MULTIDIMENSIONAL LINEAR MODELS OF INDUCTION HEATING PROCESSES

Most mathematical models of induction heating tend to be handled with a combination of the following assumptions (Section 2.3):

- Neglecting nonlinearity by averaging process parameters on appropriate temperature intervals;
- Concentrating on considering of regular-shaped workpieces (i.e., plate, cylinder, rectangle slab, parallelepiped, sphere);
- Simplifying the description of the geometric input data for the induction heating system;
- Taking into account the nonuniformity of temperature distribution only along one or two coordinates (reducing the three-dimensional temperature field to one-dimensional or two-dimensional forms).

The linear one-dimensional model as a basic model of induction heating of regular-shaped bodies has been considered in previous chapters. This simplified model describes one-dimensional processes of heat transfer along the thickness of an infinitely long plate or along the radius of an infinite length cylinder. It has been assumed that temperature nonuniformity along other coordinates can be neglected. The last assumption does not lead to substantial errors in some conventional induction heating applications because of inessential nonuniformity of heat source distribution along the workpiece length and relatively small heat loss from butt ends of the heated body.

However, these assumptions do not provide an accurate enough description of electromagnetic and temperature fields in many practical applications. In particular, electromagnetic end and edge effects and the field distortion caused by them are primarily responsible for nonuniform temperature profiles in cylindrical, and rectangular workpieces. Due to the great importance of these effects in induction heating applications, multidimensional models of heating processes should be used.[15,16] For the majority of practical problems, the temperature distribution within the heated cylindrical billet can be considered accurately enough as an axially symmetrical temperature field.

On the other hand, the temperature distribution along the length of the slab can be assumed as uniform. Therefore, depending upon the particular induction heating process, two-dimensional mathematical models can be effectively used to describe nonuniform temperature distribution of the heated workpiece (1) along axial and radial directions of a cylinder of finite length and (2) in the plane of the cross-section of the heated slab that can be considered as an infinitely long plate.

This chapter discusses the time-optimal control problem utilizing two-dimensional models of induction heating processes. An ability to extend obtained results to solution of other optimization problems will be investigated here as well as consideration of all essential features of real-life processes.

4.5.1 Linear Two-Dimensional Model of the Induction Heating Process

Let us consider the general optimal control problem assuming that complete information is available with respect to the heating process and imposed requirements. Let us also assume that there are no disturbances affecting the heating system and the time of workpiece transfer to hot working equipment can be neglected. The function $\theta(l,y,\varphi)$ will be treated as an output-controlled function of the process that describes nonuniform temperature distribution in axial/radial directions of a cylindrical billet of finite length or within a cross-section of the rectangular-shaped workpiece (rectangular slab of infinite length). The following two-dimensional linear equation of heat transfer will be considered further as a basic mathematical model of the induction heating process[1]:

$$\frac{\partial\theta(l,y,\varphi)}{\partial\varphi}=\frac{\partial^2\theta(l,y,\varphi)}{\partial l^2}+\frac{\Gamma}{l}\frac{\partial\theta(l,y,\varphi)}{\partial l}+\beta^2\frac{\partial^2\theta(l,y,\varphi)}{\partial y^2}+W_d(\xi,l,y)u(\varphi),$$

$$0<l<1;\ 0<y<1;\ 0<\varphi\le\varphi^0 \tag{4.72}$$

Equation (4.72) is written with respect to relative (dimensionless) units.

Initial and boundary conditions of the second kind can be formulated as:

$$\theta(l,y,0)=\theta_0(l,y);\ l\in[0,1];\ y\in[0,1]; \tag{4.73}$$

$$\frac{\partial\theta(1,y,\varphi)}{\partial l}=q(y,\varphi);\ \frac{\partial\theta(0,y,\varphi)}{\partial l}=0;$$

$$\beta\frac{\partial\theta(l,1,\varphi)}{\partial y}=q_{T1}(l,\varphi);\ \beta\frac{\partial\theta(l,0,\varphi)}{\partial y}=-\Gamma q_{T0}(l,\varphi) \tag{4.74}$$

$$l\in[0,1];\ y\in[0,1];\ 0<\varphi\le\varphi^0.$$

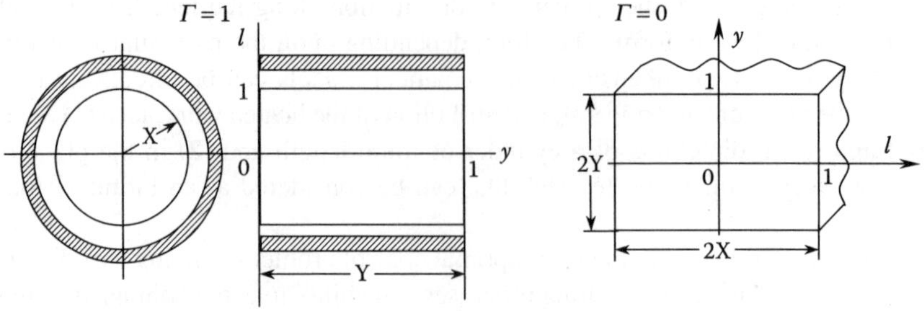

FIGURE 4.37 Geometrical model of workpieces for simulation by two-dimensional equations of heat transfer.

For the majority of induction heating problems, boundary conditions include convection heat losses. In this case, instead of Equation (4.74), the boundary conditions of the third kind can be expressed as

$$\frac{\partial\theta(1,y,\varphi)}{\partial l} = Bi\left(\theta_a(\varphi) - \theta(1,y,\varphi)\right); \frac{\partial\theta(0,y,\varphi)}{\partial l} = 0;$$

$$\frac{\partial\theta(l,1,\varphi)}{\partial y} = Bi_1\left(\theta_{a1}(\varphi) - \theta(l,1,\varphi)\right); -\frac{\partial\theta(l,0,\varphi)}{\partial y} = \Gamma Bi_0\left(\theta_{a0}(\varphi) - \theta(l,0,\varphi)\right) \tag{4.75}$$

In Equation (4.72) through Equation (4.75) (Figure 4.37):

$\Gamma = 0$ or $\Gamma = 1$ for the rectangular slab of infinite length or cylindrical billet of finite length, respectively.

φ is a relative time.

l and y are relative values of spatial coordinates with respect to X and Y, where X is a half of the largest cross-section side for the rectangular-shaped body or the radius of the cylinder.

Y is a half of the smaller cross-section side or the length of the cylinder.

$W_d(\xi,l,y)$ is a function that describes spatial distribution of internal heat source density induced by eddy currents per unit time in a unit volume.

ξ is a specific parameter that can be defined similarly to the one-dimensional problem as $\xi = X\sqrt{2}/\delta$.

q, q_{T1}, and q_{T0} are values of heat losses from a surface of the heated body at $l = 1$, $y = 1$, and $y = 0$, respectively.

Bi, Bi_1, and Bi_0 are Biot numbers for $l = 1$, $y = 1$, and $y = 0$, respectively.

θ_a, θ_{a1}, and θ_{a0} denote relative values of ambient temperatures at $l = 1$, $y = 1$, and $y = 0$, respectively.

Factor β is a specific geometric parameter that can be calculated as:

$$\beta = X/Y. \tag{4.76}$$

Other relative values can be calculated from corresponding formulas similar to those used in Chapter 1. The value of X can be considered as a basic spatial coordinate.

Equation (4.72) through Equation (4.75) describe two-dimensional temperature distribution $\theta(l,y,\varphi)$ at any time and at any point with coordinates l and y. The heating process is caused by internally generated heating power represented by function $W_d(\xi,l,y)u(\varphi)$. The initial temperature state $\theta_0(l,y)$ in Equation (4.73) refers to the fixed temperature distribution within the workpiece at time $\varphi = 0$.

The boundary conditions in Equation (4.74) or Equation (4.75) for $\Gamma = 1$ determine the value of heat losses from the butt-end and lateral surfaces of the billet and allow consideration of the cylindrical billet as an axially symmetrical body ($l = 0$ is the axis of symmetry). In the general case, heat losses from a surface of the heated body can be specified by values $q(y,\varphi)$, $q_{T1}(l,\varphi)$, and $q_{T0}(l,\varphi)$ in Equation (4.74). For heat transfer by convection, the boundary conditions include Biot numbers Bi, Bi_1, and Bi_0. Appropriate ambient temperatures in Equation (4.75) will be equal respectively to θ_a, θ_{a1}, and θ_{a0}. The center of cross-section should be taken as the coordinate origin in simulation of the heating process for the rectangular slab ($\Gamma = 0$). Therefore, it is beneficial to take advantage of symmetry and consider only the temperature field distribution within a quarter of the workpiece cross-section.

According to the symmetry condition, the temperature gradient in Equation (4.74) will be equal to zero at $l = 1$ and $y = 0$ (Figure 4.37). Therefore, an expression, $\beta > 1$, remains valid for Equation (4.76) because the value of X represents a length of the largest half-side of the rectangular cross-section. Time-dependent heating power $u(\varphi)$ is chosen as a lumped control input constrained by Equation (2.7).

The relatively complicated two-dimensional electromagnetic problem should be solved to determine the function $W_d(\xi,l,y)$ in Equation (4.72).[15,16] Approximate analytical expressions can be used for simulation of function $W_d(\xi,l,y)$. At the same time, numerical models allow representing the function $W_d(\xi,l,y)$ as a two-dimensional mesh function. Similarly to Equation (1.36), solution of the problem in Equation (4.72) under boundary conditions in Equation (4.73) and Equation (4.74) or Equation (4.73) and Equation (4.75) can be written in the form of the Duhamel integral as[1]:

$$\theta(l,y,\varphi) = \Phi(l,y,\varphi) + \int_0^{\varphi} \frac{\partial\Lambda(l,y,\varphi-\tau)}{\partial\varphi} u(\tau)\,d\tau\,. \tag{4.77}$$

Functions $\Phi(l,y,\varphi)$ and $\Lambda(l,y,\varphi)$ can be calculated taking into consideration appropriate boundary conditions of the second (Φ_2,Λ_2) and the third (Φ_3,Λ_3) kinds as[1]:

$$\Phi_2(l,y,\varphi) = (\Gamma+1)\theta_{000} + (\Gamma+1)\int_0^{\varphi} g_{00}^{*(2)}(\tau)\,d\tau +$$

$$+2\sum_{r=1}^{\infty}\frac{K(\mu_r l)}{K^2(\mu_r)}\cdot\left[\theta_{0r0}e^{-\mu_r^2\varphi} + \int_0^{\varphi} g_{r0}^{*(2)}(\tau)e^{-\mu_r^2(\varphi-\tau)}d\tau\right]+$$

$$+2(\Gamma+1)\sum_{n=1}^{\infty}\left[\theta_{00n}e^{-\beta^2\lambda_n^2\varphi} + \int_0^{\varphi} g_{0n}^{*(2)}(\tau)e^{-\beta^2\lambda_n^2(\varphi-\tau)}d\tau\right]\cdot K^*(\lambda_n y)+$$

$$+2\sum_{r=1}^{\infty}\sum_{n=1}^{\infty}\frac{K(\mu_r l)K^*(\lambda_n y)}{K^2(\mu_r)}\left[\theta_{0rn}e^{-\left(\mu_r^2+\beta^2\lambda_n^2\right)\varphi} + \int_0^{\varphi} g_{rn}^{*(2)}(\tau)e^{-\left(\mu_r^2+\beta^2\lambda_n^2\right)(\varphi-\tau)}d\tau\right]; \tag{4.78}$$

$$\Phi_3(l,y,\varphi) = \sum_{r=1}^{\infty}\sum_{n=1}^{\infty}D_{rn}\left[\theta_{0rn}e^{-\left(\mu_r^2+\beta^2\lambda_n^2\right)\varphi} + \int_0^{\varphi} g_{rn}^{*(3)}(\tau)e^{-\left(\mu_r^2+\beta^2\lambda_n^2\right)(\varphi-\tau)}d\tau\right]\times$$
$$\times K(\mu_r l)K^*(\lambda_n y) \tag{4.79}$$

$$\Lambda_2(l,y,\varphi) = U_{\max}(\Gamma+1)W_{00}(\xi)\varphi + 2U_{\max}\sum_{r-1}^{\infty}\frac{W_{r0}(\xi)K(\mu_r l)}{\mu_r^2K^2(\mu_r)}\cdot\left[1-e^{-\mu_r^2\varphi}\right]+$$

$$+2U_{\max}(\Gamma+1)\sum_{n=1}^{\infty}\frac{W_{0n}(\xi)K^*(l_n y)}{\beta^2\lambda_n^2}\left[1-e^{-\beta^2\lambda_n^2\varphi}\right]+$$

$$+4U_{\max}\sum_{r=1}^{\infty}\sum_{n=1}^{\infty}\frac{W_{rn}(\xi)K^*(\mu_r l)K^*(\lambda_n y)}{\left(\mu_r^2+\beta^2\lambda_n^2\right)K^2(\mu_r)}\left[1-e^{-\left(\mu_r^2+\beta^2\lambda_n^2\right)\varphi}\right]; \tag{4.80}$$

$$\Lambda_3(l,y,\varphi) = U_{\max}\sum_{r=1}^{\infty}\sum_{n=1}^{\infty}\frac{D_{rn}W_{rn}(\xi)K(\mu_r l)K^*(\lambda_n y)}{\mu_r^2+\beta^2\lambda_n^2}\left[1-e^{-\left(\mu_r^2+\beta^2\lambda_n^2\right)\varphi}\right]. \tag{4.81}$$

In Expression (4.78) through Equation (4.81):

$$D_{rn} = 4\mu_r^2\lambda_n^2\left(\lambda_n^2 + Bi_1^2\right)\Big/\Big[K^2(\mu_r)\Big(Bi^2 + \mu_r^2 + \big(1-\tilde{A}\big)Bi\Big)\times$$
$$\times\Big[\Big(\lambda_n^2 + \tilde{A}Bi_0^2\Big)\Big(\lambda_n^2 + Bi_1^2\Big) + \Big(\tilde{A}Bi_0 + Bi_1\Big)\Big(\lambda_n^2 + \tilde{A}Bi_0Bi_1\Big)\Big]\Big]; \tag{4.82}$$

$$W_{rn}(\xi) = \int_0^1\int_0^1 W_d(\xi, l, y)\, l^{\Gamma} K(\mu_r l)\, K^*(\lambda_n y)\, dl dy;\ r, n = 0,1,2,\ldots \tag{4.83}$$

$$\theta_{0rn} = \int_0^1\int_0^1 \theta_0(l, y)\, l^{\Gamma} K(\mu_r l)\, K^*(\lambda_n y)\, dl dy;\ r, n = 0,1,2,\ldots \tag{4.84}$$

$$g_{rn}^{*(2)}(\varphi) = q_n(\varphi) K(\mu_r) + \beta(-1)^n q_{T1r}(\varphi) + \Gamma\beta q_{T0r}(\varphi);\ r, n = 0,1,2\ldots \tag{4.85}$$

$$g_{rn}^{*(3)}(\varphi) = K(\mu_r)\left[\frac{Bi\theta_a(\varphi)}{\lambda_n}\left(\sin\lambda_n - \frac{\Gamma Bi_0}{\lambda_n}(\cos\lambda_n - 1)\right)\right] +$$
$$+\beta^2\frac{K_1(\mu_r)}{\mu_r}\left[Bi_1\theta_{a1}(\varphi)\left(\cos\lambda_n + \frac{\Gamma Bi_0}{\lambda_n}\sin\lambda_n\right) + \Gamma Bi_0\theta_{a0}(\varphi)\right]; \tag{4.86}$$

$r, n = 1, 2\ldots$

In Expression (4.85):

$$q_n(\varphi) = \int_0^1 q(y, \varphi)\cos(\pi n y)\, dy;\ n = 0,1,2,\ldots \tag{4.87}$$

$$q_{T\nu r}(\varphi) = \int_0^1 q_{T\nu}(l, \varphi)\, l^{\Gamma} K(\mu_r l)\, dl;\ \nu = 0,1;\ r = 0,1,2,\ldots \tag{4.88}$$

The fundamental functions $K(\mu r l)$ and $K_1(\mu_r)$ and fundamental numbers μ_r, $r = 0, 1, 2, \ldots$ in Equation (4.78) through Equation (4.88) should be calculated similarly to the procedure discussed in Chapter 1. In the case of the boundary conditions in Equation (4.74), the function $K^*(\lambda_n y)$ can be written using the following expressions:

$$K^*(\lambda_n y) = \cos(\lambda_n y),\ \lambda_n = \pi n,\ n = 0,1,2,\ \ldots. \tag{4.89}$$

Under the boundary conditions in Equation (4.75), the function $K^*(\lambda_n y)$ is calculated from the following expression:

$$K^*\left(\lambda_n y\right) = \cos\lambda_n y + \frac{\Gamma Bi_0}{\lambda_n}\sin\lambda_n y, n = 1,2,\ldots\ . \tag{4.90}$$

The numbers, λ_n, in Equation (4.90) represent roots of transcendental equations:

$$\frac{\lambda}{Bi_1 + \Gamma Bi_0} - \frac{\Gamma Bi_0 Bi_1}{\left(Bi_1 + Bi_0\right)\lambda} = ctg\lambda\ . \tag{4.91}$$

The finite number of terms of the series should always be used in given expressions. An important question arises at this point with regard to accuracy of obtained results, which depends upon accuracy of the infinite series. Experience shows that sufficient accuracy can be achieved by applying $r,\ n \le 10 \div 15$.

4.5.2 Two-Dimensional Time-Optimal Control Problem

This section discusses the time-optimal control problem with respect to the two-dimensional model of the induction heating process. The general form of this problem was discussed earlier in Section 2.8.

Let us assume that the induction heating process is properly described by Equation (4.72) through Equation (4.74). Two-dimensional temperature distribution within the heated workpiece will be treated as an output-controlled function of the process. Minimum process time (Equation 2.1) can be considered as a cost criterion. Control input function is chosen from the set of admissible controls to influence temperature distribution and dynamic behavior of the induction heating system. Time-dependent heating power $u(\varphi)$ is chosen as lumped control input constrained by Equation (2.7).

The optimal control problem can be formulated as follows. It is necessary to select such control function $u(\varphi) = u^*(\varphi)$ that provides the steering workpiece initial temperature distribution to desired temperature $\theta^* = const$ with prescribed accuracy ε (according to Equation 2.25) in minimal optimal process time φ^0. In this case, instead of Equation (2.25), the temperature distribution requirement at the end of the heating cycle can be obtained in the following form:

$$\max_{l,y\in[0,1]}\left|\theta\left(l,y,\varphi^0\right) - \theta^*\right| \le \varepsilon\ . \tag{4.92}$$

The requirement in Equation (4.92) that provides prescribed heating accuracy within the two-dimensional region (i.e., within the finite length cylinder or in the plane of cross-section of the infinite prism) has a unique feature. Instead of interval $[0,1] \ni l$ in Equation (2.25) with respect to the one-dimensional problem (Equation 2.25), it considers a two-dimensional area.

Similarly to Section 2.8, it is possible to conclude the following. Optimal algorithms of power control remain unchanged when using models of multidimensional temperature fields. The time-optimal control algorithm consists of alternating stages of heating with maximum power $u \equiv U_{max}$ (heat ON) and subsequent soaking stage under $u \equiv 0$ (heat OFF) (Figure 2.9).

For any particular process, the number of stages, N, and durations, Δ_i^0, $i = \overline{1,N}$, of those stages should be determined during subsequent calculation. The sought control function can be written in the form of Expression (2.27).

The problem is reduced to determination of such time intervals Δ_i, $i = \overline{1,N}$, of alternating heating and soaking stages that provide the requirement (Equation 4.92) for minimal possible time. Similarly to Equation (2.34), the time-optimal control problem is reduced to the following special problem of mathematical programming:

$$\begin{cases} I(\Delta) = \sum_{i=1}^{N} \Delta_i \to \min_{\Delta}; \\ \Phi(\Delta) = \max_{l,y\in[0,1]} \left|\theta(l,y,\Delta) - \theta^*\right| \le \varepsilon. \end{cases} \tag{4.93}$$

This optimization problem looks noticeably different in comparison with OCP (Equation 2.34) due to the important distinction between requirements in Equation (2.25) and Equation (4.92). Nevertheless, the optimal control problem (Equation 4.39) represents the particular case of the general mathematical programming problem (Equation 3.38).

Sought parameters Δ_i^0, $i = \overline{1,N}$ can be found based on alternance properties in Equation (3.39) and Δ_i^0, Equation (3.40) according to the computational procedure described in Section 3.7.

As was mentioned in Section 3.7, the basic difficulties arise here because spatial distribution of final temperature could be noticeably different compared to the optimization problem (Equation 2.34). These features primarily deal with the multidimensional nature of heating process models leading to the variety of possible final temperature distributions. These distributions differ from each other by location of points where maximum admissible temperature deviations are reached. Therefore, the proper shape of final temperature distribution and appropriate set of governing equations could not be easily found. Basic relations (Equation 3.39 and Equation 3.40) cannot be transformed into appropriate sets of equations with respect to sought parameters of the optimal process as was made earlier in Section 3.3 and Section 3.6 for the one-dimensional problem.

Nevertheless, it is possible to show that, if some preliminary conditions are satisfied, then the optimal control method offered in this text could be applied in its basic aspects to the variety of practical IHP optimization problems of the type in Equation (4.93). Special procedures based on the alternance method can be used for determining an optimal control algorithm $u^*(\varphi)$.

This conclusion is based on the following general reasons. The alternance properties (Equation 3.39 and Equation 3.40) remain valid. The limiting cases and analogies with the one-dimensional problem can be quite easily found based on common sense and obvious technical conditions. The suggested method is based on natural physical properties that can be used in the course of the computational procedure as well as effective means to check results.

The time-optimal problem (Equation 4.93) will be considered further as an example of the two-dimensional problem of the type in Equation (2.34). The obtained results can be extended to more complicated optimization problems as well as to optimal control problems subject to the variety of cost functions.

4.5.3 Time-Optimal Control of Induction Heating for Cylindrical Billets

Equation (4.72) through Equation (4.75) for $\Gamma = 1$ describe the two-dimensional temperature distribution $\theta(l,y,\varphi)$ within the axially symmetrical cylindrical billet of finite length. By substituting $u(\varphi)$ in the form of Equation (2.27) and taking into consideration Expression (4.77) through Expression (4.91), the expression for temperature distribution $\theta(l,y,\Delta)$ can be obtained in the explicit form as a function of spatial coordinates and optimal control parameters Δ_i, $i = \overline{1, N}$.[1] In the case of the second kind of boundary condition, the temperature distribution $\theta(l,y,\Delta)$ can be written as:

$$\theta(l,y,\Delta) = \Phi_2\left(l, y, \sum_{i=1}^{N}\Delta_i\right) + U_{\max}W_{00}(\xi)\sum_{i=1}^{N}\left[(-1)^{i+1}+1\right]\Delta_i +$$

$$+ 2U_{\max}\sum_{r=1}^{\infty}\frac{W_{r0}(\xi)J_0(\mu_r l)}{\mu_r^2 J_0^2(\mu_r)}\left[\sum_{j=1}^{N}(-1)^{j+1}\left(1-\exp\left(-\mu_r^2\sum_{m=j}^{N}\Delta_m\right)\right)\right] +$$

$$+ 4U_{\max}\sum_{n=1}^{\infty}\frac{W_{0n}(\xi)\cos(\pi n y)}{\pi^2\beta^2 n^2}\left[\sum_{j=1}^{N}(-1)^{j+1}\left(1-\exp\left(-\pi^2\beta^2 n^2\sum_{m=j}^{N}\Delta_m\right)\right)\right] +$$

$$+ 4U_{\max}\sum_{r=1}^{\infty}\sum_{n=1}^{\infty}\frac{W_{rn}(\xi)J_0(\mu_r l)\cos(\pi n y)}{\left(\mu_r^2+\pi^2\beta^2 n^2\right)J_0^2(\mu_r)}\times$$

$$\times\left[\sum_{j=1}^{N}(-1)^{j+1}\left(1-\exp\left(-\mu_r^2-\pi^2\beta^2 n^2\right)\sum_{m=j}^{N}\Delta_m\right)\right]. \tag{4.94}$$

In the case of the third kind of boundary condition, it can be calculated using the following expression:

$$\theta(l,y,\Delta)=\Phi_3\left(l,y,\sum_{i=1}^{N}\Delta_i\right)+U_{\max}\sum_{r=1}^{\infty}\sum_{n=1}^{\infty}\frac{D_{rn}J_0(\mu_r l)\left(\cos\lambda_n y+\dfrac{Bi_0}{\lambda_n}\sin\lambda_n y\right)}{\left(\mu_r^2+\beta^2\lambda_n^2\right)}\times$$

$$\times W_{rn}(\xi)\left[\sum_{j=1}^{N}(-1)^{j+1}\left(1-\exp\left(-\mu_r^2-\beta^2\lambda_n^2\right)\sum_{m=j}^{N}\Delta_i\right)\right]. \tag{4.95}$$

In contrast to one-dimensional models, nonuniformity of electromagnetic heat source distribution along the workpiece length and heat losses from the workpiece butt ends ($y = 1$ and $y = 0$) and lateral ($l = 1$) surfaces now will be taken into account.

The field distortion caused by these effects is primarily responsible for quantitative variation of radial temperature distribution within the heated cylindrical billet. However, in many practical cases such effects cause relatively small temperature gradients in the axial direction. Therefore, an existence of those effects does not lead to *qualitative* change of the shape of final radial temperature distribution in any cross-section of the cylinder.

The number of extremum points of the curve $\theta(l,y,\varphi^0)$ under $y = const$ can be determined according to the relationship in Equation (3.10). This condition directly links the number, N, of control intervals with the number of extremum points, $M_{\max}$. This fact was already discussed in Chapter 3. The area with maximum and minimum temperatures will be located at some internal points $x_j^0 = (l_j^0, y_j^0)$ within cylinder cross-sections. Axial coordinate y_j^0 specifies location of cross-sections where the points $x_j^0 = (l_j^0, y_j^0)$ should be found.

If axial temperature distribution is nonuniform, then points x_j^0 will be located in different cylinder cross-sections that represent the unique feature of the two-dimensional problem. All possible scenarios of location of points x_j^0, $j = \overline{1,R}$ could not be unambiguously set a priori. Therefore, the set of governing equations also could not be easily found for prescribed value ε in Equation (4.93). This adds complexity to solution of the two-dimensional time-optimal control problem of induction heating of cylindrical billets.

Because general equalities (Equation 3.39 and Equation 3.40) remain valid, in this case analogies with the one-dimensional problem can be used. Final temperature distribution $\theta(l,y,\Delta^0)$ and all optimal process parameters Δ_i^0 can be represented as continuous functions of ε. Under given conditions, it is possible to obtain the following results[1] (similar to Section 3.7).

1. If heating accuracy satisfies equality $\varepsilon = \varepsilon_{\min}^{(1)}$, then condition $N = 1$ will be valid. This means that time-optimal heating represents a single-stage process utilizing maximum power. We shall limit our consideration to the typical case of heating accuracy $\varepsilon_{\min}^{(1)} > \varepsilon_{\inf}$ that corresponds

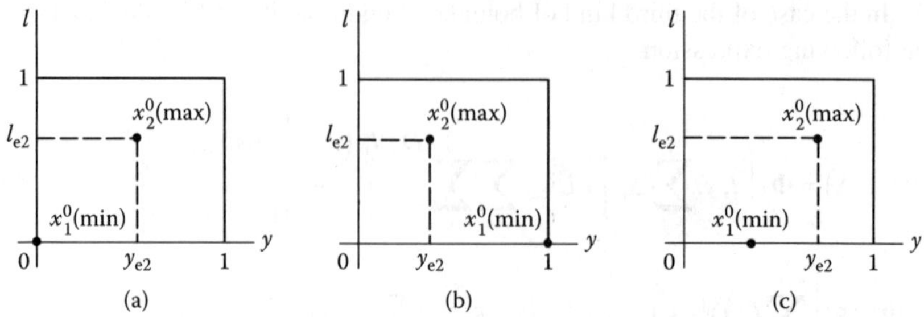

FIGURE 4.38 Variants of location of minimum and maximum temperature points at the end of time-optimal heating of finite length cylindrical billet for $\varepsilon = \varepsilon_{\min}^{(1)}$.

to substantially small heat losses from a surface of the heated body. In this case, the minimum admissible temperatures will be reached in the core of the billet at the end of the time-optimal single-stage heating cycle that provides the best heating accuracy $\varepsilon_{\min}^{(1)}$. At the same time, the region with maximum temperature will be located at a certain internal point (subsurface region) within the heated billet (see Figure 3.6 and Section 3.3). The temperature of the lateral surface (l = 1) in any cylinder cross-section remains within admissible limits. The temperature of the butt-end surface will be lower than maximum admissible level due to heat losses.

This means that, for $\varepsilon = \varepsilon_{\min}^{(1)}$, there will be only a single possible temperature distribution $\theta(l,y,\Delta^0)$ (Figure 4.38) that is characterized according to Equation (3.39) and Equation (3.40) by two extremum points $x_j^0 = (l_j^0, y_j^0)$; j = 1, 2. Both points represent maximum admissible deviations of final temperature from the required one. In this case, minimum and maximum temperatures will be reached in the core of the cylinder ($l_1^0 = 0$, y_1^0) and at some internal point ($l_2^0 = l_{e2}$, $y_2^0 = y_{e2}$), respectively.

It is feasible to determine only the set of possible variants of temperature distribution $\theta(l,y,\Delta^0)$. Every variant would correspond only to a particular case. Mainly, it is impossible to select a priori the single variant from this set for each particular case.

Minimum temperature can be positioned in three different locations on the cylinder axis. Its location primarily depends upon correlation between the value of heat loss from the butt-end surface and heat source distribution along the billet length. A relatively small coil overhang represents the typical case. In this case, the surface density of internal heat generation can be considered as uniformly distributed along the axis, y, or as reducing toward the butt-ends of the billet.[15] Two points with minimum final temperature can be located on the cyclinder axis $l = 0$ within butt-ends $y_1^0 = 0$ or $y_1^0 = 1$. Figure 4.38a and Figure 4.38b show these variants for minimum temperature points ($l_1^0 = 0$, $y_1^0 = 0$) and ($l_1^0 = 0$, $y_1^0 = 1$), respectively.

With increase of the coil overhang, the power of internal heat generation at the billet butt-ends grows in comparison with internal cross-sections of the billet.[15] Then, minimum temperature can be located at a certain point on the axis ($l_1^0 = 0$) with coordinate y that satisfies condition $0 < y_1^0 < 1$ (Figure 4.38c). Similarly to Equation (3.12), Expression (3.39) and Expression (3.40) can be transformed into the following set of equations:

$$\begin{cases} \theta\left(0, y_1^0, \Delta_1^0\right) - \theta^* = -\varepsilon_{\min}^{(1)}; \\ \theta\left(l_{e2}, y_{e2,} \Delta_1^0\right) - \theta^* = +\varepsilon_{\min}^{(1)}; \\ \dfrac{\partial\theta\left(l_{e2}, y_{e2,} \Delta_1^0\right)}{\partial l} = \dfrac{\partial\theta\left(l_{e2}, y_{e2,} \Delta_1^0\right)}{\partial y} = 0. \end{cases} \tag{4.96}$$

Observation that the temperature gradient is equal to zero at the extremum point can be used in the form of the last equality in Equation (4.96). This will allow one to define unknown coordinates l_{e2} and y_{e2}. As one can see, this equality is similar to condition $\partial\theta(l_{e2},\Delta^0)/\partial l = 0$ in Equation (3.12).

Assuming that the point of minimum temperature is located on the billet's axis, the set in Equation (4.96) should be supplemented by one of the following equations:

$$y_1^0 = 0; y_1^0 = 1; \frac{\partial\theta\left(0, y_1^0, \Delta_1^0\right)}{\partial y} = 0\,. \tag{4.97}$$

The proper choice of a single equality can be done according to the given value of the coil overhang and based on correlation between values of q_{T1} and q_{T0} (or Bi_1, θ_{a1}, and Bi_0, θ_{a0}) in Equation (4.74) or Equation (4.75). Further computation of the optimal process validates the chosen variant if minimum temperature will be reached at one of the considered points. Otherwise, the point $(0,y_1{}^0)$ should be considered as the point where temperature minimum will be located.

After substitution of $\theta(l,y,\Delta^0)$ in the form of Equation (4.94) or Equation (4.95) into the set of Equation (4.96) and Equation (4.97), this set can be solved with respect to five unknown variables: Δ_1^0, $\varepsilon_{\min}^{(1)}$, l_{e2}, y_{e2}, and y_1^0. If equalities $y_1^0 = 0$ or $y_1^0 = 1$ could be set a priori, the set (Equation 4.96) can be solved with respect to four unknown variables: Δ_1^0, $\varepsilon_{\min}^{(1)}$,

l_{e2}, and y_{e2}. At this step, the solution of the initial time-optimal problem for $N = 1$ and $\varepsilon = \varepsilon_{\min}^{(1)}$ is obtained.

2. Let us assume that prescribed value ε belongs to the range $\varepsilon_{\min}^{(2)} < \varepsilon < \varepsilon_{\min}^{(1)}$. Temperature distribution $\theta(l,y,\Delta^0)$ and vector Δ^0 can be represented as continuous functions of heating accuracy ε. If values of ε slightly differ from $\varepsilon_{\min}^{(1)}$, the following results can be obtained according to Equation (3.39) and Equation (3.40), similarly to the one-dimensional problem (see Section 3.3, Equation (3.15), and Figure 3.7):
   - The optimal control algorithm consists of two stages ($N = 2$): the heating–temperature soaking cycle.
   - Temperature distribution is similar to the case of $\varepsilon = \varepsilon_{\min}^{(1)}$ having two extremum points with minimum and maximum temperatures according to the requirement $R = N = 2$ in Equation (3.40) for $N = 2$ under condition $\varepsilon_{\min}^{(2)} < \varepsilon < \varepsilon_{\min}^{(1)}$. Similarly to Equation (3.15), the set of equations can be written as:

$$\begin{cases} \theta\left(0, y_1^0, \Delta_1^0, \Delta_2^0\right) - \theta^* = -\varepsilon; \\ \theta\left(l_{e2}, y_{e2,} \Delta_1^0, \Delta_2^0\right) - \theta^* = +\varepsilon; \\ \dfrac{\partial\theta\left(l_{e2}, y_{e2,} \Delta_1^0, \Delta_2^0\right)}{\partial l} = \dfrac{\partial\theta\left(l_{e2}, y_{e2,} \Delta_1^0, \Delta_2^0\right)}{\partial y} = 0. \end{cases} \tag{4.98}$$

After substitution of $\theta(l,y,\Delta^0)$ in the form of Equation (4.94) or Equation (4.95) into Equation (4.98) and taking into account one of the equalities in (Equation 4.97), the set of equations ought to be solved with respect to five unknown variables: parameters of optimal control Δ_1^0 and Δ_2^0 and coordinates l_{e2}, y_{e2}, and y_1^0 of points with maximum and minimum final temperatures. This leads to the final solution of initial OCP for $N = 2$. A sequence of optimization problems, which could be reduced to solution of the equation system in Equation (4.97) and Equation (4.98), should be solved for fixed values ε decreasing from $\varepsilon_{\min}^{(1)}$ until the third point, $x_3^0 = (l_3^0, y_3^0)$, with maximum temperature deviation appears for some value $\varepsilon = \varepsilon^*$ (Figure 4.39a through Figure 4.39c). According to the rule (Equation 3.40), the third point exists if the value of ε^* coincides with the value of minimax $\varepsilon_{\min}^{(2)}$. Appearance of an additional point can occur even if the expression $\varepsilon^* > \varepsilon_{\min}^{(2)}$ remains valid. This case means that temperature distribution corresponds to condition $R = N + 1 = 3$.

If the radial temperature profiles in any billet cross-section have a single temperature maximum under a two-stage control similar to the one-dimensional problem (Figure 3.7 and Figure 3.8), then a single

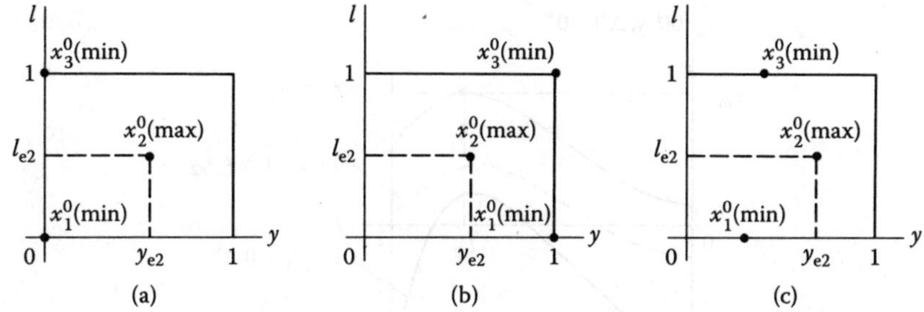

FIGURE 4.39 Variants of location of three minimum and maximum temperature points at the end of time-optimal heating of finite length cylindrical billet for $\varepsilon = \varepsilon^*$: $\varepsilon_{\min}^{(2)} \le \varepsilon^* < \varepsilon_{\min}^{(1)}$.

temperature maximum at the point (l_{e2}, y_{e2}) will be kept in all possible variants of final temperature distribution $\theta(l, y, \Delta^0)$. In this case, only minimum admissible temperature can be reached at the point x_3^0 in the core of the billet ($l_3^0 = 0$) or at the billet lateral surface ($l_3^0 = 1$). This conclusion is based on analysis of temperatures curves shown in Figure 3.7 and Figure 3.8.

Therefore, for $\varepsilon = \varepsilon^*$, the equations set in Equation (4.98) can be supplemented by the following equation:

$$\theta(l_3^0, y_3^0, \Delta_1^0, \Delta_2^0) - \theta^* = -\varepsilon^*. \tag{4.99}$$

The value of ε^* can be considered as an additional unknown variable that should be found when solving the set of Equation (4.98) and Equation (4.99). This leads to determination of optimal vector $\Delta^0(\varepsilon^*)$ and obtaining the final OCP solution for $\varepsilon = \varepsilon^*$.

3. The equality $\varepsilon^* = \varepsilon_{\min}^{(2)}$ remains valid only when the point x_3^0 is located on the surface of the billet (i.e., under $l_3^0 = 1$). In case of $l_3^0 = 0$, the heating accuracy meets condition $\varepsilon^* > \varepsilon_{\min}^{(2)}$ for the following reasons. At the end of the optimal heating process, the minimum temperature will be reached at two points, $x_1^0 = (l_1^0 = 0, y_1^0)$ and $x_3^0 = (l_3^0 = 0, y_3^0)$, of the billet's axis. During the soaking stage, temperatures at these points increase due to decreasing the maximum temperature located at point $x_2^0 = (l_{e2}, y_{e2})$. This situation, which is similar to the one-dimensional case (Figure 3.5), means that heating accuracy $\varepsilon < \varepsilon^*$ can be obtained and the expression $\varepsilon^* > \varepsilon_{\min}^{(2)}$ will be valid.

   If the minimum temperature is reached at the point $x_3^0 = (l_3^0 = 1, y_3^0)$ under $\varepsilon^* = \varepsilon_{\min}^{(2)}$, then the optimal control process consists of two stages for all $\varepsilon \in [\varepsilon_{\min}^{(2)}, \varepsilon_{\min}^{(1)}]$ and computational procedure for the time-optimal problem can be conducted according to the previously described algorithm. In the case of $\varepsilon = \varepsilon^* = \varepsilon_{\min}^{(2)}$, possible variations

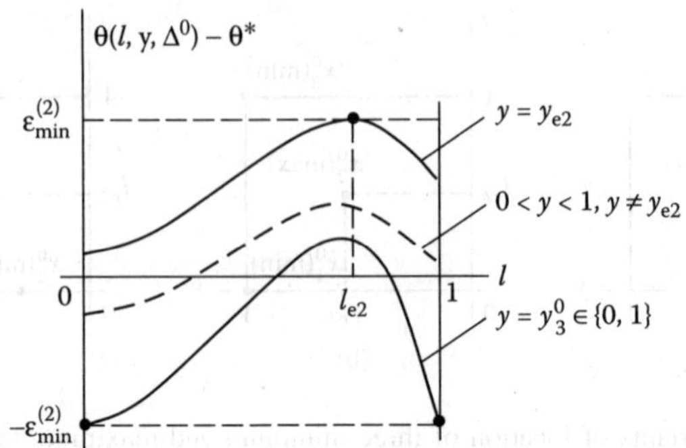

FIGURE 4.40 Variations of final temperature distribution within the finite length cylindrical billet for $\varepsilon = \varepsilon^* = \varepsilon_{\min}^{(2)}$.

of the location of points x_j^0, $j = 1, 2, 3$ are shown in Figure 4.39a and Figure 4.39b. These variations exist for the conventional case of a small coil overhang, and the expressions $y_3^0 = 0$ or $y_3^0 = 1$ will be valid. As a rule, depending upon the correlation between densities of heat loss flows q_{T1} and q_{T0}, the minimum final temperatures will be located in the cylinder core (axis) and at the cylinder surface of the one butt-end cross-section under $y_1^0 = y_3^0 = 0$ or $y_1^0 = y_3^0 = 1$. The temperature maximum point x_2^0 will be located in the internal cross-section of the billet. Figure 4.40 shows the radial temperature profiles within a cylinder cross-section where temperature takes on its extreme values. As a rule, the possible variant (Figure 4.39c) corresponds to the case of $\varepsilon^* > \varepsilon_{\min}^{(2)}$.

4. If an expression $\varepsilon^* > \varepsilon_{\min}^{(2)}$ holds true, then for $\varepsilon = \varepsilon^{**} = \varepsilon_{\min}^{(2)}$, the two-stage optimal heating cycle according to the condition in Equation (3.40) allows obtaining the final temperature distribution with three points x_j^0, $j = 1, 2, 3$. If a radial temperature distribution is identical to the one-dimensional case, then there will be only a single set of points x_j^0 (Figure 4.41):

$$x_1^0 = \left(1, y_1^0\right); x_2^0 = \left(l_{e2}, y_{e2}\right); x_3^0 = \left(1, y_3^0\right). \tag{4.100}$$

The set in Equation (4.100) differs from the case of $\varepsilon^* = \varepsilon_{\min}^{(2)}$. The final temperature distribution is characterized by reaching minimum admissible temperatures at points x_1^0 and x_3^0 that would be located at the surface of the cylinder ($l = 1$). One of these points will be located at the butt-end cross-section ($y_1^0 = 0$ or $y_1^0 = 1$) if the values of heat losses in Equation (4.74) satisfy conditions $q_{T1} < 0$ or $q_{T0} < 0$. The second point will be located in the internal cross-section of the cylinder

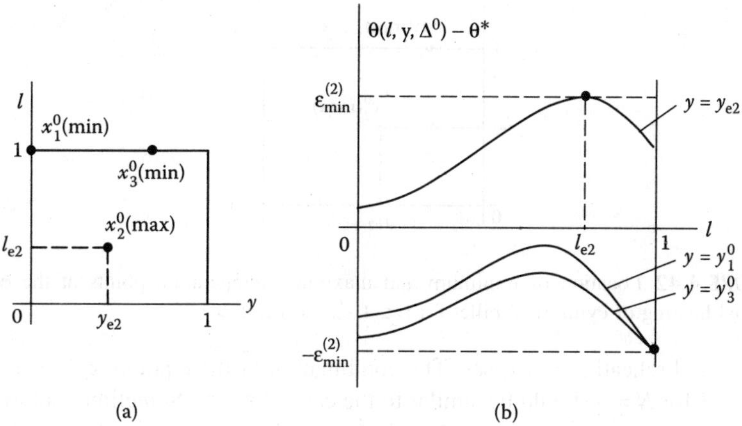

FIGURE 4.41 (a) Variants of location of minimum and maximum temperature points and (b) temperature profiles at the end of optimal heating of cylindrical billet for $\varepsilon = \varepsilon^{**} = \varepsilon_{min}^{(2)}$.

$(0 < y_3^0 < 1)$. This combination exists if the coil overhang is appreciably large. In this case, the following condition holds true:

$$\frac{\partial\theta\left(1, y_3^0, \Delta_1^0, \Delta_2^0\right)}{\partial y} = 0 . \tag{4.101}$$

The appropriate set of equations can be written as follows:

$$\begin{cases} \theta\left(1, y_1^0, \Delta_1^0, \Delta_2^0\right) - \theta^* = -\varepsilon^{**}; \\ \theta\left(1, y_3^0, \Delta_1^0, \Delta_2^0\right) - \theta^* = -\varepsilon^{**}; \\ \theta\left(l_{e2}, y_{e2}, \Delta_1^0, \Delta_2^0\right) - \theta^* = +\varepsilon^{**}; \\ \dfrac{\partial\theta\left(l_{e2}, y_{e2}, \Delta_1^0, \Delta_2^0\right)}{\partial l} = \dfrac{\partial\theta\left(l_{e2}, y_{e2}, \Delta_1^0, \Delta_2^0\right)}{\partial y} = 0. \end{cases} \tag{4.102}$$

The set in Equation (4.102) can be solved with respect to all unknown variables including Δ_1^0, Δ_2^0, and ε^{**}.

5. Let us assume that prescribed values $\varepsilon = \varepsilon_1 = \varepsilon^* - d\varepsilon < \varepsilon^*$ are slightly different from ε^* in the case of $\varepsilon^* > \varepsilon_{min}^{(2)}$. The condition (3.3) linking the number, N, of optimal control intervals with given heating accuracy would not be valid for multidimensional models of temperature fields. Therefore, according to Equation (3.40), it is proper to assume that optimal control can consist of three stages including heating, soaking,

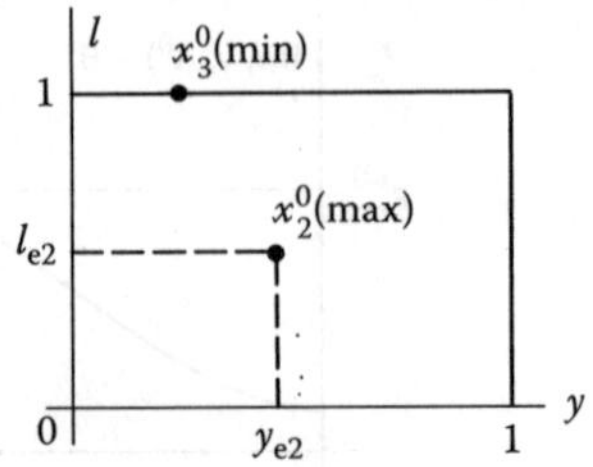

FIGURE 4.42 Location of minimum and maximum temperature points at the end of optimal heating of cylindrical billet for $\varepsilon \in (\varepsilon_{\min}^{(2)}, \varepsilon^*)$ if $\varepsilon^* > \varepsilon_{\min}^{(2)}$.

and reheating processes. The combination of three points x_j^0, $j = 1, 2, 3$ for $N = 3$ should be similar to the case of $\varepsilon = \varepsilon^*$. Sometimes solution of the corresponding set does not exist or the time required for heating becomes longer than in the case of $N = 2$. This means that for $\varepsilon = \varepsilon_1$ the optimal control process consists of two stages.

Results of computation[1] show that, for the majority of practical cases, the optimal control process consists of two stages if values of ε belong to the range $\varepsilon \in [\varepsilon_{\min}^{(2)}, \varepsilon^*]$ under $\varepsilon^* > \varepsilon_{\min}^{(2)}$. For any value of ε: $\varepsilon_{\min}^{(2)} < \varepsilon < \varepsilon^*$, the final temperature distribution $\theta(l,y,\Delta^0)$ is characterized by two points $x_2^0 = (l_{e2}, y_{e2})$ and $x_3^0 = (l_3^0, y_3^0) = (1, y_3^0)$ (Figure 4.42) from the set of three points that would correspond to the value $\varepsilon = \varepsilon^*$ in the set in Equation (4.98) and Equation (4.99) (Figure 4.39c). This combination of extreme points exists in the case of $0 < y_3^0 < 1$ when the coil overhang is relatively large. However, in this case, the set of points x_j^0 for $\varepsilon = \varepsilon^{**}$ (Figure 4.41) replaces the combinations of points shown in Figure 4.39a through Figure 4.39c.

For values of ε that satisfy the condition $\varepsilon_{\min}^{(2)} < \varepsilon < \varepsilon^*$, the following set of equations can be written:

$$\begin{cases} \theta\left(l_{e2}, y_{e2}, \Delta_1^0, \Delta_2^0\right) - \theta^* = +\varepsilon; \\ \theta\left(1, y_3^0, \Delta_1^0, \Delta_2^0\right) - \theta^* = -\varepsilon; \\ \dfrac{\partial\theta\left(l_{e2}, y_{e2}, \Delta_1^0, \Delta_2^0\right)}{\partial l} = \dfrac{\partial\theta\left(l_{e2}, y_{e2}, \Delta_1^0, \Delta_2^0\right)}{\partial y} = 0. \end{cases} \tag{4.103}$$

Assuming that $x_3^0 = (1, y_3^0)$, the set in Equation (4.103) can be solved with respect to optimal control parameters Δ_1^0 and Δ_2^0, and unknown variables l_{e2} and y_{e2}.

6. To determine an optimal control algorithm for cylindrical billet heating, the following computational procedure can be suggested for all values

of $\varepsilon \in [\varepsilon^{(2)}_{\min}, \varepsilon^{(1)}_{\min}]$. As value ε decreases from $\varepsilon^{(1)}_{\min}$ to $\varepsilon^{(2)}_{\min}$, there might be one of two possible profiles of the final temperature distribution depending upon distribution of heating power along the heater length. In the case of a relatively small coil overhang, the condition $\varepsilon^* = \varepsilon^{(2)}_{\min}$ will be valid. The temperature distribution (Figure 4.38a and Figure 4.38b) can be changed as shown in Figure 4.39a and Figure 4.39b. The appropriate set of equations (Equation 4.98 and Equation 4.99) ought to be solved instead of the set in Equation (4.96) through Equation (4.98). If the coil overhang is large enough, then the inequality $\varepsilon^* > \varepsilon^{(2)}_{\min}$ holds true. This case is represented by the sequential variations of temperature distribution in Figure 4.38c, Figure 4.39c, Figure 4.42, and Figure 4.41a. The appropriate sets of equations in Equation (4.96), Equation (4.98), Equation (4.99), Equation (4.103), and Equation (4.102) can be written. Particular scenarios could be revealed by solving a sequence of optimal control problems for the series of fixed values ε descending by sufficiently small steps from initial value $\varepsilon^{(1)}_{\min}$ to value $\varepsilon^{(2)}_{\min}$. If $\varepsilon = \varepsilon_0$ and given value $\varepsilon_0 \in [\varepsilon^{(2)}_{\min}, \varepsilon^{(1)}_{\min}]$, then solution of the initial optimal control problem can be found during the appropriate step of the described computation procedure.

7. The preceding provided analysis proves that optimal control parameters depend on coil overhang h. Given expressions for $W_d(\xi,l,y)$ in Equation (4.72) include function $W_d^*(\xi,l,y,h)$, which describes a longitudinal distribution of electromagnetic heat sources as function of coil overhang h. The temperature field $\theta(l,y,\Delta,h)$ in the form of Equation (4.94) or Equation (4.95) can be considered as a function of h (because coefficients W_{rn}, $r, n = 0, 1, 2, \ldots$ depend on h). Therefore, the variable h can be considered as one of the unknown parameters of the optimal process. Solution of the optimal control problem allows finding the vector of optimal control parameters Δ^0 and the value of coil overhang h^* that corresponds to the best heating accuracy $\varepsilon^{(2)}_{\min\min}$ attainable under two-stage control:

$$\varepsilon^{(2)}_{\min\min} = \min_h \varepsilon^{(2)}_{\min}(h). \tag{4.104}$$

As the number of unknown parameters is increased by one, then according to Equation (3.39) and Equation (3.40), the number of points x_j^0 increases at least by one in contrast to the case when values of h differ from value h^*.

In the case of $\varepsilon = \varepsilon^{(2)}_{\min}$, two possible sets of extreme points x_j^0 exist: for $\varepsilon^{(2)}_{\min} = \varepsilon^*$ (Figure 4.39a and Figure 4.39b) and for $\varepsilon^{(2)}_{\min} = \varepsilon^{**}$ (Figure 4.41). The combination of these sets corresponds to the case of $h = h^*$ (Figure 4.43).

Instead of Equation (4.102), the following set of equations can be written with respect to four points $x_j^0, j = \overline{1,4}$:

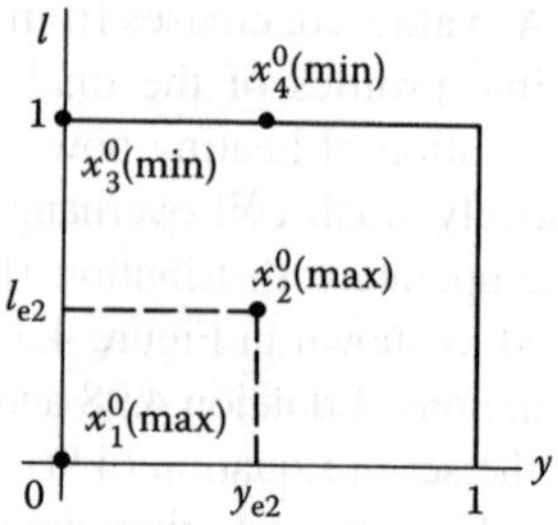

FIGURE 4.43 Location of minimum and maximum temperature points within cylindrical billet for value of coil overhang that corresponds to the best heating accuracy under two-stage optimal control.

$$\begin{cases} \theta\left(0, y_1^0, \Delta_1^0, \Delta_2^0, h^*\right) - \theta^* = -\varepsilon_{\min\min}^{(2)}; \\ \theta\left(l_{e2}, y_{e2}, \Delta_1^0, \Delta_2^0, h^*\right) - \theta^* = \varepsilon_{\min\min}^{(2)}; \\ \theta\left(1, y_3^0, \Delta_1^0, \Delta_2^0, h^*\right) - \theta^* = -\varepsilon_{\min\min}^{(2)}; \\ \theta\left(1, y_4^0, \Delta_1^0, \Delta_2^0, h^*\right) - \theta^* = -\varepsilon_{\min\min}^{(2)}; \\ \dfrac{\partial\theta\left(l_{e2}, y_{e2}, \Delta_1^0, \Delta_2^0, h^*\right)}{\partial l} = \dfrac{\partial\theta\left(l_{e2}, y_{e2}, \Delta_1^0, \Delta_2^0, h^*\right)}{\partial y} = 0. \end{cases} \tag{4.105}$$

The minimum admissible temperature is reached at three points. The set in Equation (4.105) can be solved with respect to six unknown variables: the value of optimal coil overhang h^*, "global" minimax $\varepsilon_{\min\min}^{(2)}$ and optimal control parameters Δ_1^0, Δ_2^0, l_{e2}, and y_{e2}. It is possible to show that condition $\varepsilon_{\min}^{(2)} = \varepsilon^*$ is valid under $h < h^*$ and the expression $\varepsilon_{\min}^{(2)} = \varepsilon^{**}$ is true under $h > h^*$. In both cases, the condition $\varepsilon_{\min}^{(2)} > \varepsilon_{\min\min}^{(2)}$ holds true for all values $h \neq h^*$.

8. In the case of $\varepsilon_0 < \varepsilon_{\min}^{(2)}$, the optimal control problem should be solved using the computational technique described in Section 3.7. Using additional information with respect to temperature distribution that is based on physical nature of induction heating processes, it is possible to further simplify the computation process by applying analogies with the conventional case of $\varepsilon_0 < \varepsilon_{\min}^{(2)}$ for the one-dimensional problem.

In the practicable cases the optimal control process consists of one, two, or three stages, and the condition in Equation (3.3) directly links the number of control stages with preset value of ε.[1] Table 4.1 shows some results of computation of optimal cylindrical billets heating for the initial data: $\xi = 4$; $\beta = 0.214$; $\theta^* = 0$; $U_{max} = 1$. Similarly to the one-dimensional model, heat source distribution along the billet length is assumed to be uniform.[1] Initial and boundary conditions (Equation 4.73 and Equation 4.75) are considered under the following assumption:

$$\theta_0(l,y) = \theta_a(\varphi) = \theta_{al}(\varphi) = \theta_{a0}(\varphi) \equiv \theta_0 = -0.5 = const.$$

This case corresponds to the condition $\varepsilon^* = \varepsilon_{\min}^{(2)}$.

Computation results (Table 4.1) show that, due to considering heat losses from billet butt-ends, the heating accuracy decreases in comparison with the one-dimensional problem. The last line of Table 4.1 shows results for one-dimensional OCP solution (assuming that butt-end heat losses are neglected $Bi_0 = Bi_1 = 0$). More complicated problems of time-optimal control can be solved using the computational technique described in Section 4.1 through Section 4.3. Similarly to the one-dimensional case (Section 4.1), the final temperature distribution is not changed if technological restraints are taken into account.

Control input functions u^σ and u^θ can be found similarly to the technique described in Section 4.1.2. Instead of Expression (4.9) and Expression (4.13), they can be approximated as weighted sum of the constant component and two exponential summands.[1] In the case of $\varepsilon \geq \tilde{\varepsilon}_{\min}^{(2)} = \varepsilon_{\inf}$ and $\alpha_T < 1$ in Equation (4.30), the governing set of equations remains unchanged. This conclusion is based on analogies with the one-dimensional time-optimal problem subject to the transfer stage (Section 4.2).

Similarly to Equation (4.41) and Equation (4.42) for the conventional case of $\varepsilon = \tilde{\varepsilon}_{\min}^{(2)} = \varepsilon_{\inf}$ (assuming $\alpha_T = 1$ in Equation 4.30), it is possible to obtain the following set of equations:

$$\begin{cases} \theta_T\left(1, y_1^0, \Delta_1^0, \Delta_T^0\right) - \theta^* = -\tilde{\varepsilon}_{\min}^{(2)}; \\ \theta\left(l_{e2}, y_{e2}, \Delta_1^0, \Delta_T^0\right) - \theta^* = +\tilde{\varepsilon}_{\min}^{(2)}; \\ \dfrac{\partial \theta_T\left(l_{e2}, y_{e2}, \Delta_1^0, \Delta_T^0\right)}{\partial l} = \dfrac{\partial \theta_T\left(l_{e2}, y_{e2}, \Delta_1^0, \Delta_T^0\right)}{\partial y} = 0. \end{cases} \tag{4.106}$$

The first equation of the set in Equation (4.106) is written with respect to minimum temperature at the point $(1, y_1^0)$ on the cylinder lateral surface under $l_{e2} \geq 0$ (Figure 4.42). The set in (Equation 4.106) can be solved using additional information with respect to the axial coordinate of the temperature minimum

TABLE 4.1
Optimal Process Parameters for Heating of Cylindrical Billet

			N = 1					*N* = 2						
Bi	**Bi_0**	**Bi_1**	y_1^0	y_{e2}	l_{e2}	Δ_1^0	$\varepsilon_{min}^{(1)}$	$y_1^0\ y_3^0$	y_{e2}	l_{e2}	Δ_1^0	Δ_2^0	φ_{min}^0	$\varepsilon_{min}^{(2)}$
0.7	1.4	1.4	0	0.5	0.860	0.364	0.095	0	0.5	0.647	0.372	0.0421	0.4141	0.0484
0.7	0.7	0.7	0	0.5	0.861	0.356	0.086	0	0.5	0.650	0.361	0.0423	0.4033	0.0360
0.7	0	0	—	—	0.889	0.347	0.077	—	—	0.661	0.349	0.0426	0.3916	0.0237

point y_1^0. Similarly to Equation (4.105), it is possible to obtain the system of equations that corresponds to coil overhang value $h = h^*$ in the time-optimal control problem, taking into consideration the transfer stage.

In this case, additional point x_3^0 (compare with Equation 4.106) appears because the minimum temperature can be reached at one of two points on the cylinder lateral surface: in the internal cross-section ($0 < y_1^0 < 1$) or in the butt-end cross-section ($y_1^0 = 0$ or $y_1^0 = 1$). These variations of location of temperature minimum will be similar to the variant shown in Figure 4.41. The combination of both scenarios allows obtaining the following set of equations:

$$\begin{cases} \theta_T\left(1, y_1^0, \Delta_1^0, \Delta_T^0, h^*\right) - \theta^* = -\tilde{\varepsilon}_{\min\min}^{(2)}; \\ \theta_T\left(1, y_3^0, \Delta_1^0, \Delta_T^0, h^*\right) - \theta^* = -\tilde{\varepsilon}_{\min\min}^{(2)}; \\ \theta_T\left(l_{e2}, y_{e2}, \Delta_1^0, \Delta_T^0, h^*\right) - \theta^* = +\tilde{\varepsilon}_{\min\min}^{(2)}; \\ \dfrac{\partial\theta_T\left(l_{e2}, y_{e2,}\Delta_1^0, \Delta_T^0, h^*\right)}{\partial l} = \dfrac{\partial\theta_T\left(l_{e2}, y_{e2,}\Delta_1^0, \Delta_T^0, h^*\right)}{\partial y} = 0, \end{cases} \tag{4.107}$$

where $y_1^0 = 0$ or $y_1^0 = 1$; $0 < y_3^0 < 1$. Expression for $\theta_T(l,y,\Delta_1^0,\Delta_T)$ can be found similarly to Equation (4.34) through Equation (4.36). This expression will include fourfold series instead of twofold series in Equation (4.34).

The technique described in Section 4.3 allows taking into account incomplete information with respect to the heating system under control. In real-life production situations, we often face incompleteness of source information with respect to the induction heating system. Incompleteness of information is caused, first of all, by imperfection of our knowledge about certain features of the heating system. This includes but is not limited to uncertainties of electromagnetic and thermal-physical properties of heated billets, inductor geometry, power supply limitations, presence of harmonics, etc.

On another hand, there are uncertainties with respect to particular conditions of the technological process. Initial temperature distribution of the heated workpiece and/or the value of heat losses from its surface to the surrounding area could be regarded as other examples of incompletely defined operating characteristics of the heating process (Section 4.3). It is assumed that these disturbances can be taken into account while solving optimal-control problems. For this purpose, each equation of the governing set should be considered for appropriate particular values of ill-defined characteristics. These values should be chosen within possible variation intervals.

Let us consider as an example a two-dimensional problem of time-optimal control of induction heating of aluminum alloy cylindrical billets.[1] Figure 4.44 represents some results of computation for $\varepsilon = \tilde{\varepsilon}_{\min}^{(2)}$.

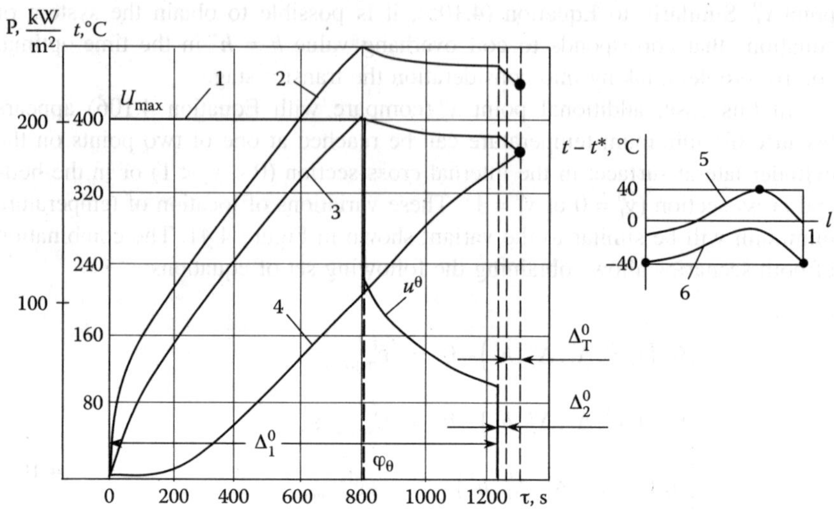

FIGURE 4.44 Optimal control, time–temperature history, and final temperature profiles in the two-dimensional problem of time-optimal control for induction heating of D19 aluminum alloy cylindrical billets (diameter is 240 mm; length is 835 mm; required temperature is 400°C). $\Gamma = 1$; $\theta_0 \in [-0.50;-0.44]$; $Bi_h \in [0.015;0.050]$; $Bi_T \in [0.10;0.12]$; $\xi = 20$; $\theta_{adm} = 0.09$; $\theta^* = 0$; $\Delta_T{}^0 = 50$ s; $\varepsilon_0 = \bar{\bar{\varepsilon}}^{(2)}_{\min}$; $h = 1.75$ mm. 1: optimal control $u^\theta(\varphi) = U_{\max}\{0.24 + 0.264 \exp[-8.3(\varphi - \varphi_\theta)] + 0.002 \exp[-4.4(\varphi - \varphi_\theta)]\}$; 2: temperature at point $l = 1$; $y = 0.5$; under $\theta_0 = -0.44$; $Bi_h = 0.015$; $Bi_T = 0.10$; 3: temperature at point $l = 1$; $y = 0$; under $\theta_0 = -0.5$; $Bi_h = 0.05$; $Bi_T = 0.12$; 4: temperature at point $l = 0$; $y = 0$; under $\theta_0 = -0.5$; $Bi_h = 0.05$; $Bi_T = 0.12$; 5: final temperature profile at $y = 0.5$; under $\theta_0 = -0.44$; $Bi_h = 0.015$; $Bi_T = 0.10$; 6: final temperature profile at $y = 0$; $\theta_0 = -0.5$; under $Bi_h = 0.050$; $Bi_T = 0.12$.

Calculations were performed taking into account constraint on maximum temperature and constraint imposed on transportation time. Incomplete information with respect to initial temperature and level of heat losses (during heating and transfer stages) has been considered in order to have a proper OCP formulation. Similarly to the one-dimensional problem, it was assumed that distribution of heat source power could be considered as uniform along the billet length.

Computations of the optimal process have been conducted with respect to constant surface heat losses that can be estimated by relative values Bi_h (during the heating stage) and Bi_T (during the transfer stage). The preset value $\Delta_T = \Delta_T^0$ should satisfy the condition $\Delta_T^0 < \Delta_{T2}^*$ and, therefore, the inequality $\alpha_T < 1$ remains valid. Under incomplete information with respect to parameters θ_0, Bi_h, and Bi_T, it is assumed that the maximum temperature is reached for $Bi_h = Bi_{h\min}$, $Bi_T = Bi_{T\min}$, and $\theta_0 = \theta_{0\max}$; minimum temperature is reached for $Bi_h = Bi_{h\max}$, $Bi_T = Bi_{T\max}$, and $\theta_0 = \theta_{0\min}$.

In the case of $\varepsilon = \bar{\bar{\varepsilon}}^{(2)}_{\min}$, $\alpha_T < 1$ computational sets keep the form of Equation (4.98), Equation (4.99) or Equation (4.102) with substitution $\theta_T(l_j^0, y_j^0, \Delta_1^0, \Delta_2^0, \Delta_T^0)$ instead of $\theta(l_j^0, y_j^0, \Delta_1^0, \Delta_2^0, \Delta_T^0)$. The function $\theta_T(l_j^0, y_j^0, \Delta_1^0, \Delta_2^0, \Delta_T^0)$ can be calculated at each of points (l_j^0, y_j^0) for the set of values $(Bi_{hj}^0, Bi_{Tj}^0, \theta_{0j}^0)$ chosen in the same manner. Figure 4.44 shows that, similarly to Figure 4.39a (or Figure 4.39b), the final temperature distribution is characterized by combination of points x_j^0.

In this case, it is possible to obtain $\bar{\tilde{\varepsilon}}_{\min}^{(2)} = \varepsilon^* = 40°C$. According to denotations discussed in Section 4.2 and Section 4.3, the upper indexes "~" and "–" indicate that solution of the optimal control problem has been found in the class of two-stage control, taking into account the condition $\Delta_T \neq 0$ and uncertainty in the definition of values Bi, Bi_T, and θ_0. The optimal power control algorithm can be written in the form of Expression (4.3) for $N = 2$. In contrast to the case of Equation (4.9), the control input $u^\theta(\varphi)$ can be described by the sum of two exponential summands.

4.5.4 Time-Optimal Control of Induction Heating of Rectangular-Shaped Workpieces

Slabs, plates, and blooms of various shapes and thicknesses may be heated and later rolled down to the final desired size. Some of these are cooled and later reheated for rolling or forming; others are reheated as a part of the continuous process to facilitate the subsequent rolling or forming operation.[1] This section is devoted to discussion of optimal induction heating of rectangular-shaped workpieces.

Solution of the two-dimensional time-optimal control problem for heating of rectangular-shaped workpieces is noticeably different compared to previously described cases. The condition $\Gamma = 0$ should be accepted in Equation (4.72) through Equation (4.75). In this case, direct analogies with one-dimensional models cannot be applied.[16,17] The two-dimensional mathematical model (Equation 4.72 through Equation 4.75) describes temperature distribution in the plane of the cross-section of rectangular-shaped workpieces, assuming uniform temperature distribution along the workpiece length.

This means that inductor load can be approximately considered as a rectangular prism of infinite length (Figure 4.37). Similarly to the problem for heating of finite length cylindrical billets, the time-optimal control problem can be reduced to the form of Equation (4.93). The complicated shape of final temperature distribution in the slab cross-section adds complexity to the solution of OCP.

4.5.4.1 Surface Heat-Generating Sources

Internally generated heating power is the most significant process parameter that affects temperature distribution (Section 2.4). Proper control of the heating process involves appropriate choice of induced power density. Complication of the control process occurs due to difficulties in providing desired spatial distribution of internal heat sources. It could be impractical to realize certain power distribution along the length of the induction coil. Many of the existing means to control power distribution face strict limitations or merely cannot produce a particular spatial distribution of internal heat sources when heating certain metals.

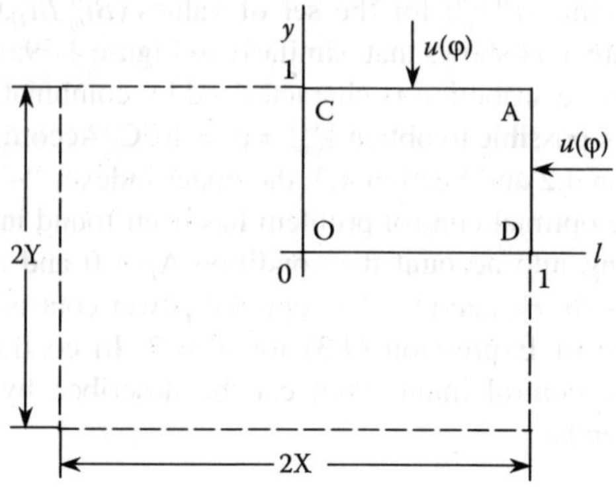

FIGURE 4.45 Geometrical model of infinite length rectangular slab for simulation by two-dimensional equations of heat transfer with surface heat-generating sources.

Straightforward solution of the time-optimal problem faces major difficulties when spatial distribution of internal heat sources within the workpiece is represented by the function $W_d(\xi,l,y)$ of arbitrary form. To simplify the problem, let us examine the process of heating where heat sources are located in the surface. In this case, internal sources can be replaced by an external surface heat flow. The distribution of this flow along the slab perimeter is assumed to be uniform.

The density of external surface heat flow $u(\varphi)$ can be chosen as a control input (Figure 4.45).The function $W_d(\xi,l,y)$ that describes heat source distribution depends upon current penetration depth ξ. Because the value of ξ increases, representation of the real distribution, $W_d(\xi,l,y)$, in the form of surface heat-generating flow becomes more accurate. As ξ tends to infinity, the approximate distribution approaches the actual one.[17] Let us assume the following initial data: $\theta_0(l,y) \equiv \theta_0 = const$; $q(y, \varphi) = q_{T1}(l,\varphi) \equiv q_0 = const < 0$ in Equation (4.73) and Equation (4.74).

Then, for $\Gamma = 0$, the homogeneous heat conduction equation can be written as[1]:

$$\frac{\partial\theta(l,y,\varphi)}{\partial\varphi} = \frac{\partial^2\theta(l,y,\varphi)}{\partial l^2} + \beta^2\frac{\partial^2\theta(l,y,\varphi)}{\partial y^2};$$

$$0 < l < 1;\ 0 < y < 1;\ 0 < \varphi \le \varphi^0; \tag{4.108}$$

$$\theta(l,y,0) = \theta_0; \tag{4.109}$$

$$\frac{\partial\theta(1,y,\varphi)}{\partial l} = u(\varphi) + q_0;\quad \beta\frac{\partial\theta(l,1,\varphi)}{\partial y} = u(\varphi) + q_0;$$

$$\frac{\partial\theta(0,y,\varphi)}{\partial l} = \frac{\partial\theta(l,0,\varphi)}{\partial y}. \tag{4.110}$$

The control input $u(\varphi)$ is included into the boundary conditions (Equation 4.110).

The optimal control by density of heat flow from the surface has the same form of Equation (2.27) as the control by density of internal heat sources. The control function is constrained by the condition in Equation (2.7).

According to the boundary condition (Equation 4.110), the optimal heating process consists of alternating stages of heating with heat flow density $U_{max} + q_0 < U_{max}$ and subsequent soaking under heat losses $q_0 < 0$. In all practicable cases it can be assumed that $|q_0| < U_{max}$ and $U_{max} + q_0 > 0$. The number, $N \geq 1$, of stages can be uniquely defined based on given heating accuracy ε and it increases with decrease of ε. Therefore, the shape of the optimal control algorithm is known, but the number of stages, N, and durations, $\Delta_1, \Delta_2, \ldots, \Delta_N$, of those stages remain unknown.

The expression for temperature field $\theta(l,y,\Delta)$ can be obtained in the explicit form as a function of spatial coordinates and optimal control parameters Δ_i, $i = \overline{1, N}$[1]:

$$\theta(l,y,\Delta) = \theta_0 + (1+\beta)\sum_{m=1}^{N}\left[q_0 + \frac{U_{max}}{2}\left(1+(-1)^{m+1}\right)\right]\Delta_m - \frac{1}{2}\left(\frac{1}{3} - l^2\right) \times$$

$$\times\left[q_0 + \frac{U_{max}}{2}\left(1+(-1)^{N+1}\right)\right] - \frac{1}{2\beta}\left(\frac{1}{3} - y^2\right)\left[q_0 + \frac{U_{max}}{2}\left(1+(-1)^{N+1}\right)\right] +$$

$$+2\sum_{r=1}^{\infty}(-1)^{r+1}\cdot\left[(U_{max}+q_0)\exp\left(-\pi^2 r^2\sum_{m=1}^{N}\Delta_m\right) + U_{max}\sum_{j=2}^{N}(-1)^{j+1}\times\right.$$

$$\left.\times\exp\left(-\pi^2 r^2\sum_{m=j}^{N}\Delta_m\right)\right]\frac{\cos\pi r l}{\pi^2 r^2} + \frac{2}{\beta}\sum_{n=1}^{\infty}(-1)^{n+1}\times$$

$$\times\left[(U_{max}+q_0)\exp\left(-\pi^2 n^2\beta^2\sum_{m=1}^{N}\Delta_m\right) + U_{max}\sum_{j=2}^{N}(-1)^{j+1}\times\right.$$

$$\left.\times\exp\left(-\pi^2 n^2\beta^2\sum_{m=j}^{N}\Delta_m\right)\right]\frac{\cos\pi n y}{\pi^2 n^2}. \tag{4.111}$$

The optimal control problem is reduced to searching for number, N, and optimal values of parameters Δ_i^0, $i = \overline{1, N}$ that uniquely specify the optimal control input.

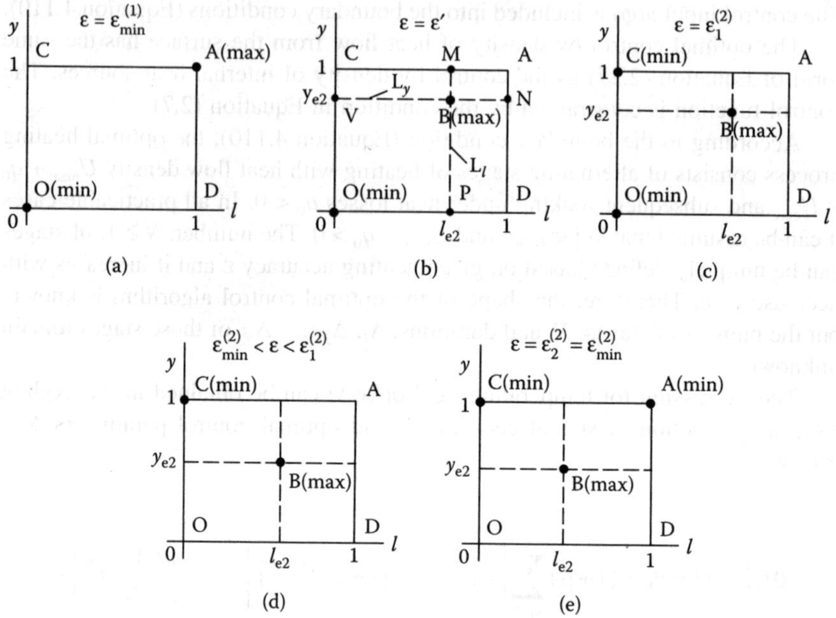

FIGURE 4.46 Variants of location of minimum and maximum temperature points within rectangular slab cross-section at the end of single-stage and two-stage optimal control by power of surface heat-generating sources for $\varepsilon \in [\,\varepsilon_{\min}^{(2)}, \varepsilon_{\min}^{(1)}\,]$.

Problems of this type can be solved for the conventional case of $\varepsilon \in [\,\varepsilon_{\min}^{(3)}, \varepsilon_{\min}^{(1)}\,]$.[1,18,19]

Optimal single-stage control. In the simplest case of $\varepsilon = \varepsilon_{\min}^{(1)}$ similar to heating of the cylindrical billet (Section 4.5.3), one can assume that $N = 1$. Under maximum heat flow from the slab surface, the maximum and minimum final temperatures can be reached only at two points $x_1^0 = (0,0)$ and $x_2^0 = (1,1)$, respectively: in the vertices $O(l = 0, y = 0)$ and $A(l = 1, y = 1)$ of rectangular cross-section OCAD (Figure 4.46a). With regard to Equation (3.39) and Equation (3.40) for $\varepsilon = \varepsilon_{\min}^{(1)}$ under $R = N + 1 = 2$, at the end of the time-optimal process, maximum deviations of final temperature from the required one, θ^*, will be reached at these points with different signs. As a result, the basic system of correlations (Equation 3.39) can be transformed into the appropriate set of two equations:

$$\begin{cases} \theta\left(1,1,\Delta_1^0\right) - \theta^* = \varepsilon_{\min}^{(1)}; \\ \theta\left(0,0,\Delta_1^0\right) - \theta^* = -\varepsilon_{\min}^{(1)}. \end{cases} \tag{4.112}$$

The set in Equation (4.112) can be solved with respect to two unknown variables, $\Delta_1{}^0$ and $\varepsilon_{\min}^{(1)}$, by substituting the expression for temperature field $\theta(l,y,\Delta)$ in the form of Equation (4.111).

Optimal two-stage control. Let us consider the case of $\varepsilon = \varepsilon' = \varepsilon_{\min}^{(1)} - d\varepsilon$, where $d\varepsilon > 0$ is a sufficiently small value. The continuous nature of the final temperature distribution $\theta(l,y,\Delta^0)$ as a function of ε and inequality $N \geq 2$ for $\varepsilon' < \varepsilon_{\min}^{(1)}$ allows establishing the following property of optimal final temperature profile $\theta(l,y,\Delta^0(\varepsilon'))$. The number, R, of points $x_j{}^0$ in Equation (3.39) and Equation (3.40), where the maximum admissible absolute deviations of final temperature are reached, is equal to number, N, of optimal control intervals, i.e., $R = N = 2$ under $\varepsilon = \varepsilon'$.

Therefore, in the case of $\varepsilon = \varepsilon'$, the time-optimal control consists of two stages; at the end of heating, maximum admissible deviations are reached at two points: $x_1^0 = (l_1^0, y_1^0)$ and $x_2^0 = (l_2^0, y_2^0)$. This conclusion remains valid for all values of ε decreasing from $\varepsilon_{\min}^{(1)}$ until, for some value $\varepsilon = \varepsilon_1^{(2)}$, the third point, x_3^0, with maximum temperature deviation appears in the prism cross-section. According to the condition in Equation (3.40), three points $x_j{}^0$ always exist for the value $\varepsilon = \varepsilon_{\min}^{(2)}$; therefore, the condition $\varepsilon_1^{(2)} \geq \varepsilon_{\min}^{(2)}$ will be valid. A similar situation has been observed earlier when considering optimal heating of cylindrical billets (Section 4.5.3).

Variations of locations of points $x_j{}^0$, $j = 1, 2, 3$ could be set from the study of final temperature distribution $\theta(l,y,\Delta)$ under two-stage control. The shape of temperature field $\theta(l,y,\Delta^0)$ can be uniquely determined by values of temperature gradients, $\partial\theta(l,y,\Delta_1,\Delta_2)/\partial l$, $\partial\theta(l,y,\Delta_1,\Delta_2)/\partial y$, within the cross-section of the prism.

When passing from the first control stage under $u(\varphi) \equiv U_{\max} + q_0 > 0$ to the second one under $u(\varphi) \equiv q_0 < 0$, the temperature gradients on the border of the cross-section (on the lines $l = 1$ and $y = 1$) change their signs. This means that only two lines, L_l and L_y, appear in the plane of rectangle OCAD, for which the following conditions remain valid: $\partial\theta(l,y,\Delta_1,\Delta_2)/\partial l = 0$ and $\partial\theta(l,y,\Delta_1,\Delta_2)/\partial y = 0$, respectively (Figure 4.46).

It is important to underline that these conditions remain valid for the points that belong to the rectangle's sides $y = 0$ and $l = 0$. There is a general analogy to the previously described one-dimensional case where not more than one extremum point (i.e., the point with zero gradient) appears if the number of control stages increases by one (Section 3.2). As it follows from Equation (4.111), the expression for final temperature distribution can be represented as a sum, and each summand depends only on one of the spatial coordinates:

$$\theta(l,y,\Delta) = \theta_1(l,\Delta) + \frac{1}{\beta}\theta_2(y,\beta^2\Delta), \tag{4.113}$$

where $\beta^2\Delta = (\beta^2\Delta_1, \beta^2\Delta_2, \ldots, \beta^2\Delta_N)$.

With regard to the control input (Equation 2.27) applied on borders $l = 1$ and $y = 1$, functions $\theta_1(l,\Delta)$ and $\theta_2(y,\beta^2\Delta)$ (in relative units) describe one-dimensional temperature fields within infinitely long plates of thickness $2X$ and $2Y$, respectively.

According to Equation (4.113), the following conditions are met:

$$\frac{\partial\theta(l,y,\Delta)}{\partial l}=\frac{\partial\theta_1(l,\Delta)}{\partial l}\ \textit{for all } y\in[0,1];$$
$$\frac{\partial\theta(l,y,\Delta)}{\partial y}=\frac{1}{\beta}\frac{\partial\theta_2\left(y,\beta^2\Delta\right)}{\partial y}\ \textit{for all } l\in[0,1]. \tag{4.114}$$

This implies that lines L_l and L_y represents straight lines in the plane of the rectangular cross-section of the prism. They are parallel to coordinate axes y and l, respectively (Figure 4.46b through Figure 4.46e). During the first control interval under maximum heating power, the temperature gradients $\partial\theta/\partial l$ and $\partial\theta/\partial y$ are positive within the whole prism cross-section — i.e., at any point of rectangle *OCAD*.

When passing on to the second interval, straight lines L_l and L_y of zero gradient appear on borders $l = 1$ and $y = 1$ and then they move towards coordinate axes $l = 0$ and $y = 0$ (Figure 4.46b through Figure 4.46e), leaving behind the negative gradients zone adjoining surfaces $l = 1$ and $y = 1$. If only one straight line, L_l, as well as only one line, L_y, exists in the plane of rectangle *OCAD*, then signs of gradients will be the same except on these lines. Therefore, gradients remain positive in the areas prior to these lines in the same way as during the first control interval. As a result, straight lines L_l and L_y divide rectangle *OCAD* into regions of positive and negative gradients. This allows determining their signs at any point of this rectangle in the end of $\partial\theta/\partial l$ two-stage optimal control process (Figure 4.46b).

For the second interval, the following conditions can be obtained:

$$\frac{\partial\theta(l,y,\Delta)}{\partial l}\geq 0 \text{ for rectangle OCMP;}$$
$$\frac{\partial\theta(l,y,\Delta)}{\partial y}\geq 0 \text{ for rectangle OVND;}$$
$$\frac{\partial\theta(l,y,\Delta)}{\partial l}\leq 0 \text{ for rectangle PMAD;} \tag{4.115}$$
$$\frac{\partial\theta(l,y,\Delta)}{\partial y}\leq 0 \text{ for rectangle VCAN.}$$

According to boundary conditions in Equation (4.110), the temperature gradient will be equal to zero only on the lines L_l and L_y and on the sides $l = 0$ and $y = 0$.

The single internal point $B(l_{e2},y_{e2})$ of the maximum final temperature represents a cross point of the lines L_l and L_y (Figure 4.46b through Figure 4.46e). Minimum final temperature will be reached in one of vertices $O(l = 0, y = 0)$, $C(l = 0, y = 1)$, $A(l = 1, y = 1)$, or $D(l = 1, y = 0)$ or in several vertices of the rectangle *OCAD*.

Described analysis allows one to define the set of points $\{B, O, C, A, D\}$ that can be considered under $N = 2$ as "potential candidates" to be considered as points $x_j^0(l_j^0,y_j^0)$ in Expression (3.39) and Expression (3.40). Proper choice of points x_1^0 and x_2^0 can be performed for the case of $\varepsilon_1^{(2)} < \varepsilon < \varepsilon_{\min}^{(1)}$. Because the final temperature distribution, $\theta(l,y,\Delta^0(\varepsilon'))$, is slightly different from $\theta(l,y,\Delta^0(\varepsilon_{\min}^{(1)}))$, it is possible to assume that the set of points x_1^0 and x_2^0 for $\varepsilon = \varepsilon'$ will be similar to the appropriate set for $\varepsilon = \varepsilon_{\min}^{(1)}$.

For $\varepsilon = \varepsilon_{\min}^{(1)}$, one can obtain the following points: $x_1^0 = (0,0)$ and $= x_2^0 = (1,1)$ (Figure 4.46a). Therefore, for $\varepsilon = \varepsilon'$, the rectangle's vertex O and point B can be considered as points of minimum and maximum final temperature, respectively (Figure 4.46b). This means that extremum points have the following coordinates: $x_1^0 = (0,0)$, $x_2^0 = (l_{e2},y_{e2})$. General relations (Equation 3.39) lead to the following system:

$$\begin{cases} \theta\left(0,0,\Delta_1^0,\Delta_2^0\right)-\theta^* = -\varepsilon'; \\ \theta\left(l_{e2},y_{e2},\Delta_1^0,\Delta_2^0\right)-\theta^* = +\varepsilon'; \\ \dfrac{\partial\theta\left(l_{e2},y_{e2},\Delta_1^0,\Delta_2^0\right)}{\partial l} = \dfrac{\partial\theta\left(l_{e2},y_{e2},\Delta_1^0,\Delta_2^0\right)}{\partial y} = 0. \end{cases} \tag{4.116}$$

For preset value of ε', the set in Equation (4.116) can be solved with respect to four unknown variables, Δ_1^0, Δ_2^0, l_{e2}, and y_{e2}, after substitution of $\theta(l,y,\Delta^0)$ in the form of Equation (4.111).

The set in Equation (4.116) remains valid for all values of ε decreasing from $\varepsilon_{\min}^{(1)}$ to certain value $\varepsilon = \varepsilon_1^{(2)}$, for which the third point, x_3^0, appears in the prism cross-section. As shown previously, it is reasonable to assume that only rectangle vertices C, A, and D should be considered as points where minimum admissible final temperature would take place.

It can be shown that under conditions $\beta > 1$ in Equation (4.76) and for $\varepsilon = \varepsilon_1^{(2)}$, the only vertex $C(l = 0, y = 1)$ represents point x_3^0 located in the middle of the short side of the prism cross-section (Figure 4.45 and Figure 4.46c).[1,18]

The following analysis of the variety of possible scenarios leads to this conclusion. Let us assume that the vertex D represents the point x_3^0. Using the equality in Equation (4.113), it can be shown that the condition in Equation (4.92) will be violated in the vertex C. On the other hand, assuming that the vertex A represents point x_3^0, it is possible to conclude that minimum admissible temperature would be reached at all vertices O, C, A, and D. This conclusion leads to contradiction under $\beta > 1$.[1,18]

As a result, for $\varepsilon = \varepsilon_1^{(2)}$ under $N = 2$, $R = N + 1 = 3$, basic correlations (Equation 3.39 and Equation 3.40) can be rewritten as the following set of equations (Figure 4.46c):

$$\begin{cases} \theta\left(0,0,\Delta_1^0,\Delta_2^0\right)-\theta^* = -\varepsilon_1^{(2)}; \\ \theta\left(l_{e2},y_{e2},\Delta_1^0,\Delta_2^0\right)-\theta^* = +\varepsilon_1^{(2)}; \\ \theta\left(0,1,\Delta_1^0,\Delta_2^0\right)-\theta^* = -\varepsilon_1^{(2)}; \\ \dfrac{\partial\theta\left(l_{e2},y_{e2},\Delta_1^0,\Delta_2^0\right)}{\partial l} = \dfrac{\partial\theta\left(l_{e2},y_{e2},\Delta_1^0,\Delta_2^0\right)}{\partial y} = 0. \end{cases} \tag{4.117}$$

This system of five equations should be solved with respect to all unknown variables, including optimal control parameters Δ_1^0 and Δ_2^0, coordinates l_{e2} and y_{e2} of the maximum temperature point B, and the value of $\varepsilon_1^{(2)}$. If obtained value $\varepsilon_1^{(2)}$ coincides with the utmost heating accuracy for two-stage control $\varepsilon_{\min}^{(2)}$, the previously described analysis can be used when solving the time-optimal control problem for all values $\varepsilon \in [\varepsilon_{\min}^{(2)}, \varepsilon_{\min}^{(1)}]$. If the expression $\varepsilon_1^{(2)} > \varepsilon_{\min}^{(2)}$ is valid, then optimal control processes should be considered for values of ε within the range $\varepsilon \in [\varepsilon_{\min}^{(2)}, \varepsilon_1^{(2)}]$.

One can prove[1] that, in this case, the optimal control process consists of two stages for all $\varepsilon \in [\varepsilon_{\min}^{(2)}, \varepsilon_{\min}^{(1)}]$. In such cases, the correlation (Equation 3.3) remains valid for the two-dimensional problem. Similar analysis allows one to obtain the following results.[1]

1. If the value of ε belongs to the range $\varepsilon_{\min}^{(2)} < \varepsilon < \varepsilon_1^{(2)}$, then it is possible to assume that $N = R = 2$. Similarly to the case of $\varepsilon = \varepsilon_1^{(2)}$, the maximum and minimum final temperatures are reached at the point $B(l_{e2}, y_{e2})$ and in the rectangle vertex $C(0,1)$, respectively. Temperature in the prism center $O(0,0)$ will not reach its utmost admissible value. This means that condition $|\theta(0, 0, \Delta_1^0, \Delta_2^0) - \theta^*| < \varepsilon$ is valid (Figure 4.46d). The following set of equations can be written with respect to four unknown variables Δ_1^0, Δ_2^0, l_{e2}, and y_{e2} that uniquely determine optimal heating mode for preset values of ε:

$$\begin{cases} \theta\left(0,1,\Delta_1^0,\Delta_2^0\right)-\theta^* = -\varepsilon; \\ \theta\left(l_{e2},y_{e2},\Delta_1^0,\Delta_2^0\right)-\theta^* = +\varepsilon; \\ \dfrac{\partial\theta\left(l_{e2},y_{e2},\Delta_1^0,\Delta_2^0\right)}{\partial l} = \dfrac{\partial\theta\left(l_{e2},y_{e2},\Delta_1^0,\Delta_2^0\right)}{\partial y} = 0. \end{cases} \tag{4.118}$$

2. Minimum admissible temperature is reached at one additional point — the vertex $A(1,1)$ of the prism cross-section (as well as at points B and C of maximum deviations) when heating accuracy ε decreases until the certain level $\varepsilon_2^{(2)} < \varepsilon_1^{(2)}$ that coincides with minimax value $\varepsilon_{\min}^{(2)}$ (Figure 4.46e). With respect to Equation (3.39), the set of equations can be obtained for $N = 2$, $R = N + 1 = 3$ as follows:

$$\begin{cases} \theta\left(0,1,\Delta_1^0,\Delta_2^0\right) - \theta^* = -\varepsilon_2^{(2)}; \\ \theta\left(l_{e2}, y_{e2}, \Delta_1^0, \Delta_2^0\right) - \theta^* = +\varepsilon_2^{(2)}; \\ \theta\left(1,1,\Delta_1^0,\Delta_2^0\right) - \theta^* = -\varepsilon_2^{(2)}; \\ \dfrac{\partial\theta\left(l_{e2}, y_{e2,}\Delta_1^0,\Delta_2^0\right)}{\partial l} = \dfrac{\partial\theta\left(l_{e2}, y_{e2,}\Delta_1^0,\Delta_2^0\right)}{\partial y} = 0. \end{cases} \tag{4.119}$$

This system should be solved with respect to unknown variables Δ_1^0, Δ_2^0, l_{e2}, and y_{e2} and minimax value $\varepsilon_2^{(2)}$. Obtained results fully define the set of points $x_j^0, j = \overline{1,R}$ and appropriate system of equations. At this step, the initial time-optimal control problem is solved with regard to two possible variants of the transition from $\varepsilon_{\min}^{(1)}$ to $\varepsilon_{\min}^{(2)}$ for all values $\varepsilon \in [\, \varepsilon_{\min}^{(2)}, \varepsilon_{\min}^{(1)} \,]$.

Optimal three-stage control. For prescribed heating accuracy $\varepsilon \in [\, \varepsilon_{\min}^{(3)}, \varepsilon_{\min}^{(2)})$, the condition $N \geq 3$ can be obtained regarding the number of optimal control intervals. First of all, let us assume that $N = 3$, according to Equation (3.3). Similarly to a case of a two-stage control, it is possible to investigate signs of temperature gradients with respect to final temperature distribution $\theta(l,y,\Delta_1^0,\Delta_2^0,\Delta_3^0)$.

As shown in Rapoport,[1,19] due to specifics of boundary conditions in Equation (4.110), temperature gradients on the borders $l = 1$ and $y = 1$ of rectangle *OCAD* (Figure 4.45) change their signs from "+" to "–" when passing from the first control stage to the second stage. These gradients change their signs from "–" to "+" when passing from the second to the third control stage.

In both cases, at the moment of transition from one control stage to the following one, there will be only two moving straight lines, L_l and L_y, that are parallel to coordinate axes y and l. Temperature gradients $\partial\theta/\partial l$ and $\partial\theta/\partial y$ will be equal to zero along these lines as well as along the axes $y = 0$ and $l = 0$ (Figure 4.47a).

Similarly to Equation (4.115), signs of temperature gradients at any point within the prism cross-section can be uniquely specified at the end of the optimal control process. Therefore, the following conditions can be obtained:

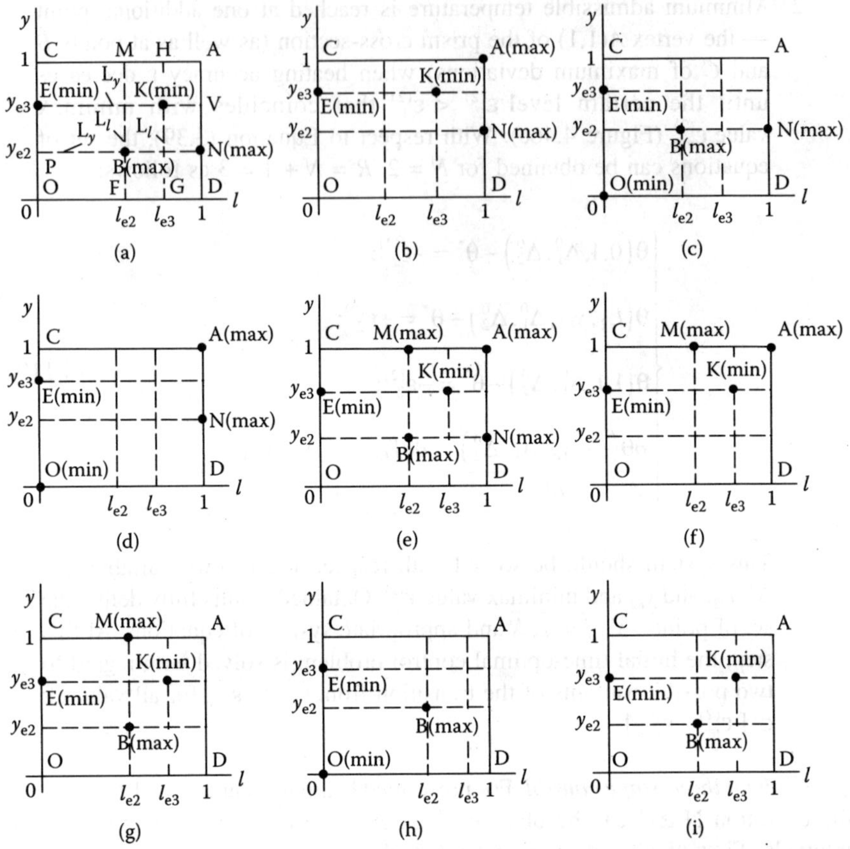

FIGURE 4.47 Variants of location of minimum and maximum temperature points within rectangular prism cross-section at the end of three-stage optimal control by power of surface heat-generating sources for (a–g) $\varepsilon = \varepsilon_{min}^{(3)}$ and (h, i) $\varepsilon_{min}^{(3)} < \varepsilon < \varepsilon_{min}^{(2)}$.

$$
\begin{aligned}
&\frac{\partial\theta\left(l, y, \Delta_1^0, \Delta_2^0, \Delta_3^0\right)}{\partial l} \geq 0 \text{ for rectangles OCMF and GHAD;}\\
&\frac{\partial\theta\left(l, y, \Delta_1^0, \Delta_2^0, \Delta_3^0\right)}{\partial y} \geq 0 \text{ for rectangles OPND and ECAV;}\\
&\frac{\partial\theta\left(l, y, \Delta_1^0, \Delta_2^0, \Delta_3^0\right)}{\partial l} \leq 0 \text{ for rectangle FMHG;}\\
&\frac{\partial\theta\left(l, y, \Delta_1^0, \Delta_2^0, \Delta_3^0\right)}{\partial y} \leq 0 \text{ for rectangle PEVN.}
\end{aligned}
\tag{4.120}
$$

Obtained correlations allow revealing the profile of the final temperature distribution under three-stage control. As it follows from Equation (4.120), during the first and the third control stages, the extremum points of minimum temperature (O and E, K, respectively) appear on the concurrence of lines L_l and L_y (including coordinate axes $l = 0$ and $y = 0$). During the second control stage, the extremum point B of maximum temperature appears. The maximum temperature can also be reached at points M, A, and N along the perimeter of the rectangle; the minimum temperature could appear at points O, E, and K.

Therefore, the complete set $\{O, E, \text{K}, B, M, A, N\}$ of seven points considered as "potential candidates" for points x_j^0 is determined. In the next step, it would be necessary to choose particular profiles of final temperature distribution according to the preset value of ε.

The final temperature distribution can be represented in the form of Expression (4.113). Analysis provided in Rapoport[1,19] shows that there might be six combinations with maximum number $R = 4$ of points x_j^0 within the prism cross-section. The number, R, exceeds by one the number of optimal control stages, $N = 3$. Figure 4.47a through Figure 4.47d, and Figure 4.47f through Figure 4.47g, show all possible profiles. The first four profiles represent the condition $1 < \beta < \beta^*$ in Equation (4.76). The last two profiles satisfy the condition $\beta > \beta^*$, where the value β^* corresponds to the case of six points x_j^0 (Figure 4.47e). According to the basic condition in Equation (3.40), one of these profiles should conform to the case of $\varepsilon = \varepsilon_{\min}^{(3)}$. In each particular problem, such a profile can be chosen by trial solution of the corresponding set of equations. For the given combination of four points x_j^0, $j = \overline{1, 4}$ the set of equations can be written as:

$$\begin{cases} \theta\left(l_1^0, y_1^0, \Delta_1^0, \Delta_2^0, \Delta_3^0\right) - \theta^* = -\varepsilon_k^{(3)}; \\ \theta\left(l_2^0, y_2^0, \Delta_1^0, \Delta_2^0, \Delta_3^0\right) - \theta^* = -\varepsilon_k^{(3)}; \\ \theta\left(l_3^0, y_3^0, \Delta_1^0, \Delta_2^0, \Delta_3^0\right) - \theta^* = \varepsilon_k^{(3)}; \\ \theta\left(l_4^0, y_4^0, \Delta_1^0, \Delta_2^0, \Delta_3^0\right) - \theta^* = \varepsilon_k^{(3)}; \\ \dfrac{\partial\theta\left(l_{e2}, y_{e2,}\Delta_1^0, \Delta_2^0, \Delta_3^0\right)}{\partial l} = \dfrac{\theta\left(l_{e2}, y_{e2,}\Delta_1^0, \Delta_2^0, \Delta_3^0\right)}{\partial y} = 0; \\ \dfrac{\partial\theta\left(l_{e3}, y_{e3,}\Delta_1^0, \Delta_2^0, \Delta_3^0\right)}{\partial l} = \dfrac{\theta\left(l_{e3}, y_{e3,}\Delta_1^0, \Delta_2^0, \Delta_3^0\right)}{\partial y} = 0. \end{cases} \tag{4.121}$$

Here, for the first scenario (Figure 4.47a):

$$l_1^0 = 0, y_1^0 = y_{e3}, l_2^0 = l_{e3}, y_2^0 = y_{e3}, l_3^0 = l_{e2}, y_3^0 = y_{e2}, l_4^0 = 1, y_4^0 = y_{e2}$$

for the second scenario (Figure 4.47b):

$$l_1^0 = 0, y_1^0 = y_{e3}, l_2^0 = l_{e3}, y_2^0 = y_{e3}, l_3^0 = 1, y_3^0 = 1, l_4^0 = 1, y_4^0 = y_{e2}$$

for the third scenario (Figure 4.47c):

$$l_1^0 = 0, y_1^0 = y_{e3}, l_2^0 = 0, y_2^0 = 0, l_3^0 = l_{e2}, y_3^0 = y_{e2}, l_4^0 = 1, y_4^0 = y_{e2}$$

for the fourth scenario (Figure 4.47d):

$$l_1^0 = 0, y_1^0 = y_{e3}, l_2^0 = 0, y_2^0 = 0, l_3^0 = 1, y_3^0 = 1, l_4^0 = 1, y_4^0 = y_{e2}$$

for the fifth scenario (Figure 4.47f):

$$l_1^0 = 0, y_1^0 = y_{e3}, l_2^0 = l_{e3}, y_2^0 = y_{e3}, l_3^0 = 1, y_3^0 = 1, l_4^0 = l_{e2}, y_4^0 = 1$$

and for the sixth scenario (Figure 4.47g):

$$l_1^0 = 0, y_1^0 = y_{e3}, l_2^0 = l_{e3}, y_2^0 = y_{e3}, l_3^0 = l_{e2}, y_3^0 = y_{e2}, l_4^0 = l_{e2}, y_4^0 = 1$$

Each of the systems in Equation (4.121) can be solved with respect to eight unknown variables: durations of optimal control stages Δ_1^0, Δ_2^0, and Δ_3^0, coordinates l_{e2}, y_{e2}, l_{e3}, and y_{e3} of extremum points B and K, and value of heating accuracy $\varepsilon_k^{(3)}$, $k \in \{\overline{1,6}\}$ for each kth scenario. The least value of all $\varepsilon_k^{(3)}$, $k = \{\overline{1,6}\}$ represents the minimax value $\varepsilon_{\min}^{(3)}$.

The first four scenarios should be considered in the course of solving optimal control problems (Equation 4.121) for values of β increasing by small steps from $\beta = 1$ down to the preset value $\beta_0 > 1$. If the value of β becomes equal to β^* (under condition $\beta^* < \beta_0$), the fifth and the sixth variants should be taken into account.

In each specific situation, further analysis is necessary to determine the real number, N, of optimal control stages and particular combinations of points x_j^0 for all values $\varepsilon \in [\varepsilon_{\min}^{(3)}, \varepsilon_{\min}^{(2)})$. Similarly to the case of $\varepsilon = \varepsilon' = \varepsilon_{\min}^{(1)} - d\varepsilon$, regarding two-stage control the following conclusions can be obtained based on alternance properties in Equation (3.39) and Equation (3.40) and the continuous nature of final temperature distribution $\theta(l,y,\Delta^0)$ as a function of ε.

If preset values of $\varepsilon = \varepsilon'' = \varepsilon_{\min}^{(2)} - d\varepsilon$, $d\varepsilon > 0$ slightly differ from $\varepsilon_{\min}^{(2)}$, then the optimal control would consist of three stages ($N = 3$) and optimal temperature distribution with three extremum points x_j^0, $j = 1, 2, 3$ will be "close" to the prior distribution in the case of $\varepsilon = \varepsilon_{\min}^{(2)}$.

According to Equation (4.120) and depending on two possible scenarios of value assignment for minimax $\varepsilon_{\min}^{(2)}$: $\varepsilon_{\min}^{(2)} = \varepsilon_1^{(2)}$ or $\varepsilon_{\min}^{(2)} = \varepsilon_2^{(2)}$ (Figure 4.46c or Figure 4.46e), only points *O, B* , *E* or *E, B,* K (Figure 4.47h and Figure 4.47i) can be considered as points $x_j^0, j = 1, 2, 3$. Similarly to the case of two-stage control, this conclusion remains valid as value of ε decreases to the lower limit, $\varepsilon = \varepsilon^*$, for which the fourth point x_4^0 of maximum temperature deviation would appear. Obviously, the value ε^* coincides with one of values $\varepsilon_k^{(3)}, k \in \{1,6\}$. If comparative analysis leads to the equality $\varepsilon^* = \varepsilon_{\min}^{(3)}$, then the optimal control would consist of three piecewise constant stages for all values of ε: $\varepsilon_{\min}^{(3)} \le \varepsilon < \varepsilon_{\min}^{(2)}$.

Computation results show[1] that the OCP solution leads to three-stage control in practicable cases of $\varepsilon_{\min}^{(2)} = \varepsilon_2^{(2)}$, and $\varepsilon_{\min}^{(3)} = \varepsilon_1^{(3)}$, or $\varepsilon_{\min}^{(3)} = \varepsilon_6^{(3)}$.

For all values of ε: $\varepsilon_{\min}^{(3)} < \varepsilon < \varepsilon_{\min}^{(2)}$, the point $B(l_{e2}, y_{e2})$ of maximum temperature and minimum temperature points $E(0, y_{e3})$, $K(l_{e3}, y_{e3})$ should be considered as $x_j^0, j = 1, 2, 3$ (Figure 4.47i). Therefore, the following set of equations can be obtained:

$$\begin{cases} \theta\left(0, y_{e3}, \Delta_1^0, \Delta_2^0, \Delta_3^0\right) - \theta^* = -\varepsilon; \\ \theta\left(l_{e3}, y_{e3}, \Delta_1^0, \Delta_2^0, \Delta_3^0\right) - \theta^* = -\varepsilon; \\ \theta\left(l_{e2}, y_{e2}, \Delta_1^0, \Delta_2^0, \Delta_3^0\right) - \theta^* = \varepsilon; \\ \dfrac{\partial\theta\left(l_{e2}, y_{e2}, \Delta_1^0, \Delta_2^0, \Delta_3^0\right)}{\partial l} = \dfrac{\theta\left(l_{e2}, y_{e2}, \Delta_1^0, \Delta_2^0, \Delta_3^0\right)}{\partial y} = 0; \\ \dfrac{\partial\theta\left(l_{e3}, y_{e3}, \Delta_1^0, \Delta_2^0, \Delta_3^0\right)}{\partial l} = \dfrac{\theta\left(l_{e3}, y_{e3}, \Delta_1^0, \Delta_2^0, \Delta_3^0\right)}{\partial y} = 0. \end{cases} \tag{4.122}$$

For known values of ε, this system can be solved with respect to seven unknown variables: Δ_1^0, Δ_2^0, Δ_3^0, l_{e2}, y_{e2}, l_{e3}, and y_{e3}.

If the condition $\varepsilon_{\min}^{(3)} = \varepsilon_1^{(3)}$ still remains valid, then for $\varepsilon = \varepsilon_{\min}^{(3)}$, the prior combination of points x_j^0 can be supplemented by the second point $N(1, y_{e2})$ with maximum temperature (Figure 4.47a). The maximum temperature point $M(l_{e2}, 1)$ appears if the equality $\varepsilon_{\min}^{(3)} = \varepsilon_6^{(3)}$ is true (Figure 4.47g). The following analysis is necessary to determine the number, *N*, within interval [$\varepsilon_{\min}^{(3)}$,ε^*) under condition $\varepsilon^* > \varepsilon_{\min}^{(3)}$.

Computational technique for optimal heating process. To determine time-optimal control for induction heating of an infinite length rectangular prism, the following simple computational procedure is suggested under the assumption that heat losses from the surface cannot be considered as negligible.

1. At the beginning, the sets of Equation (4.112), Equation (4.117), and Equation (4.119) can be solved. This leads to computation of a priori unknown values $\varepsilon_{\min}^{(1)}$, $\varepsilon_1^{(2)}$, and $\varepsilon_2^{(2)}$ as well as corresponding parameters of the optimal process. Then, minimax value $\varepsilon_{\min}^{(2)}$ can be determined as the least of the values $\varepsilon_1^{(2)}$ and $\varepsilon_2^{(2)}$.
2. The optimal control problem is already solved at this stage if the prescribed numerical value of $\varepsilon = \varepsilon_0$ coincides with the obtained numerical value of minimax $\varepsilon_{\min}^{(1)}$: $\varepsilon = \varepsilon_0 = \varepsilon_{\min}^{(1)}$. The initial problem is also solved for the cases $\varepsilon_0 = \varepsilon_1^{(2)}$ or $\varepsilon_0 = \varepsilon_2^{(2)}$.
3. Computations should be continued if the expression $\varepsilon_{\min}^{(2)} < \varepsilon_0 < \varepsilon_{\min}^{(1)}$ is valid. If the value of heating accuracy satisfies the condition $\varepsilon_{\min}^{(2)} \le \varepsilon_1^{(2)} < \varepsilon_0$, then it would still be necessary to solve the set of Equation (4.116) under assumption $\varepsilon' = \varepsilon_0$. In the case of $\varepsilon_{\min}^{(2)} = \varepsilon_2^{(2)} < \varepsilon_0 < \varepsilon_1^{(2)}$, the set in Equation (4.118) should be solved.
4. If the expression $\varepsilon_0 < \varepsilon_{\min}^{(2)}$ is true, then the OCP solution that provides heating accuracy $\varepsilon_{\min}^{(3)}$ should be found among optimal control functions with three piecewise constant intervals ($N = 3$). A sequence of optimization problems that could be reduced to solution of six possible sets of the type in Equation (4.121) should be solved and the least value $\varepsilon_k^{(3)}$ should be chosen as the minimax value $\varepsilon_{\min}^{(3)}$.

For the majority of practical applications, the expression $\varepsilon_{\min}^{(3)} \le \varepsilon_0 < \varepsilon_{\min}^{(2)}$ remains valid. Then the number, N, and vector, Δ^0, can be found for all values $\varepsilon_0 \in (\varepsilon_{\min}^{(3)}, \varepsilon_{\min}^{(2)})$ according to the computational procedure described earlier.

As was mentioned earlier, in typical cases of $\varepsilon_{\min}^{(2)} = \varepsilon_2^{(2)}$, $\varepsilon_{\min}^{(3)} = \varepsilon_1^{(3)}$, or $\varepsilon_{\min}^{(3)} = \varepsilon_6^{(3)}$, the following results can be obtained:

- The expression $N = 3$ is true for all values of ε_0: $\varepsilon_{\min}^{(3)} \le \varepsilon_0 < \varepsilon_{\min}^{(2)}$.
- In the case of $\varepsilon_0 = \varepsilon_{\min}^{(3)}$, the problem is solved as a result of the previous step.
- In the case of $\varepsilon_{\min}^{(3)} < \varepsilon_0 < \varepsilon_{\min}^{(2)}$, the problem can be reduced to the solution of the equations set in Equation (4.122).

All sets of equations can be solved after substitution of the expression for final temperature field $\theta(l, y, \Delta_0)$ in the form of Equation (4.111). Required number $N = 10 \ldots 15$ of series terms in the set of equations can be utilized without considerable difficulties using modern computers.

The described computational algorithm allows solving the optimal control problems for any required values of q_0, β, and $\theta_0 < \theta^*$ in the whole range of heating accuracy variation $\varepsilon_0 \in [\varepsilon_{\min}^{(3)}, \varepsilon_{\min}^{(1)}]$. Figure 4.48a and Figure 4.48b represent some results of computation. Using the system of dimensionless units discussed in Chapter 1, it is possible to obtain $\theta^* = 0$.[1,18] The examples in Figure 4.48a and Figure 4.48b show that minimax values $\varepsilon_{\min}^{(2)} = \varepsilon_1^{(2)}$ and $\varepsilon_{\min}^{(2)} = \varepsilon_2^{(2)}$ can be obtained respectively within the whole range of β variation.

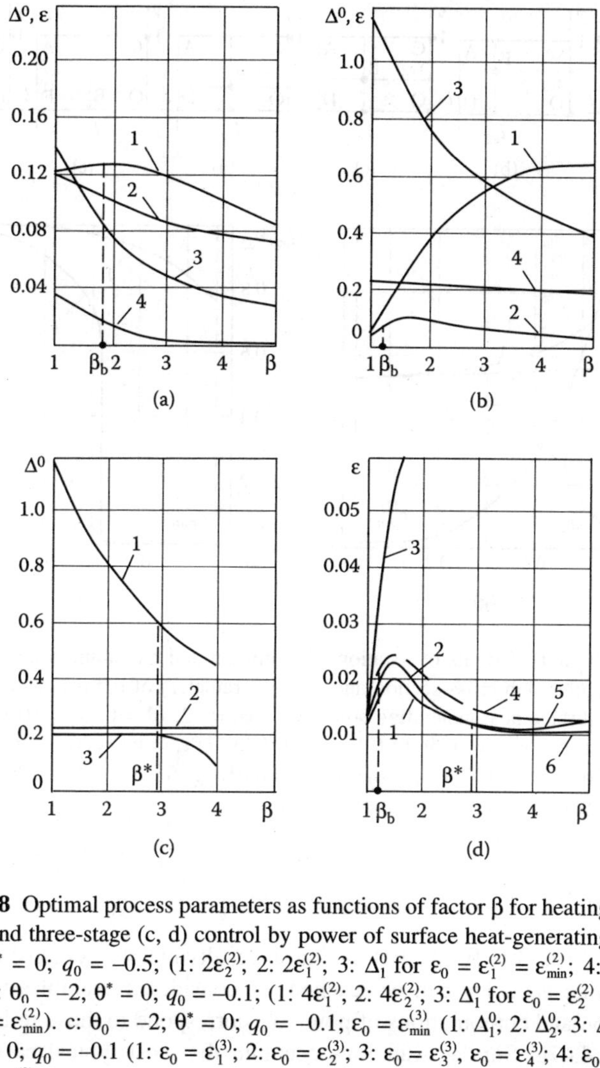

FIGURE 4.48 Optimal process parameters as functions of factor β for heating under two-stage (a, b) and three-stage (c, d) control by power of surface heat-generating sources. a: $\theta_0 = -0.1$; $\theta^* = 0$; $q_0 = -0.5$; (1: $2\varepsilon_2^{(2)}$; 2: $2\varepsilon_1^{(2)}$; 3: Δ_1^0 for $\varepsilon_0 = \varepsilon_1^{(2)} = \varepsilon_{\min}^{(2)}$; 4: Δ_2^0 for $\varepsilon_0 = \varepsilon_1^{(2)} = \varepsilon_{\min}^{(2)}$). b: $\theta_0 = -2$; $\theta^* = 0$; $q_0 = -0.1$; (1: $4\varepsilon_1^{(2)}$; 2: $4\varepsilon_2^{(2)}$; 3: Δ_1^0 for $\varepsilon_0 = \varepsilon_2^{(2)} = \varepsilon_{\min}^{(2)}$; 4: Δ_2^0 for $\varepsilon_0 = \varepsilon_2^{(2)} = \varepsilon_{\min}^{(2)}$). c: $\theta_0 = -2$; $\theta^* = 0$; $q_0 = -0.1$; $\varepsilon_0 = \varepsilon_{\min}^{(3)}$ (1: Δ_1^0; 2: Δ_2^0; 3: $\Delta_3^0 \times 10^3$). d: $\theta_0 = -2$; $\theta^* = 0$; $q_0 = -0.1$ (1: $\varepsilon_0 = \varepsilon_1^{(3)}$; 2: $\varepsilon_0 = \varepsilon_2^{(3)}$; 3: $\varepsilon_0 = \varepsilon_3^{(3)}$, $\varepsilon_0 = \varepsilon_4^{(3)}$; 4: $\varepsilon_0 = \varepsilon_{\min}^{(2)}$; 5: $\varepsilon_0 = \varepsilon_5^{(3)}$; 6: $\varepsilon_0 = \varepsilon_6^{(3)}$).

The computations demonstrate that the most typical case of $\varepsilon_{\min}^{(2)} = \varepsilon_2^{(2)}$ arises due to relatively small heat losses from the surface of the heated workpiece. The case of $\varepsilon_{\min}^{(2)} = \varepsilon_1^{(2)}$ takes place if the value $|q_0|$ increases significantly and if the difference $\theta^* - \theta_0$ diminishes. If the expression $\beta \geq \beta_b$ is valid for $\varepsilon_1^{(2)} = \varepsilon_{\min}^{(2)}$ or $\varepsilon_2^{(2)} = \varepsilon_{\min}^{(2)}$ (Figure 4.48a and Figure 4.48b), then the point B that represents the maximum temperature will be located on the side $y = 0$ of rectangle $OCAD$; therefore, $y_{e2} = 0$ (Figure 4.49).

Figure 4.48c and Figure 4.48d represent optimal heating process parameters in the case of $\varepsilon_0 = \varepsilon_{\min}^{(3)}$. As can be seen, the equality $\varepsilon_{\min}^{(3)} = \varepsilon_1^{(3)}$ is true if $\beta < \beta^*$,

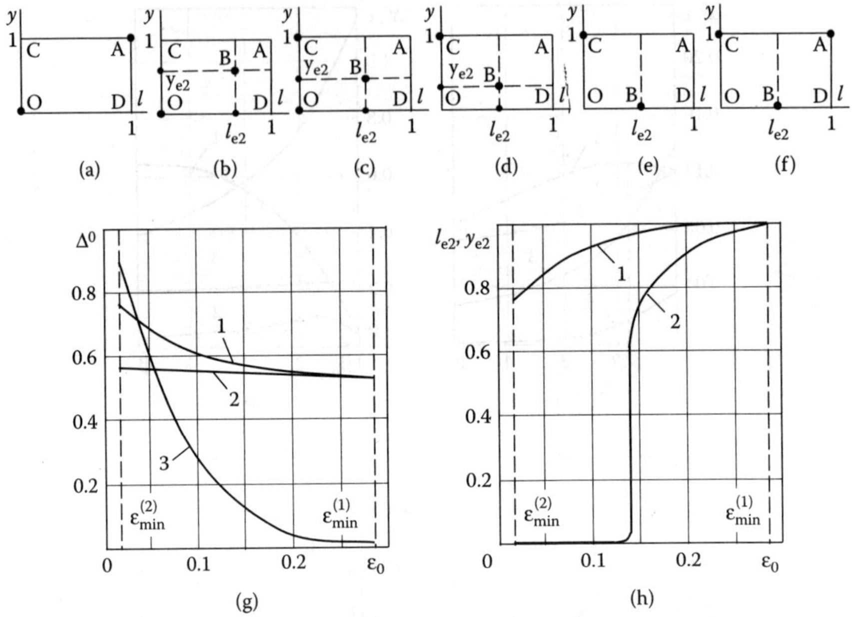

FIGURE 4.49 (a–f) Variants of location of minimum and maximum temperature points within rectangular slab cross-section and (g, h) parameters of optimal two-stage control by power of surface heat-generating sources for $\theta_0 = -2$; $\theta^* = 0$; $q_0 = -0.1$; $\beta = 3$; $\varepsilon \in [\varepsilon_{min}^{(2)}, \varepsilon_{min}^{(1)}]$. a: $\varepsilon_0 = \varepsilon_{min}^{(1)} = 0.29$; b: $0.14 < \varepsilon_0 < 0.29$. c: $\varepsilon_0 = \varepsilon_1^{(2)} = 0.14$. d: $0.125 < \varepsilon_0 < 0.140$. e: $0.016 < \varepsilon_0 \leq 0.125$. f: $\varepsilon_0 = \varepsilon_2^{(2)} = \varepsilon_{min}^{(2)} = 0.016$. g: for Δ^0. 1: $\Delta_1^0 + \Delta_2^0$; 2: Δ_1^0; 3: $5\Delta_2^0$. h: For coordinates of point (l_{e2}, y_{e2}): 1: l_{e2}, 2: y_{e2}.

and $\varepsilon_{min}^{(3)} = \varepsilon_6^{(3)}$ if $\beta > \beta^*$. Therefore, these results clearly demonstrate that the utmost heating accuracy can be reached for considered examples under three-stage control, and combinations of points x_j^0 correspond to those in Figure 4.47a and Figure 4.47g. Similar situations occur in the most typical cases of initial temperatures and levels of heat losses.

Figure 4.48d shows the results of computation that are provided for the case of substantially small heat loss, q_0 — for example, 10% of maximum heating power. Three-stage control can provide required heating accuracy, if the difference does not exceed 20% from heating accuracy under two-stage control. Optimal duration of reheating stage Δ_3^0 (the third stage of the control process) represents only a small fraction of the previous soaking stage (the second stage of control process Δ_2^0). The advantage of three-stage control in comparison with the two-stage one becomes more substantial with surface heat loss increase. For example, if value q_0 increases up to 50%, then heating accuracy under three-stage control becomes three times better.

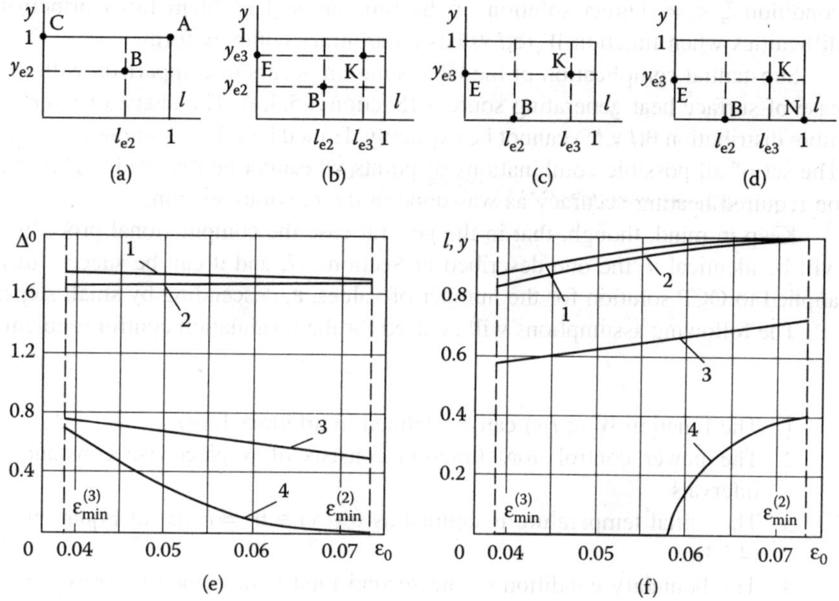

FIGURE 4.50 (a–d) Variants of location of minimum and maximum temperature points within rectangular slab cross-section and (e, f) parameters of optimal three-stage control by power of surface heat-generating sources for $\theta_0 = -2$; $\theta^* = 0$; $q_0 = -0.5$; $\beta = 1.5$; $\varepsilon \in [\varepsilon^{(3)}_{min}, \varepsilon^{(2)}_{min}]$. a: $\varepsilon_0 = \varepsilon_2^{(2)} = \varepsilon^{(2)}_{min} = 0.073$. b: $0.058 < \varepsilon_0 < 0.073$. c: $0.039 < \varepsilon_0 \leq 0.058$. d: $\varepsilon_0 = \varepsilon_1^{(3)} = \varepsilon^{(3)}_{min} = 0.039$. e: Δ^0. 1: φ^0_{min}; 2: Δ^0_1; 3: $\Delta^0_2 \times 10$; 4: $\Delta^0_3 \times 10^2$. f: For coordinates of points (l_{e2}, y_{e2}), (l_{e3}, y_{e3}). 1: l_{e3}; 2: y_{e3}; 3: l_{e2}, 4: y_{e2}.

Let us consider examples of optimal control process computations for values of ε_0 that belong to the ranges $[\varepsilon^{(2)}_{min}, \varepsilon^{(1)}_{min}]$ and $[\varepsilon^{(3)}_{min}, \varepsilon^{(2)}_{min}]$. Figure 4.49 and Figure 4.50 show results for $\varepsilon^{(2)}_{min} = \varepsilon_2^{(2)} < \varepsilon_1^{(2)} < \varepsilon^{(1)}_{min}$ and $\varepsilon^{(3)}_{min} = \varepsilon_1^{(3)}$ (under $\varepsilon^{(2)}_{min} = \varepsilon_2^{(2)}$), respectively, when the condition $\beta > \beta_b$ holds true. Combinations of points x_j^0 are changed sequentially as the value of ε decreases from $\varepsilon^{(1)}_{min}$ to $\varepsilon^{(2)}_{min}$ and from $\varepsilon^{(2)}_{min}$ to $\varepsilon^{(3)}_{min}$ (Figure 4.47 and Figure 4.48). This means that, in the course of the computational procedure, there is a necessity to replace one of the considered sets of equations.

4.5.4.2 Optimization of Internal Source Heating

Proper control of the heating process involves appropriate choice of such important process parameters as internally generated heating power. Two-dimensional mathematical models of induction heating of rectangular-shaped solids should take into account a spatial distribution of internal heat sources within the slab cross-section that is represented by the function $W_d(\xi, l, y)$ in Equation (4.72) under condition $\xi < \infty$. Direct solution of the time-optimal problem faces principal difficulties when function $W_d(\xi,l,y)$ has an arbitrary complex form.

Substantial complication of the OCP solution occurs in comparison with the case of surface heat-generating sources (Section 4.5.4.1). The shape of temperature distribution $\theta(l,y,\Delta^0)$ cannot be explicitly defined based on "common sense." The set of all possible combinations of points x_j^0 cannot be revealed depending on required heating accuracy as was done in the previous section.

Keep in mind, though, that in the general case the computational procedure will be identical to the one described in Section 3.7, and it can be successfully applied to OCP solution for the number of values, ε, descending by small steps.

The following assumptions will be used for the formulation control problem:

1. The function $W_d(\xi,l,y)$ can be defined in arbitrary form.
2. The power control $u(\varphi)$ function consists of N piecewise constant intervals.
3. The initial temperature is defined as $\theta_0(l,y) \equiv \theta_0 = const$ in Equation (4.73).
4. The boundary condition of the second kind is described by $q(y,\varphi) = q_{T1}(l,\varphi) \equiv q_0 = const$ in Equation (4.74).

Then, similarly to Equation (4.94) and Equation (4.111), the following solution of the heat conduction equation (Equation 4.72 through Equation 4.74) can be written for $\Gamma = 0$[1,17]:

$$\theta(l,y,\Delta) = \theta_0 + q_0(1+\beta)\sum_{m=1}^{N}\Delta_m - \frac{q_0}{2}\left(\frac{1}{3}-l^2\right) - \frac{q_0}{2\beta}\left(\frac{1}{3}-y^2\right) +$$

$$+ 2q_0\sum_{r=1}^{\infty}\frac{(-1)^{r+1}}{\pi^2 r^2}\cdot\exp\left(-\pi^2 r^2\sum_{m=1}^{N}\Delta_m\right)\cos\pi r l + \frac{2q_0}{\beta}\sum_{n=1}^{\infty}\frac{(-1)^{n+1}}{\pi^2 n^2}\cdot$$

$$\exp\left(-\pi^2 n^2\beta^2\sum_{m=1}^{N}\Delta_m\right)\cos\pi n y + W_{00}(\xi)U_{\max}\sum_{m=1}^{N}\frac{1+(-1)^{m+1}}{2}\Delta_m +$$

$$+ 2U_{\max}\sum_{r=1}^{\infty}W_{r0}(\xi)\frac{\cos\pi r l}{\pi^2 r^2}\left[\sum_{j=1}^{N}(-1)^{j+1}\cdot\left(1-\exp\left(-\pi^2 r^2\sum_{m=j}^{N}\Delta_m\right)\right)\right]+$$

$$+\frac{2}{\beta^2}U_{\max}\sum_{n=1}^{\infty}W_{0n}(\xi)\frac{\cos\pi n y}{\pi^2 n^2}\left[\sum_{j=1}^{N}(-1)^{j+1}\cdot\left(1-\exp\left(-\pi^2 n^2\beta^2\sum_{m=j}^{N}\Delta_m\right)\right)\right]+$$

$$+ 4U_{\max}\sum_{r=1}^{\infty}\sum_{n=1}^{\infty}W_{rn}(\xi)\frac{\cos\pi r l\cdot\cos\pi n y}{\pi^2\left(r^2+\beta^2 n^2\right)}\times$$

$$\times\left[\sum_{j=1}^{N}(-1)^{j+1}\cdot\left(1-\exp\left(-\pi^2\left(r^2+\beta^2 n^2\right)\sum_{m=j}^{N}\Delta_m\right)\right)\right]. \tag{4.123}$$

The sufficiently large number of terms of series is always used in Equation (4.123). The given expression should be substituted in governing sets of equations on appropriate steps of computational procedure. All values of $W_{rn}(\xi)$ r, $n = 0, 1, 2, \ldots$ computed by Equation (4.83) can be determined using analytical approximations for $W_d(\xi,l,y)$.[16,17]

Nevertheless, for the most practical cases, it is possible to simplify the computational procedure significantly, taking into account the usual physical concepts with respect to spatial temperature distribution at the end of the most typical heating modes.[1,17] Optimization of induction heating processes with the highly pronounced eddy current "skin" effect is quite similar to the previously described OCP for heating by surface heat-generating sources.

The simplest case takes place when required heating accuracy $\varepsilon = \varepsilon_{\min}^{(1)}$ can be reached by applying typical single-stage heating mode under maximum power ($N = 1$, $u^*(\varphi) \equiv U_{\max}$). The resulting final temperature distribution similar to (Figure. 4.46) is characterized by a single internal point of temperature maximum B and by the set of vertices O, C, D, A, at least one of which represents the point of minimum admissible temperature (depending on values of ξ, β, and q_0).[1,17]

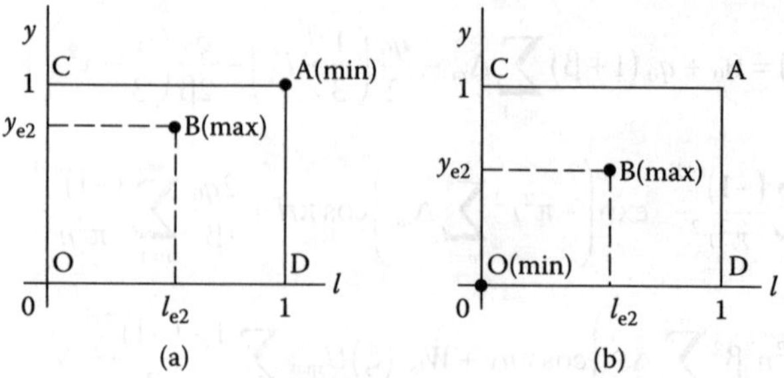

FIGURE 4.51 Variants of location of minimum and maximum temperature points within rectangular slab cross-section at the end of internal source heating with constant power for $\frac{\beta\xi}{\sqrt{2}} \le 1.5$ (a) and $\frac{\beta\xi}{\sqrt{2}} > 1.5$ (b).

In the real-life process of induction slab heating (in particular, heating of nonmagnetic materials), the parameter ξ takes on limiting values in expressions for $W_d(\xi,l,y)$ in Equation (4.72). These values differ significantly from each other. If the parameter ξ is substantially small, meeting the condition $\frac{\beta\xi}{\sqrt{2}} \le 1.5$, then the rectangle corner A turns out to be the least heated area.[1,16,17] Only the points A and B can be considered in this case as x_1^0 and x_2^0 according to the properties in Equation (3.39) and Equation (3.40) for $\varepsilon = \varepsilon_{\min}^{(1)}$ (Figure 4.51a).

The set of governing equations resulting from Equation (3.39) and Equation (3.40) for $\varepsilon = \varepsilon_{\min}^{(1)}$ under $N = 1$ can be written as follows:

$$\begin{cases} \theta\left(l_{e2}, y_{e2}, \Delta_1^0\right) - \theta^* = \varepsilon_{\min}^{(1)}; \\ \theta\left(1,1,\Delta_1^0\right) - \theta^* = -\varepsilon_{\min}^{(1)}; \\ \dfrac{\partial\theta\left(l_{e2}, y_{e2}, \Delta_1^0\right)}{\partial l} = \dfrac{\partial\theta\left(l_{e2}, y_{e2}, \Delta_1^0\right)}{\partial y} = 0. \end{cases} \tag{4.124}$$

After substituting $\theta(l,y,\Delta^0)$ in the form of Equation (4.123) under $N = 1$, this system can be solved with respect to unknown variables Δ^0, $\varepsilon_{\min}^{(1)}$, l_{e2}, and y_{e2}. Such a situation arises, for example, when heating titanium alloys slabs before rolling (the applied frequency is 50 Hz).[1,16,17]

As a rule, the value $\varepsilon_{\min}^{(1)}$ represents the utmost attainable accuracy that cannot be further improved by increasing the number, N, of constancy interval of function $u^*(\varphi)$.[1,17] This means that the expression $\varepsilon_{\min}^{(1)} = \varepsilon_{\inf}$ will be valid. Therefore, if the required heating accuracy ε_0 coincides with minimax value $\varepsilon_{\min}^{(1)}$, then the time-optimal control problem is reduced to solving the set of equations in Equation

(4.124). For preset values of heating accuracy meeting the condition $\varepsilon_0 < \varepsilon_{\min}^{(1)} = \varepsilon_{\inf}$, the problem of heating with required accuracy proves to be unsolvable.

When the value of ξ is large enough, the minimum final temperature will be reached in the core of the slab[1,16,17] that is similar to the case of heating by surface heat-generating sources (Figure 4.51b). Then, the following system of the equations can be obtained instead of Equation (4.124):

$$\begin{cases} \theta\left(l_{e2}, y_{e2}, \Delta_1^0\right) - \theta^* = \varepsilon_{\min}^{(1)}; \\ \theta\left(0, 0, \Delta_1^0\right) - \theta^* = -\varepsilon_{\min}^{(1)}; \\ \dfrac{\partial\theta\left(l_{e2}, y_{e2}, \Delta_1^0\right)}{\partial l} = \dfrac{\partial\theta\left(l_{e2}, y_{e2}, \Delta_1^0\right)}{\partial y} = 0. \end{cases} \tag{4.125}$$

The final temperature distribution differs from the case of $\xi = \infty$ for $\varepsilon = \varepsilon_{\min}^{(1)}$ (Figure 4.46a). Here, the maximum temperature is reached in the internal point B instead of vertex A. As a result, temperature gradients $\partial\theta/\partial l$ and $\partial\theta/\partial y$ will be negative within the whole plane of the slab cross-section, adjoining to external outline $l = 1$, $y = 1$. The temperature distribution profile becomes similar to the case of two-stage control under $\xi = \infty$ for $\varepsilon = \varepsilon'$, $\varepsilon_1^{(2)} < \varepsilon' < \varepsilon_{\min}^{(1)}$ (Figure 4.46b). However, temperature gradients will be equal to zero along the curves L_l and L_y that cannot be considered as straight lines.

In this case, the inequality $\varepsilon_{\min}^{(1)} > \varepsilon_{\inf}$ is always true, and heating accuracy can be improved by increasing the number, N, of optimal control stages. The considered case arises, in particular, when heating aluminum or aluminum alloy slabs (applied frequency is 50 Hz). These processes usually correspond to the condition $\frac{\beta\xi}{\sqrt{2}} >> 1.5$.[1,16,17]

Because the final temperature distribution $\theta(l,y,\Delta^0)$ and all optimal process parameters Δ^0 can be represented as continuous functions of ξ, it is possible to use analogies with results of OCP solution for the case of surface heat-generating sources ($\xi = \infty$). For all values of ε within the range $[\varepsilon_{\min}^{(2)}, \varepsilon_{\min}^{(1)}]$, the spatial final temperature distribution and governing sets of equations are kept identical to the case of $\xi = \infty$. As a rule, the range $[\varepsilon_{\min}^{(2)}, \varepsilon_{\min}^{(1)}]$ covers all practical requirements with regard to the heating accuracy for aluminum workpieces. Expression (4.123) for final temperature distribution should be used in all systems of the type in Equation (4.116), Equation (4.117), and Equation (4.119) instead of Equation (4.111).

Figure 4.52 through Figure 4.55 represent some results of computation for the time-optimal heating of rectangular-shaped bodies based on the two-dimensional model (Equation 4.123). The set of initial data corresponds to the real-life

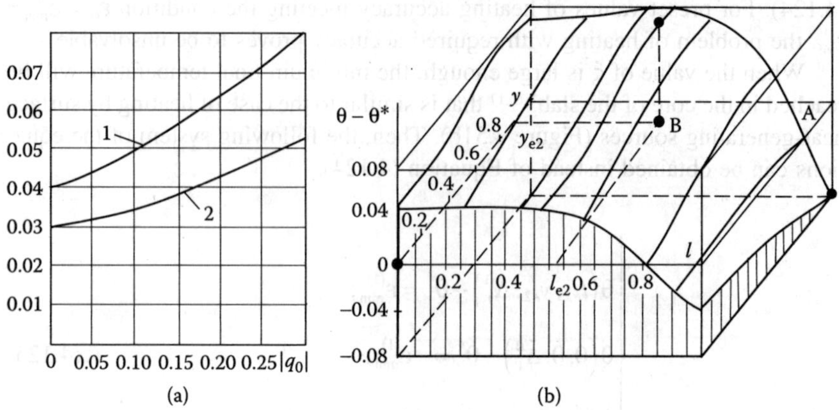

FIGURE 4.52 (a) Process parameters of time-optimal heating of rectangular slab and (b) final temperature distribution for $\varepsilon_0 = \varepsilon_{min}^{(1)} = \varepsilon_{inf}$. a: $\theta_0 = -0.2$; $\theta^* = 0$; $\xi = 11.3$; $\beta = 5.7$; $U_{max} = 1$; 1: $\varepsilon_{min}^{(1)}$; 2: Δ_1^0. b: $q_0 = -0.3$; $\varepsilon_0 = \varepsilon_{min}^{(1)} = 0.08$; $\Delta_1^0 = 0.053$.

process of induction heating utilizing line frequency (50 Hz).[1,16] Analytical approximations for heat sources $W_d(\xi, l, y)$ were in the calculations.[17]

Figure 4.52 shows the temperature profile at the end of heating and optimal process parameters as functions of heat losses under condition of $\varepsilon_{min}^{(1)} = \varepsilon_{inf}$ when heating titanium slabs. Figure 4.53 shows the results of computations of optimal process when heating accuracy belongs to the range $[\varepsilon_{min}^{(2)}, \varepsilon_{min}^{(1)}]$, under condition of $\varepsilon_{min}^{(1)} > \varepsilon_{inf}$ and large enough values of ξ, that is typical for heating aluminum alloy slabs. Presented combinations of points x_j^0 within the slab cross-section are identical to the case of $\xi = \infty$ (Figure 4.46b through Figure 4.46e, and Figure 4.49). Similarly to the case of infinite value of ξ, preset value $\varepsilon = \varepsilon_0$ decreases from $\varepsilon_{min}^{(1)}$ to $\varepsilon_{min}^{(2)}$; combinations of points x_j^0 are changed. Thus, the governing sets of equations remain unchanged as applied to the case of $\xi = \infty$.

Figure 4.54 shows optimal spatial temperature distributions within the slab cross-section for typical values $\varepsilon = \varepsilon_1^{(2)}$ and $\varepsilon = \varepsilon_2^{(2)} = \varepsilon_{min}^{(2)}$. As an example, let us consider optimization of induction heating of AMG alloy slabs prior to the hot forming operation. The size of the slab is 280 × 1540 × 2300 mm^3; applied frequency is 50 Hz; maximal surface density of heating power is about 82.5 kW/m^2. Figure 4.55 represents results of computations in comparison with experimental data.[1]

4.5.4.3 Exploration of Three-Dimensional Optimization Problems for Induction Heating

Suppose a slab is placed in the initially uniform magnetic field. If the slab length and width are much larger than its thickness, then the electromagnetic field within the slab can be viewed as an area consisting of three zones: central part, transverse edge effect area, and longitudinal end effect area (Figure 1.3). In the central part

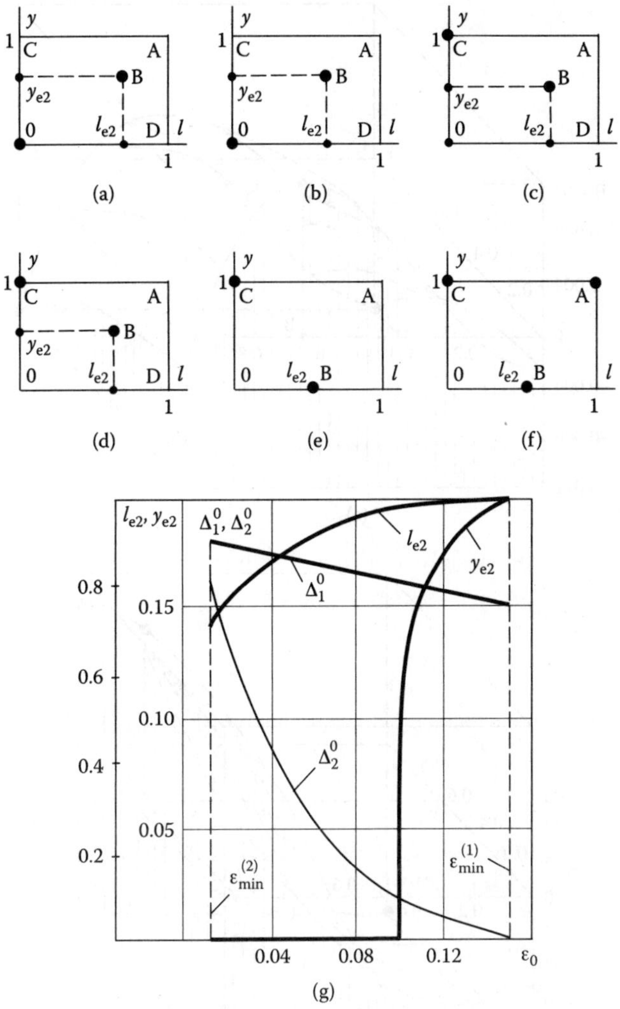

FIGURE 4.53 (a–f) Variants of location of minimum and maximum temperature points and (g) optimal process parameters for time-optimal heating of rectangular prism for $\varepsilon_{min}^{(1)} > \varepsilon_{inf}$ ($\theta_0 = -0.95$; $\theta^* = 0$; $U_{max} = 1$; $q_0 = -0.1$; $\xi = 52.3$; $\beta = 5.5$; $\varepsilon_0 \in [\varepsilon_{min}^{(2)}, \varepsilon_{min}^{(1)})$). a: $\varepsilon_0 = \varepsilon_{min}^{(1)} = 0.15$. b: $0.104 < \varepsilon_0 < 0.150$. c: $\varepsilon_0 = \varepsilon_1^{(2)} = 0.104$. d: $0.0984 < \varepsilon_0 < 0.1040$. e: $0.012 < \varepsilon_0 \leq 0.0984$. f: $\varepsilon_0 = \varepsilon_2^{(2)} = \varepsilon_{min}^{(2)} = 0.012$.

of the slab, the emf distribution corresponds to the field in the infinite plate. Basically, end and edge effects have a two-dimensional space distribution excluding only the zone of three-dimensional corners where the field is three dimensional representing a mixture of the end and edge effects.

The electromagnetic edge effect is typically negligible when heating cylindrical workpieces. However, this effect can play an essential role for many practical applications such as induction heating of slabs before rolling. In these

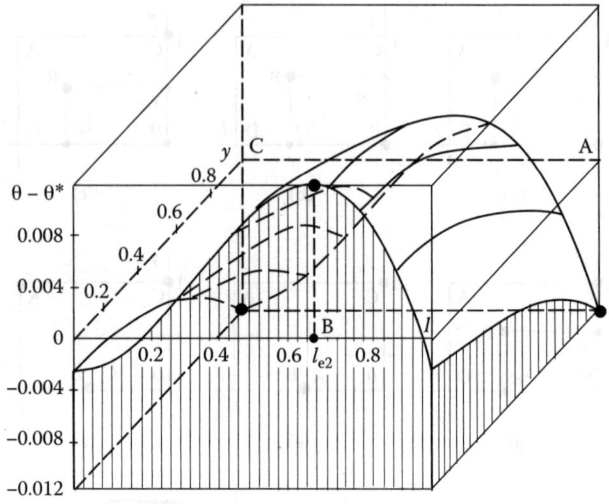

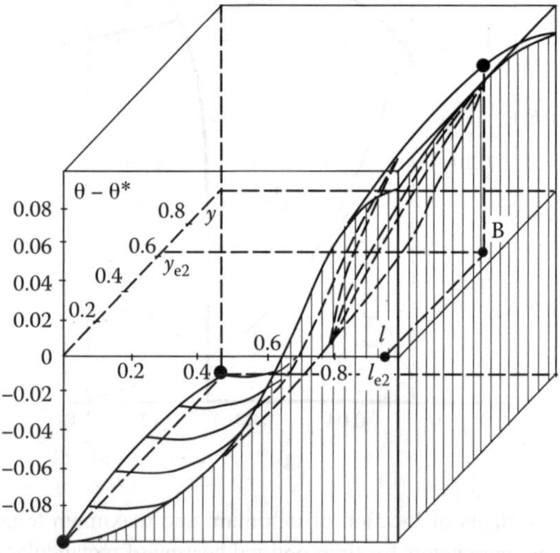

FIGURE 4.54 Spatial temperature distribution within slab cross-section at the end of time-optimal heating process ($\theta_0 = -0.95$; $\theta^* = 0$; $q_0 = -0.1$; $U_{max} = 1$; $\xi = 52.3$; $\beta = 5.5$. a: $\varepsilon_0 = \varepsilon_2^{(2)} = \varepsilon_{min}^{(2)} = 0.012$; $\Delta_1^0 = 0.182$; $\Delta_2^0 = 0.169$. b: $\varepsilon_0 = \varepsilon_1^{(2)} = 0.104 > \varepsilon_{min}^{(2)}$; $\Delta_1^0 = 0.159$; $\Delta_2^0 = 0.007$.

cases, three-dimensional mathematical models properly describe the technological process.[15]

Similarly to Section 4.1, it is possible to conclude that the optimal control method can be applied regardless of the complexity of the models. The general

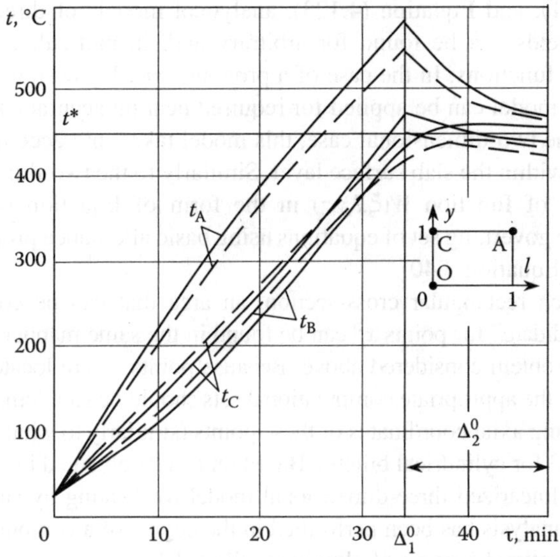

FIGURE 4.55 Time-optimal slab heating up to 450°C for $\varepsilon_{min}^{(2)} < \varepsilon_0 = 22.5°C < \varepsilon_{min}^{(1)}$ (dashed lines — computations for ($\theta_0 = -0.95$; $\theta^* = 0$; $q_0 = -0.1$; $\xi = 52.0$; $\beta = 5.5$; λ = 140 W/(m·°C); $a = 49 \cdot 10^6$ m²/s).

shape of the optimal control and computational technique remains unchanged when using three-dimensional models as well as in the case of transition from the one-dimensional to the two-dimensional problem.

However, it is important to be aware of some essential features of three-dimensional optimization problems. One of these features deals with difficulties that arise when modeling three-dimensional electromagnetic fields. Analysis becomes more complicated when we must consider temperature distribution in three-dimensional space.

To overcome difficulties of modeling, it is possible to use approximate descriptions of the temperature field that provide not absolutely accurate, but acceptable, engineering solutions. For example, similarly to Equation (4.72) through Equation (4.75), a linear three-dimensional heterogeneous equation of heat conduction can be applied. The function that describes internal heat source distribution within slab volume can be written as follows[1]:

$$W(\xi,l,y,z) = W_d(\xi,l,y)W^*(z). \tag{4.126}$$

Here, $W_d(\xi,l,y)$ is a function that describes the spatial distribution of internal heat source density in the two-dimensional problem (Equation 4.72 through Equation 4.75); function $W^*(z)$ takes into account spatial distribution of internal heat source density along the longitudinal coordinate z.

Analytical approximations of functions $W_d(\xi,l,y)$ and $W^*(z)$ can be used, for example, as shown in Kolomejtseva.[20] Similarly to Equation (4.77) through

Equation (4.91), and Equation (4.123), analytical models of three-dimensional temperature fields can be found for arbitrary and, in particular, stepwise heat power control functions. In the case of a pronounced eddy current "skin" effect, the simplified model can be applied for required heating accuracy $\varepsilon = \varepsilon_0 \geq \varepsilon_{\min}^{(2)}$. Similarly to the two-dimensional case, this model takes into account heat-generating sources within the slab surface layer. Similarly to the two-dimensional case, representation of function $W(\xi,l,y,z)$ in the form of Equation (4.126) allows developing the governing set of equations using basic alternance properties (Equation 3.39 and Equation 3.40).

Within each rectangular cross-section, an area that can be considered as a potential "candidate" for points x_j^0 can be found in the same manner as in the two-dimensional problem considered above. Because points x_j^0 are located in different cross-sections, the appropriate computational sets should be supplemented by equations determining axial coordinates of these points (similarly to what was discussed in Section 4.5.3 for cylindrical billets). Based on results obtained in Section 4.5.4.1 and using the linearized three-dimensional model for heating by surface sources, the described analysis has been performed in the course of a computational procedure for time-optimal heating of aluminum alloy slabs.[1]

4.6 OPTIMAL CONTROL FOR COMPLICATED MODELS OF THE INDUCTION HEATING PROCESS

4.6.1 Overview

Generally speaking, space–time temperature distribution within the heated workpiece is described by a highly complex mathematical model of interrelated Maxwell and Fourier equations for electromagnetic and temperature fields, respectively (Section 1.2.1). The inherently nonlinear nature of the induction heating process implies nonlinearity of the model that takes into consideration important features of induction heating processes, including end and edge effects, nonlinear material properties, nonlinear nature of boundary conditions, etc.

The solution of nonlinear interrelated electrothermal equations by rather cumbersome numerical methods represents a separate topic. In this case, the set of equations cannot be solved analytically but approximate solutions can be found with the accuracy that a properly designed and applied model provides. Development of modern computers leads to widespread use of effective numerical methods offering maximal advantages of computation of interrelated electromagnetic and heat transfer problems.[1,15]

This section will detail how the alternance method can be extended onto the wide range of optimization problems for complicated numerical models. It is important to clarify whether the shape of optimal control algorithms for the simplified IHP models can be transferred into these more complex cases. Generally speaking, the nonlinear nature of the induction heating process affects the choice of optimal control algorithm. To obtain optimal solution, it is necessary to apply an exact statement to each particular case of the optimal control problem.

Nevertheless, detailed analysis shows[1] that, in most cases, suitable practical solutions found for linear models remain valid for nonlinear numerical models.

It is possible to prove mathematically[1] that the time-optimal control consists of alternating stages of heating with maximum power $u \equiv U_{max}$ (heat ON) and subsequent soaking under $u \equiv 0$ (heat OFF) cycles regardless of nonlinearity, dimensionality, and other specifics of applied models. Consequently, optimal control $u^*(\varphi)$ can be written in the form of a stepwise function (Equation 2.27) (Figure 2.8). Additional constraints lead to a number of essential features and complications of optimal control in the same manner as for linear models (Section 4.1).

Similar conclusions can be drawn for the wider range of optimal control problems, using piecewise linear approximation of equations describing the induction system under control. As a rule, such approximation provides temperature field description by a set of heat conduction equations for each time interval. Accuracy of approximation is fair when the number of time steps is large enough.[1,7,21] Then, optimal controls on appropriate time intervals can be determined for corresponding linear models. The linkage of control functions on separate intervals into the general control algorithm leads to an optimal solution identical to optimal control for the linear single-equation model of the heating process.

Because the optimal control is kept unchanged, the shape of final spatial temperature distribution is kept identical as well. In other words, similar-by-shape temperature distributions can be considered as outputs obtained in response to similar-by-form optimal control inputs. This conclusion remains correct for heating process models of different complexity.

As a result, it is possible to conclude that, in the most practical cases, general qualitative features of optimal processes discussed earlier for the simplified IHP models remain valid for complex mathematical models. This means that the alternance method can be successfully applied for a wide variety of specific IHP optimization problems for complicated heating models.

Similarly to OCP solution for linear models, the shape of the optimal control algorithm is known, but the number, N, and the values $\Delta_1, \Delta_2, \ldots, \Delta_N$ remain unknown. The task of vector $\Delta^0 = (\Delta_1^0, \Delta_2^0, \ldots, \Delta^0_N)$ search is reduced to the problem of mathematical programming (Equation 3.38) that can be solved on the base of general correlations in Equation (3.39) and Equation (3.40). The sets of governing equations are similar to the linear case, but the solving procedure differs significantly from the cases described earlier because it is impossible to obtain an exact expression for the temperature field as a function of vector Δ. Here, approximate analytical or numerical methods are applicable to simulate the heating process. Both techniques will be used later on the base of proposed method for two typical time-optimal control problems.

4.6.2 Approximate Method for Computation of the Optimal Induction Heating Process for Ferromagnetic Billets

The heating processes for ferromagnetic billets are widely used in numerous advanced technologies; however, they are the most complex because of their nonlinearity. Nonlinear effects due to temperature dependency of magnetic permeability result in the transition between ferromagnetic and paramagnetic phases of heated material at the Curie point.[7,15,21] These phenomena lead to significant changes in intensity and spatial distribution of internal heat sources during the heating process. Additionally, temperature dependencies of electromagnetic and thermophysical characteristics affect the temperature field significantly.

In a number of typical cases, the process of ferromagnetic billet heating can be represented with acceptable-in-practice accuracy as a piecewise linear process.[1,7,21] The heating process can be divided into "cold," "intermediate," and "hot" stages. Within each stage, a temperature field can be represented by the transfer equation with corresponding boundary conditions. The values of thermophysical characteristics of the heated material, spatial distribution of internal heat sources, and constraints on control inputs will be noticeably different during these stages.[1]

During the cold stage, the heated materials exhibit magnetic properties. In the intermediate stage, at temperatures below the Curie point, the magnetic moments are partially aligned within the magnetic domain in ferromagnetic materials. As the temperature increases from below the Curie point, thermal fluctuations increasingly destroy this alignment, until the net magnetization becomes zero at and above the Curie point. On the hot stage above the Curie point, the material is purely paramagnetic.

It is possible to show[1,7] that the temperature distribution during the heating process can be described by linear equation with respect to relative units under boundary conditions of the second kind in the form of Equation (1.27) through Equation (1.35) for the one-dimensional case or Equation (4.72) through Equation (4.74) for the two-dimensional case. At transition from stage to stage, the utmost maximum values of heating power U_{max} and functions $W(\xi,l)$ (or $W(\xi,l,y)$) will be changed in the stepwise manner[1] that reflects stepwise changes in billet magnetic properties.

The induction heating process will be adequately represented by the described mathematical model under boundary conditions of the third kind (Equation 1.34 or Equation 4.75), if the heat loss level (the value of the Biot number) remains constant during the whole process.[1,7]

Similarly to the preceding case, the induction heating process for the nonmagnetic slabs can be described by a piecewise linear model when other nonlinear thermophysical properties should be taken into account. The resulting linearized representation of the heating process allows one to extend an optimal control method (developed for linear models) to nonlinear optimization problems.

As was shown earlier, the time-optimal heating power control consists of alternating stages of heating with maximum power $u \equiv U_{max}$ (heat ON) and subsequent soaking under $u \equiv 0$ (heat OFF) cycles. The number, $N \geq 1$, of stages is defined uniquely by given heating accuracy ε and increases with decreasing ε. Therefore, the shape of the optimal control algorithm is known, but the number of stages, N, and durations, $\Delta_1, \Delta_2, \ldots, \Delta_N$, of those stages remains unknown. The expression for final temperature distribution $\theta(l, \Delta)$ can be written in the explicit form as a function of spatial coordinate l and vector Δ similarly to Equation (2.28), Equation (4.15), and Equation (4.34). However, here the necessity arises to examine a positional relationship between borders of linearity intervals and optimal control intervals. As a rule, this additional problem can be solved based on physical features of optimal induction heating processes or using the successive approximations method.

Let us consider a typical problem of a time-optimal process for static induction heating of ferromagnetic billets to the temperature above the Curie point. The temperature should be constrained by a certain maximum level. Similarly to the case of the one-dimensional linear model in Equation (4.3), the typical algorithm of time-optimal heat power control can be obtained in the following form if the constraint on maximum temperature is taken into account:

$$u^*(\varphi) = \begin{cases} U_{\max 1}, 0 < \varphi < \varphi_{\max 1}; \\ U_{\max 2}, \varphi_{\max 1} < \varphi < \varphi_{\max 2}; \\ U_{\max 3}, \varphi_{\max 2} < \varphi < \varphi_{\theta}; \\ u^{\theta}(\varphi), \varphi_{\theta} < \varphi < \Delta_1^0; \\ \frac{U_{\max 3}}{2}\left[1+(-1)^{j+1}\right], \varphi_{j-1} < \varphi < \varphi_j, j = 2, 3, \ldots, N. \end{cases} \tag{4.127}$$

Because constraint on thermal stress is neglected, the maximum heating power can be applied until the moment φ_θ, when the temperature reaches its limitation. In contrast to the algorithm in Equation (4.3), the maximum admissible level of heating power U_{max} will be changed in a stepwise manner over transition from the cold stage ($0 < \varphi < \varphi_{max1}$) to the intermediate stage ($\varphi_{max1} < \varphi < \varphi_{max2}$) and then to the hot stage ($\varphi > \varphi_{max2}$). During each of these stages, U_{max} will be equal to one of preset constant values U_{max1}, U_{max2}, and U_{max3}, respectively.[7,21] This interval is over at a certain moment $\varphi = \varphi_\theta$ when temperature maximum becomes equal to admissible value θ_{adm}. The hot stage also includes time interval ($\varphi_\theta, \Delta_1^0$) of temperature holding at the maximum level and the next alternating stages of heating with maximum power U_{max3} and temperature soaking.

During the interval $(\varphi_\theta, \Delta_1^0)$, temperature maximum will be held at the maximum admissible level by applying control input $u^\theta(\varphi)$. The holding control $u^\theta(\varphi)$ can be described approximately by Expression (4.13) under boundary conditions of the third kind for constant value Bi in Equation (1.34). After substitution of $u^\theta(\varphi)$ in the form of Equation (4.13) into Equation (4.15) and Equation (4.16) it is possible to obtain the expression for final temperature distribution $\theta(l,\Delta)$. By substituting control input of the type in Equation (4.127) into solution of general heat transfer Equation (1.36) and calculating appropriate integrals, it is possible to obtain the final temperature distribution in the following form with respect to the simplified one-dimensional model[1]:

$$\theta(l,\Delta) = \theta_0 + \sum_{n=1}^{\infty} \frac{2K(\mu_n l)}{\left[\mu_n^2 + Bi^2 + (1-\Gamma)Bi\right]K^2(\mu_n)} \times$$
$$\times\Bigg\{\Bigg[\gamma_1 W_n^{(1)}\left(e^{\mu_n^2\varphi_{\max 1}} - 1\right) + \gamma_2 W_n^{(2)}\left(e^{\mu_n^2\varphi_{\max 2}} - e^{\mu_n^2\varphi_{\max 1}}\right) +$$
$$+W_n^{(3)}\left(e^{\mu_n^2\varphi_\theta} - e^{\mu_n^2\varphi_{\max 2}}\right) + W_n^{(3)} a_\theta \left(e^{\mu_n^2\Delta_1} - e^{\mu_n^2\varphi_\theta}\right) + \tag{4.128}$$
$$+\frac{W_n^{(3)}\mu_n^2 b_\theta}{\mu_n^2 - \beta_\theta}\left(e^{\mu_n^2\Delta_1 - \beta_\theta(\Delta_1-\varphi_\theta)} - e^{\mu_n^2\varphi_\theta}\right)\Bigg] e^{-\mu_n^2\sum_{m=1}^{N}\Delta_m}$$
$$+\sum_{j=3}^{N} W_n^{(3)}(-1)^{j+1}\left(1 - e^{-\mu_n^2\sum_{m=j}^{N}\Delta_m}\right)\Bigg\}.$$

Relative units can be calculated by Equation (1.28), Equation (1.29), Equation (1.33), and Equation (1.35) with regard to parameters of the hot stage under maximum specific power of internal heat sources $P_{\max} = P_{\max 3}$. The following denotations are used in Expression (4.128):

1. $W_n^{(\nu)}$, $\nu = 1, 2, 3$ represent coefficients $W_n(\xi)$ calculated by the formula in Equation (1.45) for known functions $W(\xi,l)$ during the cold ($\nu = 1$), intermediate ($\nu = 2$), and hot ($\nu = 3$) stages, respectively.[7,21]
2. Factors γ_1 and γ_2 can be defined by correlations:

$$\gamma_1 = \frac{\lambda_3}{\lambda_1} U_{\max 1}; \gamma_2 = \frac{\lambda_3}{\lambda_2} U_{\max 2}\,. \tag{4.129}$$

3. The values $U_{\max 1}$ and $U_{\max 2}$, can be computed from formulas:

TABLE 4.2
Computation of Time-Optimal Control for Induction Heating of Steel Cylindrical Billets

	Initial Data		
Heating Stage	**λ,W/(m·°C)**	**a·10⁴,m²/sec**	**P_{0max},кW/m²**
Cold	48.1	0.093	4463
Intermediate	34.3	0.066	3214
Hot	30.5	0.055	1965

$$U_{\max 1} = \frac{P_{\max 1}}{P_{\max 3}}; U_{\max 2} = \frac{P_{\max 2}}{P_{\max 3}}; U_{\max 3} = 1 , \tag{4.130}$$

where λ_v, $P_{\max v}$, $v = 1, 2, 3$ are given values of thermal conductivity and maximum specific power of internal heat sources during appropriate stages of the heating process. The value φ_θ can be found using the computational technique described in Section 4.1.2. Times $\varphi_{\max 1}$ and $\varphi_{\max 2}$ represent the borders of appropriate linearity intervals.[7,21]

If required heating accuracy is preset, the considered optimal control problem can be reduced to search for a vector of unknown parameters Δ^0. The problem can be solved using a technique that differs from the case of linear models (Section 4.1.2) only by the fact that Expression (4.128) should be used for $\theta(l,\Delta^0)$ in the equations sets.

As an example, let us consider a typical problem of time-optimal control of the induction heating process for steel cylindrical billets. Computations have been performed according to the previously described technique with respect to initial data presented in Table 4.2 under $X = 0.02$ m; $t_0 = 20°C$; $t^* = 1200°C$; $f = 50$ Hz; and $Bi = 0.1$. Expression for final temperature distribution $\theta(l,\Delta^0)$ in the form of Equation (4.128) was used in sets of equations. This example shows the optimal control mode when it is necessary to provide maximum heating accuracy $\varepsilon = \varepsilon^{(2)}_{\min}$. Figure 4.56 clearly shows that the difference between control algorithms in Equation (4.127) and Equation (4.3) lies in stepwise variation of maximum power during the first control interval until the maximum temperature reaches its utmost admissible value. In this case, the shape of final temperature distribution along the billet radius and its properties remain unchanged in comparison with the case of linear models.

4.6.3 Optimal Control for Numerical Models of Induction Heating Processes

Highly effective numerical methods are widely and successfully used in the computation of interrelated electromagnetic and heat transfer problems of different complexity. Modern electroheat numerical computation techniques allow

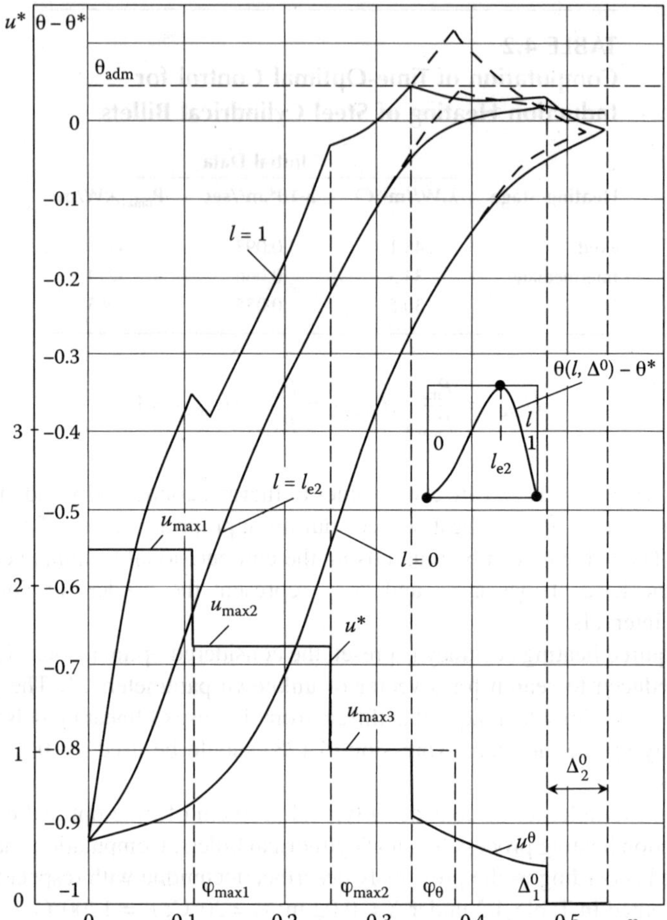

FIGURE 4.56 Control input and time–temperature history during time-optimal control of piecewise linear model of steel billet heating: $\Gamma = 1$; $\xi = 2.82$; $\theta_0 = -0.92$; $\theta^* = 0$; $\theta_{adm} = 0.039$; $Bi = 0.10$; $\gamma_1 = 1.44$; $\gamma_2 = 1.46$; $u^\theta(\varphi) = 0.094 + 0.452\exp[-9.3(\varphi - \varphi_\theta)]$; $\varepsilon_0 = \varepsilon_{min}^{(2)}$; solid lines: with constraint on θ_{max}; dashed lines: without constraints.

obtaining acceptable engineering solutions that will satisfy requirements of modern technology from a practical standpoint. Many numerical modeling methods exist or are under development. The modern electrothermal numerical models based on complex numerical solution of electromagnetic and heat problems become the most commonly used universal tools for investigation and design of a variety of induction heating installations.[15,21,22]

The alternance method can be successfully applied to a wide variety of optimization problems for numerical IHP models; however, in comparison with previously discussed problems, the number of essential distinctions appears. On the one hand, numerical modeling does not allow one to obtain analytical expressions for temperature fields; therefore, the final temperature distribution $t(x,\Delta)$ cannot be represented in the explicit form as a function of spatial coordinates and vector of unknown parameters.

On the other hand, according to the alternance method, solution of equations sets can be obtained by an iterative technique (see Section 3.4) that requires repeated calls for the numerical model. This can result in an increasing amount of required computations that can prove to be inadmissible. It is more effective to find an extremum of the auxiliary function $I_P(\Delta)$ within the space of unknown parameters Δ_i, $i = \overline{1,N}$.[1,15] The auxiliary problem is equivalent to the initial one if the auxiliary function $I_P(\Delta)$ is chosen in such a way that its global minimum is equal to zero for the optimal values of sought parameters Δ_i, $i = \overline{1,N}$.

In this case a function $I_P(\Delta)$ can be considered as the sum of linear combinations of final temperatures $t(x_j^0,\Delta)$. Variables x_j^0 represent coordinates of points where maximum admissible deviations of final temperature from the desired one are reached. These combinations can be obtained as a result of the algebraic addition of the left-hand sides of the equations in the equations set. All of these combinations tend to zero under optimal values of the vector $\Delta = \Delta^0$. For particular application, the appropriate set of equations should be chosen and solved. The proper choice of the set of equations can be performed using additional information with respect to temperature distribution during the optimal processes subject to simplified linear models of the controlled heating process.

To calculate temperatures $t(x_j^0,\Delta)$, it is necessary to define the expression $t(x_j^0,\tau)$ for temperature at the point x_j^0 during process time τ under control input uniquely specified by the vector Δ. If the function $t(x_j^0,\tau)$ is found, then, for any vector Δ, it is possible to obtain the expression $t(x_j^0,\Delta) = t(x_j^0,\tau^0)$, where the value of τ^0 is equal to total heating process time. The problem of definition $t(x_j^0,\tau)$ can be solved using numerical models. The orthogonal mesh discretizes the area of modeling into the certain array of nodes. The numerical models allow one to calculate the maximum t_{max} or minimum t_{min} temperatures at nodes of the spatial mesh in any required time point. The computational procedure often becomes simpler if additional information with regard to location of points x_j^0 is employed.

The described computational procedure can be reduced to the definition of dependencies $t_{max}(\tau)$ and $t_{min}(\tau)$, and $t_{max}(\Delta)$ and $t_{min}(\Delta)$, respectively, instead of $t(x_j^0,\tau)$ and $t(x_j^0,\Delta)$. Search of function $I_P(\Delta)$ minimum is performed using known numerical methods. Reduction of required computations is provided if simple analytical approximations of $t_{max}(\tau)$ and $t_{min}(\tau)$ are used to diminish the number of calls for numerical models.[15] Approximations of functions $t_{max}(\tau)$ and $t_{min}(\tau)$ can be written in the polynomial form according to the method of least squares. Then the computational procedure represents the sequence of optimization problems that could be reduced to minimization of function $I_p(\Delta)$.

This procedure typically exhibits fast convergence. In each iteration step, the analytical approximations are refined with respect to expressions for temperature calculated by the numerical model under Δ^0. Here, the value of Δ^0 is obtained as a result of the previous step. The described iteration procedure is repeated until values of sought parameters Δ_i^0, $i = \overline{1,N}$ will coincide satisfactorily with the values of the previous step. The initial values of Δ^0 can be chosen properly from simple analysis of possible ranges for variations of values Δ_i^0.[15]

The described approach has been performed for computation of time-optimal heating of aluminum alloy cylindrical billets before pressing (applied frequency is 50 Hz).[1,15] The two-dimensional numerical model has been used to provide combined solution of electromagnetic and heat transfer problems. The electromagnetic problem was solved by the method for integral equations, and the heat transfer problem by the finite difference method. The spatial temperature distribution $t(x^{(1)},x^{(2)},\tau)$ within the cylindrical billet evolving over time τ can be described by the following nonlinear two-dimensional heterogeneous equation of heat transfer in absolute units[1]:

$$c(t)\gamma(t)\frac{\partial t}{\partial \tau} = \frac{1}{x^{(1)}}\frac{\partial}{\partial x^{(1)}}\left(\lambda(t)x^{(1)}\frac{\partial t}{\partial x^{(1)}}\right) + \frac{\partial}{\partial x^{(2)}}\left(\lambda(t)\frac{\partial t}{\partial x^{(2)}}\right) + \\ + W\left(x^{(1)},x^{(2)},t,U\right); \quad 0 < x^{(1)} < X^{(1)}; 0 < x^{(2)} < X^{(2)}; 0 < \tau \le \tau^0 \tag{4.131}$$

with boundary conditions:

$$\left(\frac{\partial t}{\partial x^{(1)}}\right)_{x^{(1)}=0} = 0; \lambda(t)\left(\frac{\partial t}{\partial x^{(1)}}\right)_{x^{(1)}=X^{(1)}} = -\alpha(t)\left(t\left(X^{(1)},x^{(2)},\tau\right)-t_a\right); \\ \lambda(t)\left(\frac{\partial t}{\partial x^{(1)}}\right)_{x^{(2)}=0} = \alpha(t)\left(t\left(x^{(1)},0,\tau\right)-t_a\right); \lambda(t)\left(\frac{\partial t}{\partial x^{(2)}}\right)_{x^{(2)}=X^{(2)}} = \\ = -\alpha(t)\left(t\left(x^{(1)},X^{(2)},\tau\right)-t_a\right); \tag{4.132}$$

and initial conditions:

$$t\left(x^{(1)},x^{(2)},0\right) = t_0\left(x^{(1)},x^{(2)}\right); x^{(1)} \in \left[0,X^{(1)}\right]; x^{(2)} \in \left[0,X^{(2)}\right].$$

Here, $x^{(1)}$ and $x^{(2)}$ are radial and axial coordinates; $X^{(1)}$, $X^{(2)}$ are radius and length of the billet, respectively.

Voltage applied to induction coil $U(\tau)$ can be considered as a required control function constrained by Equation (2.7). The function $W(x^{(1)},x^{(2)},t,U)$ describes internal heat source distribution depending on spatial coordinates, control input, and temperature. This function can be defined by using the numerical model in the course of computation of the electric field. The inherently nonlinear nature of the induction heating process implies nonlinearity of the model in Equation (4.131) and Equation (4.132) that takes into account temperature dependencies of all electromagnetic and thermal properties of heated material: specific heat

$c(t)$, density $\gamma(t)$, thermal conductivity $\lambda(t)$, and convection surface heat transfer coefficient $\alpha(t)$. The function $W(x^{(1)},x^{(2)},t,U)$ depends on temperature as well.

The previously described method can be applied to the optimal control problem for numerical models of induction heating processes. Algorithms for time-optimal control, all established properties of final temperature distribution $t(x,\Delta^0)$, corresponding sets of equations, and computational techniques remain similar to the case of linear models of the heating process (Section 4.5.3). Only expressions for temperature distribution $t(x,\Delta^0)$ will be changed. Similarly to the one-dimensional case (Section 4.1), these conclusions are valid for the optimal control problem with technological constraints under the condition in Equation (4.17). In particular, for the case of $\varepsilon = \varepsilon_{\min}^{(2)}$, one of the sets in Equation (4.98), Equation (4.99) under $l_3^0 = 1$, or Equation (4.102) should be solved.

When solving the set in Equation (4.98) and Equation (4.99) under $l_3^0 = 1$, the cost function $I_P(\Delta)$ can be written as:

$$I_{P1}(\Delta) = (t_{\max}(\Delta_1,\Delta_2) + t_{\min 1}(\Delta_1,\Delta_2) - 2t^*)^2 + (t_{\max}(\Delta_1,\Delta_2) + t_{\min 2}(\Delta_1,\Delta_2) - 2t^*)^2. \tag{4.133}$$

For the set in Equation (4.102), one can obtain

$$I_{P2}(\Delta) = (t_{\max}(\Delta_1,\Delta_2) + t_{\min 3}(\Delta_1,\Delta_2) - 2t^*)^2 + (t_{\max}(\Delta_1,\Delta_2) + t'_{\min 2}(\Delta_1,\Delta_2) - 2t^*)^2. \tag{4.134}$$

Here:

$t_{\max}(\Delta_1,\Delta_2)$ is the maximum final temperature within the billet (which is written instead of the temperature $\theta(l_{e2},y_{e2},\Delta_1,\Delta_2) = \theta(x_2{}^0,\Delta_1,\Delta_2)$ in Equation (4.98) and Equation (4.102) and in Figure 4.39 and Figure 4.41a).

$t_{\min 1}(\Delta_1,\Delta_2)$ and $t_{\min 2}(\Delta_1,\Delta_2)$ are minimum final temperatures on the axis and at the lateral surface of the billet (compare with $\theta(0,y_1{}^0,\Delta_1,\Delta_2) = \theta(x_1{}^0,\Delta_1,\Delta_2)$ in Equation 4.98 and $\theta(1,y_3{}^0,\Delta_1,\Delta_2) = \theta(x_3{}^0,\Delta_1,\Delta_2)$ in Equation 4.99; Figure 4.39a through Figure 4.39c).

In contrast to $t_{\min 2}$, the temperature $t'_{\min 2}$ is found at the lateral surface in internal cross-sections of the billet, which is similar to $\theta(1,y_3{}^0,\Delta_1,\Delta_2) = \theta(x_3{}^0,\Delta_1,\Delta_2)$ in Equation (4.102) for the one-dimensional case (Figure 4.41a).

$t_{\min 3}$ is the minimum temperature at the lateral surface in the butt-end cross-section that is similar to the temperature $\theta(1,y_1{}^0,\Delta_1,\Delta_2) = \theta(x_1{}^0,\Delta_1,\Delta_2)$ in Equation (4.102) (Figure 4.41a).

t^* is the required final temperature.

As one can see from Equation (4.133) and Equation (4.134), the first summand of cost function $I_{P1}(\Delta)$ represents the square of the sum of the left-hand sides of the first two equations in the set in Equation (4.98); the second summand is the square of the sum of the left-hand sides of the second equation in the set in Equation (4.98) and the second equation in Equation (4.99) (for $l_3^0 = 1$). In a similar manner, the function $I_{P2}(\Delta)$ is formed by summation of the first and the third equations of the set in Equation (4.102) and then its second and third equations. Thus, functions $I_{P1}(\Delta)$ and $I_{P2}(\Delta)$ include final temperatures in all points x_j^0.

Because all equations of computational sets are satisfied under $\Delta = \Delta^0$, the equality $I_{P1}(\Delta^0) = I_{P2}(\Delta^0) = 0$ is true, and vector Δ^0 represents global minimum for both functions $I_{P1}(\Delta)$ and $I_{P2}(\Delta)$, those values are non-negative. Similarly to the previously described case, the expression for I_P can be written also for $\varepsilon_{min}^{(2)} < \varepsilon < \varepsilon_{min}^{(1)}$. In accordance with the set of equations in Equation (4.98) one can obtain:

$$I_{P3}(\Delta) = (t_{max}(\Delta_1,\Delta_2) - t^* - \varepsilon)^2 + (t_{min1}(\Delta_1,\Delta_2) - t^* + \varepsilon)^2. \tag{4.135}$$

In the case of Equation (4.103), the auxiliary function can be written:

$$I_{P4}(\Delta) = (t_{max}(\Delta_1,\Delta_2) - t^* - \varepsilon)^2 + (t'_{min2}(\Delta_1,\Delta_2) - t^* + \varepsilon)^2. \tag{4.136}$$

Here, already each summand in Equation (4.135) and Equation (4.136) could be written on the base of one equation of the appropriate set. As was mentioned earlier, these cost functions can be applied to optimal control problems with additional technological constraints if the condition in Equation (4.17) is met.

With respect to the transportation problem for $\varepsilon = \tilde{\varepsilon}_{min}^{(2)} = \varepsilon_{inf}$ under $\alpha_T = 1$ (Section 4.2), the equations set of the type in Equation (4.106) corresponds to the following cost function:

$$I_{P5}(\Delta,\Delta_T^0) = (t_{max}(\Delta_1,\Delta_T^0) + t_{min2}(\Delta_1,\Delta_T^0) - 2t^*)^2. \tag{4.137}$$

The polynomial approximations of the fourth order for time-dependent variables t_{max}, t_{min1}, t_{min2}, t'_{min2}, and t_{min3} have been used on each control interval when searching for extremum of function $I_P(\Delta_1,\Delta_2)$ by gradient methods. These approximations provide virtually complete convergence for the computational procedure after three to four iterations. The values t_{max} and t_{min} have been determined by scanning the array of temperatures at the spatial nodes of the numerical model mesh. To satisfy the constraint on maximum temperature, the inductor voltage has been chosen in such a way that maximum temperature would be very close to the utmost admissible value t_{adm}. This temperature has been calculated using the numerical model in the fixed times within the interval of temperature holding at the admissible value $t_{max} \equiv t_{adm}$.

As an example, let us consider the time-optimal process of induction heating of D1 alloy cylindrical billets before hot forming. Diameter of billet is 0.48 m, length of billet is 1 m, initial temperature is 20°C, and required final temperature

TABLE 4.3
Initial Data for Numerical Simulation

Inductor	
Length, m	1.14
Radius, m	0.282
Number of turns	50
Active resistance, Om	0.015
Maximum voltage, V	380
Billet	
Length , m	1.0
Radius, m	0.24
Initial temperature, °C	20
Required temperature, °C	460
Ambient temperature, °C	20
Maximum admissible temperature, °C	500
Time of transportation, s	180
Heat conductivity, W/(m·°C)	$115 + 0, 152t + 6{\cdot}10^{-2}\,t^2$
Thermal conductivity, m²/s	$0.46{\cdot}10^{-4} + 6{\cdot}10^{-7}t + 6{\cdot}10^{-10}\,t^2$
Electrical resistivity, Om/m	$0.51{\cdot}10^{-3} + 0.16{\cdot}10^{-4}t$
Heat transfer coefficient inside inductor W/(m²·°C)	20
Heat transfer coefficient during transportation, W/(m²·°C)	40
Coordinates of Spatial Mesh Nodes	
$x^{(1)}$, m:	0.000; 0.00534; 0.01708; 0.0300; 0.0660; 0.1020; 0.1380; 0.1740; 0.2100; 0.2179; 0.2245; 0.2300; 0.2346; 0.2384; 0.2400
$x^{(2)}$, m:	0.0000; 0.0178; 0.0571; 0.1003; 0.1478; 0.2000; 0.2667; 0.3333; 0.4000; 0.4667; 0.8522; 0.8997; 0.9429; 0.9822; 1.000

is 460°C.[1,15,23] Computations were performed for different values of symmetrical coil overhang that can be calculated by the expression

$$h = \frac{1}{2}\,(X_I - X^{(2)}),$$

where X_I is a coil length.

The two-dimensional model has been used for numerical simulation; initial data are presented in Table 4.3. Figure 4.57 and Figure 4.58 show some results of computation.

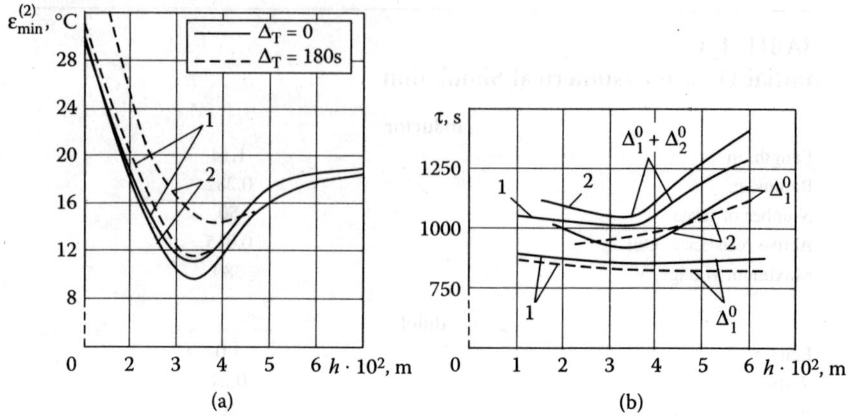

FIGURE 4.57 Time-optimal control parameters as function of coil overhang. a: $\varepsilon^{(2)}_{min}$. b: Durations of optimal control stages: 1: without constraint on maximum temperature; 2: with constraint $\theta_{max} \le 500°C$.

The optimum value of coil overhang h^* corresponds to maximum heating accuracy $\varepsilon^{(2)}_{min\,min}$ that can be found as an extremum point of the following cost function obtained with regard to the set in Equation (4.105):

$$
\begin{aligned}
I_{p6}\left(\Delta_1,\Delta_2,h^*\right) &= \left(t_{max}\left(\Delta_1,\Delta_2,h^*\right)+t_{min1}\left(\Delta_1,\Delta_2,h^*\right)-2t^*\right)^2+\\
&+\left(t_{max}\left(\Delta_1,\Delta_2,h^*\right)+t'_{min2}\left(\Delta_1,\Delta_2,h^*\right)-2t^*\right)^2+\\
&+\left(t_{max}\left(\Delta_1,\Delta_2,h^*\right)+t_{min3}\left(\Delta_1,\Delta_2,h^*\right)-2t^*\right)^2.
\end{aligned}
\tag{4.138}
$$

Minimization of this cost function leads to the optimum values of three variables: h^*, Δ_1, and Δ_2. Here, dependencies of t_{max}, t_{min1}, t_{min2}, t'_{min2}, and t_{min3} on h should be taken into account.

The numerical search procedure consists of two steps when applied to minimization of I_{P6}. During the first step, the minimum of the objective function should be defined within two-dimensional space of parameters Δ_1, and Δ_2, assuming $h = const$. During the second stage, minimization should be performed with respect to variable h under obtained values $\Delta_1{}^0(h)$ and $\Delta_2{}^0(h)$.

Therefore, the following expression can be written:

$$
\min_{h,\Delta_1,\Delta_2} I_{P6}\left(\Delta_1,\Delta_2,h\right)=\min_h\left(\min_{\Delta_1,\Delta_2} I_P\left(\Delta_1,\Delta_2,h\right)\right)=\min_h I_P\left(\Delta_1^0(h),\Delta_2^0(h)\right).
$$

Figure 4.57 shows that the value h^* corresponds to the minimum process time. This means that the following equality holds true:

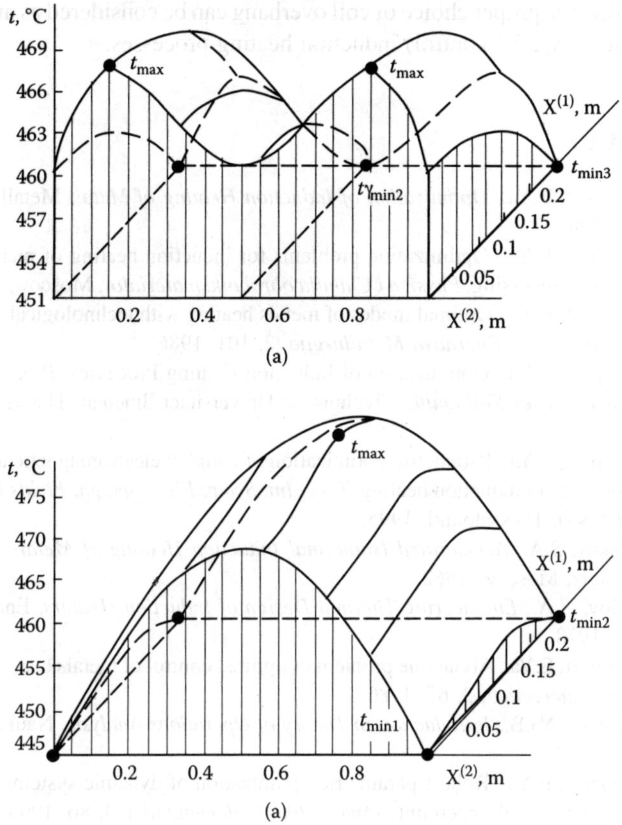

FIGURE 4.58 Final temperature distribution for $\varepsilon_0 = \varepsilon_{min}^{(2)}$. a: $h = 5 \cdot 10^{-2}$ m. b: $h = 1 \cdot 10^{-2}$ m.

$$\tau_{min}^{0}\left(h^{*}\right) = \min_{h} \tau_{min}^{0}\left(h\right).$$

Therefore, the coil overhang that is equal to the value h^* should be recommended as the optimal decision. Computational results confirm that, for $\varepsilon = \varepsilon_{min}^{(2)} = \varepsilon^*$, the function I_{P1} reaches its minimum value when condition $h < h^*$ is met. In the case of $\varepsilon = \varepsilon_{min}^{(2)} = \varepsilon^{**}$, the function I_{P2} reaches its minimum value if the value of the coil overhang is greater than h^*. To define the value of h^* in the "transportation" problem, the following objective function can be obtained using equations from the set in Equation (4.107):

$$I_{p7}\left(\Delta_1, \Delta_T^0, h^*\right) = \left(t_{max}\left(\Delta_1, \Delta_T^0, h^*\right) + t'_{min2}\left(\Delta_1, \Delta_T^0, h^*\right) - 2t^*\right)^2 + \\ + \left(t_{max}\left(\Delta_1, \Delta_T^0, h^*\right) + t'_{min3}\left(\Delta_1, \Delta_T^0, h^*\right) - 2t^*\right)^2. \quad (4.139)$$

Therefore, the proper choice of coil overhang can be considered as an effective way to control (spatial control) induction heating processes.

REFERENCES

1. Rapoport, E.Ya., *Optimization of Induction Heating of Metal,* Metallurgy, Moscow, 1993.
2. Rapoport, E.Ya., Optimization problems for induction heating of metals prior to pressure processing, *Physica i Chimija obrabotki materialov,* Moscow, 3, 54, 1984.
3. Rapoport, E.Ya., Optimal modes of metals heating with technological constraints, *Izvestija Vuzov, Chernaya Metallurgija*, 2, 101, 1986.
4. Rapoport, E.Ya., Optimization of Induction Heating Processes, *Proc. 40 Int.Wissenschaftliches Kolloquium,* Technische Universitaet Ilmenau, Thueringen, 1995, 48.
5. Rapoport, E.Ya., Parametric optimization of coupled electromagnetic and temperature fields in induction heating, *Proc. Int. Symp. Electromagn. Fields Electr. Eng. (ISEF'95)*, Thessaloniki, 1995, 319.
6. Jaitskov, S.A., *Accelerated Isothermal Induction Heating of Metals in Forging*, Mashgiz, Moscow, 1962.
7. Pavlov, N.A., *Engineering Thermal Design of Induction Heaters,* Energia, Moscow, 1978.
8. Rapoport, E.Ya., About one problem of optimal control of metals heating, *Izvestija vuzov, Energetica*, 3, 67, 1980.
9. Germejer, Yu.B., *Introduction in Theory of Operations Analysis*, Nauka, Moscow, 1971.
10. Rapoport, E.Ya., Robust parametric optimization of dynamic systems under condition of limited uncertainty, *Avtomatika i telemehanika*, 3, 86, 1995.
11. Butkovskij, A.G., Malyj, S.A., and Andreev, Yu.N., *Optimal Control of Metals Heating*, Metallurgy, Moscow, 1972.
12. Butkovskij, A.G., Malyj, S.A., and Andreev, Yu.N., *Control of Metals Heating*, Metallurgy, Moscow, 1981.
13. Bryson, A. and Yu-Chi Ho, J.R., *Applied Optimal Control: Optimization, Estimation and Control,* Waltham, MA, Toronto, London, 1969.
14. Rapoport, E.Ya., Method of optimal process computation for heat treatment of materials, *Physica i Chimija obrabotki materialov*, 5, 5, 1987.
15. Nemkov, V.S. and Demidovich, V.B., *Theory and Computation of Induction Heating Installations,* Energoatomizdat, St. Petersburg, 1988.
16. Zimin, L.S., Features of heating of rectangular shape bodies, *Primenenie tokov vysokoj chastoty v Electrotermii,* Mashinostroenie, St. Petersburg, 25, 1973.
17. Rapoport, E.Ya. and Zimin, L.S., Optimal control of slabs induction heating before rolling, *Physica i chimija obrabotki materialov*, 3, 21, 1986.
18. Rapoport, E.Ya., Optimal control in two-dimensional problems of heat transfer, *Izvestija AN SSSR. Energetica i transport*, 6, 102, 1984.
19. Rapoport, E.Ya., Optimization of two-dimensional processes of nonstationary heat transfer, *Izvestija AN SSSR. Energetica i transport*, 1, 86, 1985.

20. Kolomejtseva, M.B., Simulation of heat sources in the course of solving optimization problems in electrothermy, *Proc. Optimization Modes Complex Dynamic Objects*, Moscow Energetics Institute, 1980, 47.
21. Sluhotskij, A.E., Nemkov, V.S., Pavlov, N.A., and Bamunauer, A.V., *Induction Heating Installations,* Energoatomizdat, St. Petersburg, 1981.
22. Rudnev, V.I. et al., *Handbook of Induction Heating*, Marcel Dekker, New York, 2003.
23. Rapoport, E.Ya., Pleshivtseva, Yu.E., and Livshits, M.Yu., Alternance method in problems of induction heating processes: basic principles and experience of applications, *Proc. Int. Symp. Electromagn. Fields Electr. Eng.*, Pavia, 1999, 141.

5 Optimal Control of Progressive and Continuous Induction Heating Processes

The majority of induction billet heaters utilize a heating mode where billets are moved through a heater. This chapter will provide a study of the wide range of optimal control problems for progressive and continuous induction heating modes (Section 1.3). The *progressive heating* mode occurs when two or more heated workpieces (e.g., billets) are moved (via pusher, indexing mechanism, walking beam, etc.) through a single coil or inline multicoil induction heater. Therefore, components are sequentially heated (in a progressive manner) at certain predetermined heating stages inside the heater. With the *continuous heating* mode, the workpiece is moved in a continuous motion through one or more inline induction heating coils. This heating mode is commonly used when it is necessary to heat long components such as bars, slabs, strips, tubes, wires, blooms, and rods.

Progressive and continuous heaters are considered at steady-state operation conditions. This chapter discusses the optimal control problems for continuous and progressive heating modes based on a suggested parametric optimization method, taking into consideration the most typical real-life operating requirements.

The control inputs applied to continuous and progressive heating processes are quite different compared to those that were used in static heating. In continuous and progressive heating applications (see Section 2.4), control inputs include time-invariant, but space-dependent control functions (spatial or distributed control inputs). The spatial control function can be considered as a time-dependent optimal control scanning along the heater length or as a set of constant parameters of an induction heating system. Obviously, space-dependent control inputs represent design solutions employed at the stage of IHI design (see Section 2.4 and Section 2.8).

For progressive heating, these design solutions could also be complemented by cyclic periods of heating power variation in time, which allows performing optimal process control algorithms during each process stage. This includes between successive unloading and loading of billets and billet transportation between different inline induction coils.

Basic methods of optimal control problem solution with respect to different mathematical models will be presented as well. It will be shown that optimization results are in good qualitative agreement with results obtained earlier for static heating.

5.1 OPTIMIZATION OF CONTINUOUS HEATERS AT STEADY-STATE OPERATING CONDITIONS

5.1.1 OVERVIEW OF TYPICAL OPTIMIZATION PROBLEMS AND METHODS FOR THEIR SOLUTION

Let us consider continuous induction heating installation where billets are moved end to end through a heater at constant speed V (Section 1.3). The fixed value of V corresponds to required throughput of the technological complex "induction heater–hot working equipment."

As shown earlier, the problem of providing the heating accuracy ε at the exit of inductor having the minimum length under given constraints can be considered instead of a general time-optimal control problem for static heating (see Section 2.8). The general time-optimal problem is reduced to minimizing of the heater length when speed V of workpiece movement is prescribed. As was discussed earlier with respect to the basic mathematical model, the spatial distribution of power $u^*(y/V)$ along the heater length can be considered as optimal control input.

The time of static heating corresponds to the spatial coordinate along the inductor in the case of continuous heating. Similarly to time-optimal control (Equation 2.27), optimal heat power distribution can be obtained as a stepwise function in the form of Equation (2.32). Bursts of power represent the maximum power of inline coils that may have different length and windings and/or can be individually fed from different power supplies. Therefore, an optimal heat power distribution can be implemented by using alternating inline coils with "heat ON" and "heat OFF" that represent active and passive intervals of the lengths Δ_i^*, $i = \overline{1, N}$.

The shape of optimal control algorithm, u^*, is known, but the number, N, of alternating inline coils and their lengths remain unknown. The problem of heater length optimization can be reduced to the problem of mathematical programming in Equation (2.33).

After substitution of a variable (Equation 2.6), the optimal control function $u^*(\varphi)$ in the time-optimal problem (Equation 2.34) coincides with function $u^*(y/V)$ in the problem in Equation (2.33) that minimizes heater length. The basic one-dimensional mathematical model (Equation 1.27 through Equation 1.35) can be applied in this case (see Section 2.8). Indeed, the optimal control $u^*(y/V)$ can be properly considered as a time-dependent power control $u^*(\varphi)$ scanned along the heater length.

Therefore, after substitution (Equation 2.6), the power-control algorithm (Equation 2.32) for the problem in Equation (2.33) can be determined from the solution of the problem in Equation (2.34). This solution was described in detail in Chapter 3 and Chapter 4. In this case, the maximum heating accuracy problem (Equation 2.37) for continuous heater of the given length has the same solution as similar problems (Equation 2.35 and Equation 2.36) for static heating mode.

The problems in Equation (2.35), Equation (2.36), and Equation (2.37) can be reduced to the time-optimal problems, if the heater length does not exceed minimum length required for heating with utmost accuracy, ε_{inf} (Section 2.9.1). Only in rare cases, when the last condition is not satisfied, the problem for providing maximum accuracy have an infinite solution set.

Therefore, if spatial heat source distribution is chosen as a control function applied to the mathematical model of a heating process (Equation 1.27 through Equation 1.35), then the solution of the minimum-length problem for continuous induction heating with required accuracy is known. This conclusion holds true for the problem of obtaining the maximum heating accuracy with respect to a continuous heater of the given length.

The basic model has been obtained under a number of simplified assumptions. In particular, this model describes a temperature field under assumption that an effect of heat transfer in the axial direction by heat conduction is negligibly small (Section 2.2). To increase simulation accuracy further, two-dimensional heat transfer equations can be used.

In contrast to two-dimensional models of static heating (Section 4.5.1), continuous heaters at steady-state operating conditions can be described by stationary time-invariant equations for temperature distribution along the length and the radius of the billet.

In many cases, accurate simulation requires the use of more complicated mathematical models. In particular, the inherently nonlinear nature of the induction heating process of ferromagnetic billets implies nonlinearity of the model. This model takes into account temperature dependencies of all electromagnetic and thermal properties of the heated material.

It can be shown[1] that the previously described optimal control method can be extended to more complicated models of induction heating processes. Similarly to the time-optimal problem for static heating, optimal spatial distribution of heat power can be written in the form of Equation (2.32). All basic properties (Equation 3.39 and Equation 3.40) of final temperature distribution within the billet at the inductor exit remain unchanged regardless of complicating factors.

Similarly to the case of the one-dimensional model (Section 4.1), these conclusions are valid for optimal control problems taking into consideration technological constraints. This means that the form of optimal control $u^*(y/V)$ becomes more complicated than the control input (Equation 2.32), (similarly to the general time-optimal problem).

As experience of computation shows,[1] the shape of the final (i.e., at the inductor exit) temperature profile along the radius and appropriate sets of equations, obtained earlier for the time-optimal problem, remain the same. Only expressions that describe temperature distribution along the radius/thickness of the heated workpiece are varied. These expressions should be written in an explicit form as functions of spatial coordinates and optimal control parameters. They can be obtained after solving heat transfer equations using analytical or numerical methods.

However, mathematical models for static and continuous heating processes are not the same. It is still possible to accept as a physically based approach that the optimal algorithm of heat power control for continuous heaters can be considered as a spatial scanning of time-dependent optimal power control for static heaters. The dependences of these algorithms on axial coordinate and time have identical forms (Equation 2.32 and Equation 2.27), but their parameters can be noticeably different.

The voltages U_1, U_2, …, U_N for N coils of an induction heater can be considered as spatial control inputs for multicoil inductor design (Section 2.4 and Section 2.9); therefore, they can be treated as unknown parameters Δ_i, $i = \overline{1, N}$. The number of sections, N, and their lengths are defined by the given IHI design. In this case an optimal problem can be formulated as the problem (Equation 2.38) of maximum heating accuracy. Refined models should include different factors including gaps between coils.

Experience in solving similar problems[1–5] confirms that optimal final temperature distribution within a billet cross-section and computational sets of equations remain similar to the general time-optimal problem, regardless of the complexity of mathematical models. The most important conclusion is that the alternance method can be used in this case if the temperature distribution can be obtained as a function of spatial coordinates and voltages U_i, $i = \overline{1, N}$.

However, the choice of coil voltages as spatial control inputs leads to some unique features.[4,5] The set of optimal voltages for N coils of given lengths allows one to realize optimal heat power distribution along the multicoil continuous heater. This distribution represents time-dependent heat power control scanning along the length of the induction heater. Therefore, the heating cycle has preset number and durations of process stages defined by the given IHI design. Temperature distribution can be affected through a set of control inputs by choosing the values, U_i, for each inductor coil.

The opposite situation occurs when the heat power distribution, $u(y/V)$, is chosen as a control input. According to Equation (2.32), the optimal control u^* takes only fixed limit values, and one can modify temperature distribution within the billet by choosing the number and durations of alternating heating/soaking stages.

From the preceding discussion, it follows that optimal heat power control utilizing coil voltages provides multiposition control (Figure 2.11); the values U_i^0, $i = \overline{1, N}$ differ from their admissible limit values. The optimal control function (Equation 2.32) represents on–off control that takes only two extreme values.

The consideration of optimal voltages (U_1^0, U_2^0, …, U_N^0) as parameters for varying heat power distribution with time allows transforming (in terms of qualitative characteristics) a maximum accuracy problem (Equation 2.38) into an equivalent time-optimal problem. The "inverse" transformation from time-optimal problem to the problem on maximum accuracy has been explored in Section 2.9.1.

Let us assume that the condition in Equation (3.3) remains valid for the equivalent time-optimal problem. This condition links required accuracy ε with number of required optimal control intervals where voltage is constant. In this case, the number of control intervals is equal to the number of inductor coils. Duration of each interval corresponds to the time required for the billet to pass through each coil.

The assumption regarding a validity of the condition in Equation (3.3) is based on the fact that it remains correct in all previously considered time-optimal problems (Chapter 3 and Chapter 4). If the condition $\varepsilon_{\min}^{(S)} \le \varepsilon \le \varepsilon_{\min}^{(S-1)}$ is true, then the required number of coils (which is equal to S according to Equation 3.3) can differ from number N.

If condition $N < S$ is met, then required heating accuracy is unattainable in induction heating installation of the given design. However, this is a rare case. In typical cases, an expression $N > S$ is valid. This means that the number of coils exceeds the number of required control intervals with different values U_i^0. Obviously, this leads to the fact that $N - S$ inductor coils should have the same applied voltage. Therefore, S sections of induction heater will be under control — i.e., S control intervals with S sought parameters $U_1, U_2, \ldots, U_S$ exist. Each control interval can be considered as an inductor section composed of one or more inductor coils.

The preceding deals with the optimal control in an equivalent time-optimal problem at preset heating accuracy $\varepsilon = \varepsilon_0$ that satisfies the condition: $\varepsilon_{\min}^{(S)} \le \varepsilon_0 \le \varepsilon_{\min}^{(S-1)}$, if $N \ge S$. This type of control algorithm provides maximum heating accuracy $\tilde{\varepsilon}$ (which is equal to ε_0) for given time $\tilde{\varphi}^0$, if $\tilde{\varphi}^0 = \varphi_{\min}^0(\varepsilon_0)$ (see Section 2.9.1).

In the case of continuous heating mode, this type of control provides the maximum heating accuracy, $\varepsilon = \varepsilon_0$, at the exit of inductor of given length $\tilde{y}^0$, if an expression $\tilde{y}^0 = y_{\min}^0(\varepsilon_0)$ is true, where $y_{\min}^0(\varepsilon_0) = V\varphi_{\min}^0(\varepsilon_0)$. In the case when conditions $\tilde{\varphi}^0 > \varphi_{\min}^0(\varepsilon_{\inf})$, $\tilde{y}^0 > y_{\min}^0(\varepsilon_{\inf})$ are met, the problem of maximum heating accuracy has an infinite solution set. The same situation occurs when the function $u(y/V)$ is chosen as a control input.

If inductor length $\tilde{y}^0$, the number of sections, N, and their lengths are defined by given IHI design, we come to the problem in Equation (2.38). In this case, it is necessary to choose a priori the number $S \le N$ of controlled sections and variant of N inductor coils positioning into these sections under condition of a priori unknown relation between $\tilde{y}^0$ and $y_{\min}^0(\tilde{\varepsilon})$.

Similarly to computational techniques used in cases of static heating, the optimal control problem is reduced to a sequence of optimization problems that could be solved at $y^0 = \tilde{y}^0$ for a number of values of S increasing from $S = 1$. The value of $y_{\min}^0(\tilde{\varepsilon})$ is excluded from consideration in this case.

If $S = 1$, then all inductor sections have the same voltage, U_1. Under a single-parameter control algorithm, the shape of final temperature distribution across billet thickness turns out to be similar to the one shown in Figure 3.6. The corresponding set of equations can be written in the form of Equation (3.12).

After substitution of appropriate expressions for temperature as functions of U_1, a solution of this system allows one to define the optimal value, U_1^0, and maximum heating accuracy, $\varepsilon_{\min}^{(1)}$ (that is attainable in the class of single-parameter control inputs). If obtained value $\varepsilon_{min}^{(1)}$ satisfies technological requirements, then the initial optimal control problem proves to be solved.

If the optimization problem is formulated with respect to the billet movement speed, *V*, that is less than nominal value, then for the fixed inductor length, the optimal supply voltage U_1^0 can be decreased to the level when heating accuracy $\varepsilon_{min}^{(1)}$ becomes feasible. Usually, this situation corresponds to the case when inequality $\tilde{y}^0 > y_{min}^0(\varepsilon_{inf})$ is true — i.e., inductor length is bigger than minimum length required to obtain utmost heating accuracy $\varepsilon = \varepsilon_{inf}$.

If the value of $\varepsilon_{min}^{(1)}$ does not provide required heating uniformity, then it is necessary to consider the case of $S = 2$ regarding the inequalities in Equation (3.41). For $N > 2$, there are $N - 1$ variants of *N* coils' distribution to two controlled sections. Figure 5.1 shows an example for four coils, $N = 4$. For each of these variants, the problem is reduced to solving the set of equations (Equation 3.16) with respect to optimal supply voltages U_1^0, U_2^0 for controlled sections and appropriate value of maximum heating accuracy $\varepsilon_{min}^{(2)}$. The shape of the final temperature profile is similar to the case shown in Figure 3.8.

The further enumeration of variants (e.g., including 1 … $N - 1$ sections to the second zone close to the inductor exit) allows finding the optimal variant for $S = 2$ that provides minimum value of $\varepsilon_{min}^{(2)}$. This value represents limiting attainable accuracy of heating in the class of two-section controls. In most cases, the

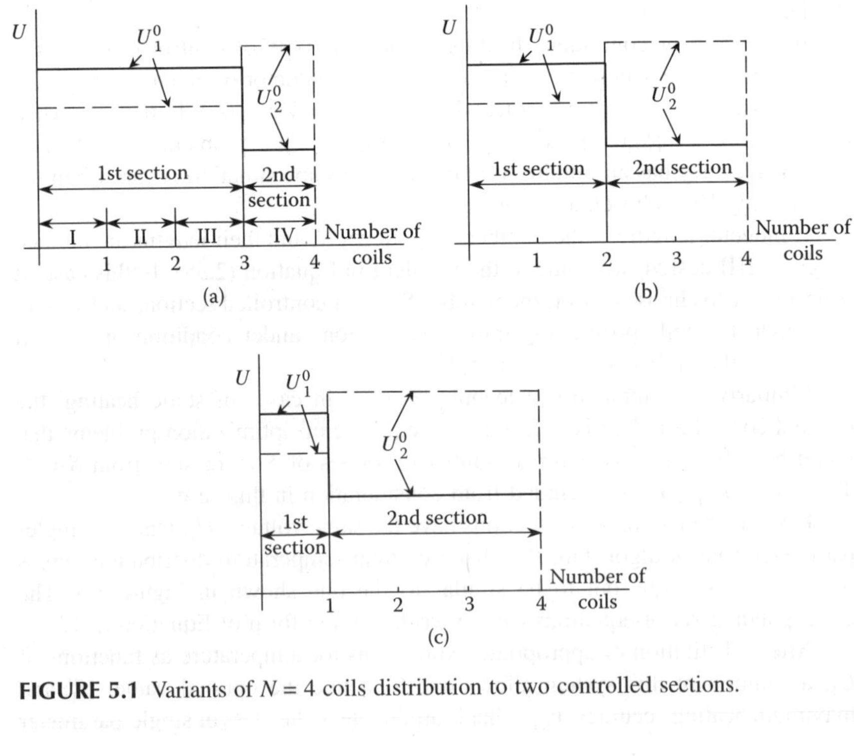

FIGURE 5.1 Variants of $N = 4$ coils distribution to two controlled sections.

obtained value, $\varepsilon_{\min}^{(2)}$, satisfies technological requirements, and the initial optimal control problem proves to be solved.

Based on common sense, it is possible to assume that, for values of $\tilde{y}^0$ close to $y_{\min}^0$ ($\varepsilon_{\min}^{(2)}$), "scanning in time" optimal programs for varying feeding voltages along inductor length approximate to the time-optimal heat power control. Therefore, for such values $\tilde{y}^0$, the feeding voltage U_1^0 of the first section usually exceeds the value U_2^0, similarly to two-stage control (Equation 2.27) (Figure 5.1, solid curves).

As the value of $\tilde{y}^0$ becomes higher than $y_{\min}^0$ ($\varepsilon_{\min}^{(2)}$), the relation between U_1^0 and U_2^0 can be varified in the opposite way (Figure 5.1, dashed curves). If the heating accuracy for $S = 2$ is not acceptable, then it is necessary to consider the case of using three controlled sections ($S = 3$); further searching for optimal values U_1^0, U_2^0, and U_3^0 can be done in a similar manner. It should be kept in mind that, under condition $N > S$, the number of variants of controlled coils positioning grows rapidly in comparison with the case of $S = 2$. Similarly to static heating mode, the heating accuracy $\varepsilon_{\min}^{(S)}$ attainable for $S = 3$ coincides with the value of utmost accuracy $\varepsilon_{\inf}$ (Section 3.3 and Section 3.6).

General considerations discussed here can be applied to different optimization problems subject to additional complicating factors (Chapter 4), as has been done for static heating optimization. Typical optimization problems will be further discussed for a variety of particular control inputs. The optimization problem with respect to continuous heating application models will be discussed later.

5.1.2 Design of Minimum Length Inductor

Let us consider the optimal design problem that minimizes heater length when speed V of workpiece movement through an induction heater is constant and a required heating accuracy should be obtained at the inductor exit.[1] Spatial distribution of power $u^*(y/V)$ along the heater length can be chosen as control input constrained by Equation (2.8). This problem can be formulated as the mathematical programming problem (Equation 2.33).

A more precise model may be used for calculation of temperature distribution, $\theta(l,y)$, taking into account the heat transfer by conduction along the radius (coordinate l) and the length (coordinate y) of the heated cylindrical billet. Using the system of relative units, the temperature distribution, $\theta(l,y)$, within the billet (that moves through the inductor at constant speed $V = const$) can be described by the linear two-dimensional stationary heat transfer equation of the following form[1]:

$$\frac{\partial^2\theta(l,y)}{\partial l^2} + \frac{1}{l}\frac{\partial\theta(l,y)}{\partial l} + \beta^2\frac{\partial^2\theta(l,y)}{\partial y^2} - \beta V\frac{\partial\theta(l,y)}{\partial y} + W_d(\xi,l,y) = 0, \tag{5.1}$$

$$0 < l < 1;\ 0 < y < y^0;$$

with the boundary conditions:

$$\frac{\partial\theta(1,y)}{\partial l} = Bi\left(\theta_a - \theta(1,y)\right); \frac{\partial\theta(0,y)}{\partial l} = 0;$$
$$\theta(l,0) = \theta^0(l); \beta\frac{\partial\theta(l,y^0)}{\partial y} = q(l) < 0. \tag{5.2}$$

Here,
$q(l)$ is a given radial distribution of heat losses' flux density at the inductor exit.
$\theta^0(l)$ is initial radial temperature distribution at the inductor enter.
y^0 is the heater length.

The speed, V, is expressed as a relation to the value a/X, where a is thermal conductivity and X is radius of the cylindrical billet. All other denotations and relative units would correspond to the ones used in the case of a two-dimensional model (Equation 4.72 through Equation 4.76) under $\Gamma = 1$.

A function, $W_d(\xi,l,y)$, in Equation (5.1) represents spatial distribution of internal heat source density induced by eddy current per unit time in a unit volume. In many cases, the function $W_d(\xi,l,y)$ for the cylindrical inductor can be written in the following simplified form (similar to Equation 4.126)[1]:

$$W_d(\xi,l,y) = W(\xi,l)\,W_1(y)\,. \tag{5.3}$$

Here, the function $W(\xi,l)$, which can be determined according to Equation (1.22), represents the one-dimensional distribution of internal heat sources along the billet radius. A distribution of internal heat sources can be varied along the heater length; therefore, the function $W_1(y)$ can be considered as a spatial control bounded by the following constraint:

$$0 \le W_1(y) \le W_{1\max},\ y \in [0, y^0]\,. \tag{5.4}$$

In the most typical technological processes, maximum admissible value ε of absolute deviation of temperature distribution $\theta(l,y^0)$ at the inductor exit from the required temperature θ^* is prescribed. It means that, at the exit of the induction heater, the temperature in any point of the billet should deviate not more than by value ε from the required temperature θ^*:

$$\max_{l\in[0,1]}\left|\,\theta(l,y^0) - \theta^*\,\right| \le \varepsilon\,. \tag{5.5}$$

The condition in Equation (5.5) is similar to that in Equation (2.26), but the expressions for $\theta(l,y^0)$ differ from ones that can be used in cases of basic mathematical models.

Let us formulate the optimal control problem with respect to a continuous heater at steady-state operating conditions as follows. It is necessary to select control function $W_1^*(y)$ restrained by Equation (5.4) that provides steering billets' initial temperature distribution to desired temperature $\theta^* = const$ with prescribed accuracy ε (according to Equation 5.5), while minimizing inductor length y^0.

It can be proved[1] that optimal heat power distribution $W_1^*(y)$ can be represented as a stepwise function in the form of Equation (2.32) (Figure 2.10):

$$W_1^*(y) = \frac{W_{1\max}}{2}\left[1+(-1)^{j+1}\right],\ y_{j-1} < y < y_j,\ j = \overline{1,N},\ y_N = y^0. \tag{5.6}$$

This means that the optimal design solution represents a heater consisting of alternating inline sections with heat ON and heat OFF power intervals (active and passive zones).

Therefore, the solution of the problem that minimizes length of continuous heater and provides required accuracy is known, but the number, N, of alternating inline sections and their lengths Δ_i, $i = \overline{1,N}$ remain unknown. Similarly to Equation (2.33), the problem that minimizes the length of the induction heater can be reduced to the problem of mathematical programming:

$$\begin{cases} I(\Delta) = \sum_{i=1}^{N} \Delta_i \to \min\limits_{\Delta}; & (5.7) \\ \Phi(\Delta) = \max\limits_{l\in[0,1]} \left| \theta(l, y^0, \Delta) - \theta^* \right| \le \varepsilon. & (5.8) \end{cases}$$

The alternance method can be successfully applied to a wide variety of specific optimization problems of the type in Equation (5.7) and Equation (5.8). In this case, previously described universal properties of final temperature distribution, sets of appropriate equations, and computational techniques remain unchanged as applied to time-optimal problems with respect to static heating.

It is important to note at this point that the expressions in governing sets that have been used in the case of optimization of static heating describe the temperature distribution at the end of the heating cycle. In the case of continuous heating mode, these expressions represent temperature distribution within billet cross-section at the inductor exit under $y = y^0$.

In contrast to the problem in Equation (2.33), there is a specific dependence of temperature $\theta(l,y^0,\Delta)$ as a function of spatial coordinates and control parameters. This dependence should be obtained after solving two-dimensional heat transfer Equation (5.1) and Equation (5.2).

Let us accept that $y^0 = 1$ with respect to relative units. Assuming that initial temperature distribution at the inductor inlet can be considered as uniform along billet radius (i.e., $\theta^0(l) \equiv \theta^0 = \theta_a$), the following expression for $\theta(l,y^0,\Delta) = \theta(l,1,\Delta)$ can be obtained[1]:

$$\theta(l,1,\Delta) = \theta^0 + \sum_{n=1}^{\infty} D_n^* \left\{ \frac{q_n}{\beta}\left(1 - e^{r_{2n}-r_{1n}}\right) + \frac{W_n(\xi)}{\beta^2} \cdot W_{1\max} \times \right.$$
$$\left. \times \sum_{j=1}^{N} (-1)^{j+1} \left[\frac{1}{r_{2n}} \left(e^{r_{2n}\sum_{m=j}^{N}\Delta_m} - 1 \right) - \frac{e^{r_{2n}-r_{1n}}}{r_{1n}} \left(e^{r_{1n}\sum_{m=j}^{N}\Delta_m} - 1 \right) \right] \right\} \cdot J_0(\mu_n l) . \tag{5.9}$$

Here,

q_n and $W_n(\xi)$ designate the coefficients of expansion of $q(l)$ and $W(\xi,l)$ in the series of Bessel function $J_0(\mu_n l)$ determined by formulas in Equation (4.88) and Equation (1.45).

The proper number μ_n is the root of Equation (1.43).

r_{1n}, r_{2n} and D_n^* can be determined as:

$$r_{1n} = \frac{1}{2\beta}\left(V + \sqrt{V^2 + 4\mu_n^2}\right); \quad r_{2n} = \frac{1}{2\beta}\left(V - \sqrt{V^2 + 4\mu_n^2}\right); \tag{5.10}$$

$$D_n^* = 2\mu_n^2 \left[\left(Bi^2 + \mu_n^2\right) J_0^2(\mu_n) \left(r_{1n} - r_{2n} e^{r_{2n}-r_{1n}}\right) \right]^{-1} . \tag{5.11}$$

Under $y^0 = 1$, it is reasonable to consider a coefficient, β, as an equivalent variable that should be optimized in the problem that minimizes the heater length. This coefficient can be expressed as a correlation (Equation 4.76) between billet radius and the heater length y^0. Lengths Δ_i of $N - 1$ sections (from the set of N sections) and the value of β can be considered as unknown variables. The length of one section (in this case, the length of last section Δ_n) can be excluded from consideration and can be found after solving the equality $\sum_{i=1}^{N} \Delta_i = 1$ using defined values $\Delta_1, \Delta_2, \ldots, \Delta_{N-1}$. Expression (5.9) represents a dependence of final radial temperature distribution $\theta(l,\beta,\Delta_1, \ldots, \Delta_{N-1})$ on sought parameters at the moment when a billet exits an inductor.

As an example, let us consider two typical cases of $\varepsilon = \varepsilon_{\min}^{(1)}$ and $\varepsilon = \varepsilon_{\min}^{(2)}$ in Equation (5.8). In the case of $\varepsilon = \varepsilon_{\min}^{(1)}$, it easy to determine the condition $N = 1$ that is similar to the time-optimal problem. According to Equation (5.6), this condition would correspond to the single-section inductor when maximum heat power is applied (which is equal to $W_{1\max}$). Maximum power is uniformly distributed along the inductor length (Figure 5.2a).

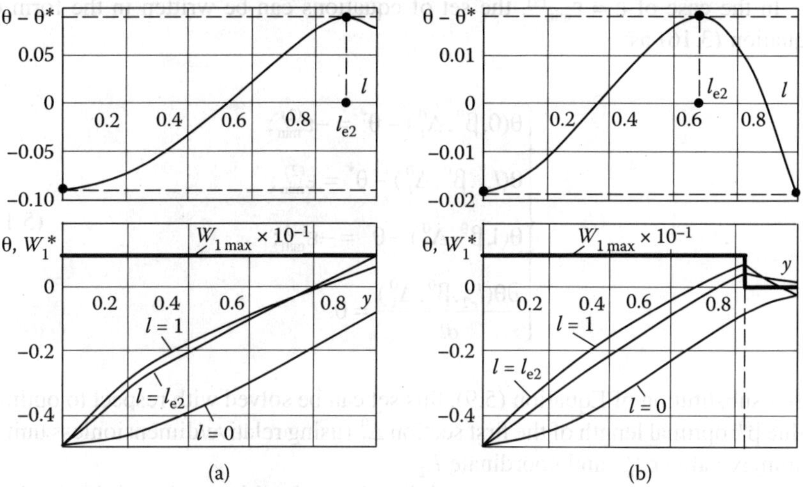

FIGURE 5.2 Temperature distribution along billet radius at the inductor exit and along the heater length for optimal spatial control of steady-state continuous induction heating ($\xi = 4$; $\theta^0 = \theta_a = -0.5$; $\theta^* = 0$; $Bi = 0.5$; $q(l) = -0.2$; $V = 100$). a: For $\varepsilon = \varepsilon^{(1)}_{\min}$; $\beta^0 = 0.031$. b: For $\varepsilon = \varepsilon^{(2)}_{\min}$; $\beta^0 = 0.026$; $\Delta^0_1 = 0.84$.

The shape of radial temperature distribution at the inductor exit (Figure 5.2a) will be identical to the shape of curve shown in Figure 3.6. Similarly to Equation (3.12), the computational set of equations can be written as follows:

$$\begin{cases} \theta(0,\beta^0) - \theta^* = -\varepsilon^{(1)}_{\min}; \\ \theta(l_{e2},\beta^0) - \theta^* = \varepsilon^{(1)}_{\min}; \\ \dfrac{\partial\theta(l_{e2},\beta^0)}{\partial l} = 0. \end{cases} \tag{5.12}$$

After substitution of Equation (5.9), this system can be solved with respect to all unknown parameters: optimal value β^0 of coefficient β, minimax value $\varepsilon^{(1)}_{\min}$, and intermediate variable l_{e2}. Solution of the set in Equation (5.12) allows determining the optimal heater length Y^0 after conversion to absolute units. Optimal heater length can be calculated using the relation $Y^0 = X/\beta^0$ with respect to actual billet radius X.

If the condition $\varepsilon = \varepsilon^{(2)}_{\min}$ is true, then one can obtain that $N = 2$. Optimal design solution leads to a heater that would consist of two sections. Maximum power $W_{1\max}$ is applied in the first heating section. The second section provides the temperature soaking without power (heat OFF): $W_1(y) \equiv 0$ (Figure 5.2b).

In the case of $\varepsilon = \varepsilon_{min}^{(2)}$, the set of equations can be written in the form of Equation (3.16) as:

$$\begin{cases} \theta(0,\beta^0,\Delta_1^0) - \theta^* = -\varepsilon_{min}^{(2)}; \\ \theta(l_{e2},\beta^0,\Delta_1^0) - \theta^* = \varepsilon_{min}^{(2)}; \\ \theta(1,\beta^0,\Delta_1^0) - \theta^* = -\varepsilon_{min}^{(2)}; \\ \dfrac{\partial\theta(l_{e2},\beta^0,\Delta_1^0)}{\partial l} = 0. \end{cases} \tag{5.13}$$

After substitution of Equation (5.9), this set can be solved with respect to optimal value β^0, optimal length of the first section Δ_1^0 (using relative/dimensionless units), minimax value $\varepsilon_{min}^{(2)}$, and coordinate l_{e2}.

Obtained value β^0 allows determining the optimal heater length Y^0. Optimal lengths of the first and second sections can be calculated as $\Delta_1^0 Y^0$ and $(1 - \Delta_1^0)Y^0$, respectively. As a matter of fact, the second section of the induction heater can be considered as "thermostat," where the heating is not required. The inductor length will be equal to the length of the first section. The problem can be reduced to the problem that minimizes the total length of the induction heating installation.

Let us consider a transfer time as total time from the moment when the heated workpiece exits an induction heater until beginning of the hot working operation. In this case, the second section can be unnecessary because the temperature will be equalized during transfer stage. This is the typical case when an expression, $\Delta_T^0 \geq \Delta_{T2}^*$, is valid and heating accuracy, $\varepsilon = \tilde{\varepsilon}_{min}^{(2)}$, will be in agreement with requirements. Here, the variables Δ_T^0, Δ_{T2}^* and $\tilde{\varepsilon}_{min}^{(2)}$ make the same sense as was discussed in Section 4.2. Time-optimal heat power control in a "transportation problem" consists of a single control interval under $u^*(\varphi) \equiv U_{max}$ (Section 4.2).

Spatial scanning of time-dependent control leads to the simplest single-section design solution with uniform maximum heat power distribution along the length of the induction heater. In contrast to the case of $N = 1$, the required uniformity of temperature distribution cannot be provided at the inductor exit, but sufficient temperature equalization, which takes place during the transfer stage of optimal duration, ensures obtaining required uniformity.

Minimum value of $\tilde{\varepsilon}_{min}^{(2)}$ can be obtained at $\Delta_T^0 = \Delta_{T2}^*$. If the time of transportation can be varied, then the duration of transfer stage meets the condition $\Delta_T^0 = \Delta_{T2}^*$ and will be considered as an optimal design solution as well.

If transfer stage Δ_T^0 is taken into account, then the computational equations set remains similar to the "transportation" time-optimal problem. According to Equation (4.40), this set keeps the form of Equation (5.13) at $\Delta_T^0 = \Delta_{T2}^*$ or can be transformed into form of Equation (4.41) or Equation (4.42) if $\Delta_T^0 > \Delta_{T2}^*$.

In the first case, the value Δ_{T2}^* would be defined after solving the set of equations. Obviously, the appropriate expressions for $\theta(l,\beta,\Delta,\Delta_T^0)$ will be changed in comparison with Equation (5.9). These expressions can be found after solving the corresponding set of heat transfer equations with regard to different levels of heat losses from the billet surface during heating stage and transfer stage (Section 4.2).

Figure 5.2 presents some computational results of determining optimal process parameters for continuous induction heater. In the typical cases of $\varepsilon = \varepsilon_{\min}^{(1)}$ and $\varepsilon = \varepsilon_{\min}^{(2)}$, the final temperature distribution can be calculated from Expression (5.9). As one can see, the optimal design solution leads to the temperature distribution along billet radius at the heater exit that is similar to temperature distribution at the end of time-optimal process (Figure 3.6 and Figure 3.8). The temperature distribution within any cross-section of the billet located inside the heater proves to be identical to the temperature distribution that would correspond to appropriate time of static heating.

Similarly to Section 4.1, the requirement on the temperature distribution during heating can be imposed. According to such a requirement, during heating stage the value of maximum temperature anywhere within the heated workpiece should not exceed a certain admissible value θ_{adm}. Therefore, instead of Expression (2.11), the temperature maximum constraint can be written as follows:

$$\theta_{\max} = \max_{l,y\in[0,1]} \theta(l, y) \leq \theta_{adm} \,. \tag{5.14}$$

As an example, let us consider an optimal heater design problem taking into consideration the condition $\theta_{adm} = \theta^*$. This condition means that temperature at any point within the billet should not be higher than required temperature θ^*. In this case, the condition $\varepsilon > \theta_{adm} - \theta^* = 0$ holds true under given heating accuracy $\varepsilon > 0$.

Similarly to optimization of static heating (Section 4.1), the utmost admissible value of heating accuracy ε coincides with $\varepsilon_{\min}^{(1)}$. Optimal heat power distribution along the length of an induction heater can be considered as a time-dependent optimal control (Equation 4.3) with scanning along the heater length. For $N = 1$, the optimal heat power distribution can be written as follows:

$$W_1^*(y) = \begin{cases} W_{1\max}, \; 0 \leq y < y_\theta; \\ W^\theta(y), \; y_\theta < y \leq y^0. \end{cases} \tag{5.15}$$

According to Equation (5.15), the heater consists of two sections. In the first section, which is y_θ long, the maximum heating power $W_{1\max}$ is applied. The second section represents a holding chamber of temperature $\theta_{\max}$ at the level $\theta_{adm} = \theta^*$ under heat power $W^\theta(y)$. Therefore, continuous induction heating installation provides accelerated heating mode (Section 4.1). Obviously, the function $W^\theta(y)$ should be written in the form similar to Expression (4.13) with respect to time-dependent function $u^\theta(\varphi)$.

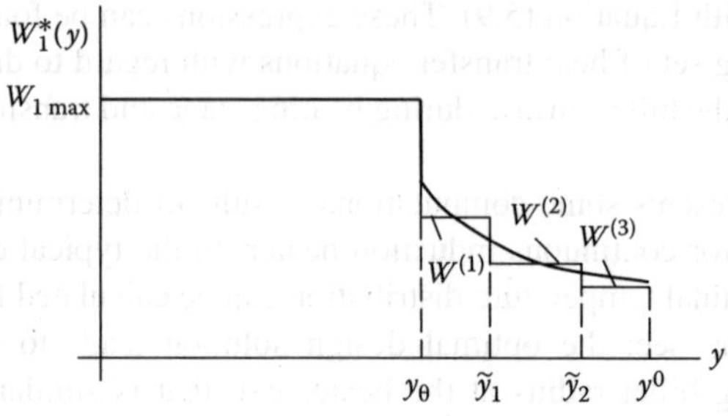

FIGURE 5.3 Piecewise constant approximation of optimal heat power distribution along the heater length.

A time-dependent control of the heat power can serve as a good example of the control input that can be relatively simply realized in the static induction heater. However, practically speaking, one can face certain problems in realization of providing nonuniform coil power distribution along the length of the second section. Therefore, it is reasonable to use the piecewise constant approximation, $W_A^\theta(y)$, of function $W^\theta(y)$ that can be relatively simply realized by providing a constant heat power $W^{(k)}$, $k = 1, 2, \ldots, \chi$, $\chi \geq 1$ within each of χ inductor sections (Figure 5.3):

$$W_A^\theta(y) = W^{(k)} = \text{const}, \quad y \in (\tilde{y}_{k-1}, \tilde{y}_k), \; k = \overline{1,\chi},$$
$$y_\theta = \tilde{y}_0 < \tilde{y}_1 < \ldots < \tilde{y}_\chi = y^0. \tag{5.16}$$

Under given number $\chi \geq 1$, the variables $\tilde{y}_k$ denote the bound point's coordinates for heater sections. The powers $W^{(k)}$ and coordinates $\tilde{y}_k$ can be considered as approximation parameters. If the temperature distribution as a function of approximation parameters is known, these parameters can be defined numerically using condition of minimum deviation of θ_{max} from θ_{adm} at points within interval $[y_\theta, y^0]$. Similarly to Equation (4.12), the coordinates (l_{max}, y_θ) of the point at which the temperature θ_{max} reaches an admissible value θ_{adm} can be found by considering the equalities:

$$\theta(l_{max}, y_\theta) = \theta_{adm} = \theta^*; \quad \frac{\partial \theta(l_{max}, y_\theta)}{\partial l} = 0 . \tag{5.17}$$

The temperature $\theta(l_{max}, y_\theta)$ should be found after solving Equation (5.1) and Equation (5.2) with substitution of control input in the form of Equation (5.15) and Equation (5.16). If we allow for relative/dimensionless units of length $y^0 =$ 1, then instead of Equation (5.9) and after replacing y^0 by coefficient β, this solution can be written as follows[1]:

$$\theta(l,\beta)=\theta^0+\sum_{n=1}^{\infty}D_n^*\left\{\frac{q_n}{\beta}\left(1-e^{r_{2n}-\eta_n}\right)+\frac{W_n(\xi)}{\beta^2}e^{r_{2n}}\times\right.$$

$$\times\left[\left(\frac{1}{r_{1n}}\left(e^{-r_{1n}y_\theta}-1\right)-\frac{1}{r_{2n}}\left(e^{-r_{2n}y_\theta}-1\right)\right)\cdot W_{1\max}\right.$$

$$\left.\left.+\sum_{k=1}^{\chi}W^{(k)}\left(\frac{1}{r_{1n}}\left(e^{-r_{1n}\tilde{y}_k}-e^{-r_{1n}\tilde{y}_{k-1}}\right)-\frac{1}{r_{2n}}\left(e^{-r_{2n}\tilde{y}_k}-e^{-r_{2n}\tilde{y}_{k-1}}\right)\right)\right]\right\}J_0(\mu_n l). \tag{5.18}$$

In the case of $\varepsilon = \varepsilon_{\min}^{(1)}$ (under defined values of y_θ and $W_A^\theta(y)$), the optimal control input (Equation 5.15) has a single unknown parameter, β, that allows determining the optimal heater length.

Similarly to Equation (4.24), the following computational set can be obtained instead of Equation (5.12):

$$\begin{cases}\theta(1,\beta^0)-\theta^* = -\varepsilon_{\min}^{(1)};\\ \theta(0,\beta^0)-\theta^* = -\varepsilon_{\min}^{(1)}.\end{cases} \tag{5.19}$$

After substitution of Equation (5.18), this set should be solved with respect to unknown variables β^0 and $\varepsilon_{\min}^{(1)}$.

However, it is impossible to find $W_1^*(y)$ and coordinates $l_{\max}$ and y_θ after separate solving of appropriate equations sets because the output $\theta(l,y)$ from a heating process is obtained in response to applied input $W_1^*(y)$ varying on the interval (y_θ, y^0). This means that, in contrast to static heating optimization, all unknown variables $l_{\max}$, y_θ, β^0, and $\varepsilon_{\min}^{(1)}$ for each given set of values $W^{(k)}, \tilde{y}_k, k=\overline{1,\chi}$ can be found by combined solution of sets of Equation (5.17) and Equation (5.19).

The values of $W^{(k)}$ and $\tilde{y}_k$ can be found by minimizing deviation of $\theta_{\max}$ from θ_{adm} at the last stage of a solution of the discussed optimal control problem.

Figure 5.4 shows the computational results for optimal mode of continuous accelerated heating that provides a holding of temperature $\theta_{\max}$ at the level $\theta_{adm} = \theta^*$. The computational procedure has been performed in the simplest case of $\chi = 1$ and $\varepsilon = \varepsilon_{\min}^{(1)} = \varepsilon_{\inf}$, according to the previously described technique. All initial data are the same as in the case shown in Figure 5.2.

Optimal temperature distribution along the radius of a cylindrical billet (Figure 5.4a) at the inductor exit replicates in shape the temperature distribution at the end of accelerated static heating (Figure 4.9). Temperature distribution along the heater length (Figure 5.4b) exhibits the acceptable accuracy of approximation of $\theta_{\max}$ to $\theta_{adm} = \theta^*$ within the interval $[y_\theta, y^0]$ in spite of the fact that uniform power distribution $W_A^\theta(y) = W^{(1)} < W_{1\max}$, $y \in (y_\theta, y^0)$ represents a crude approximation of the function $W^\theta(y)$.

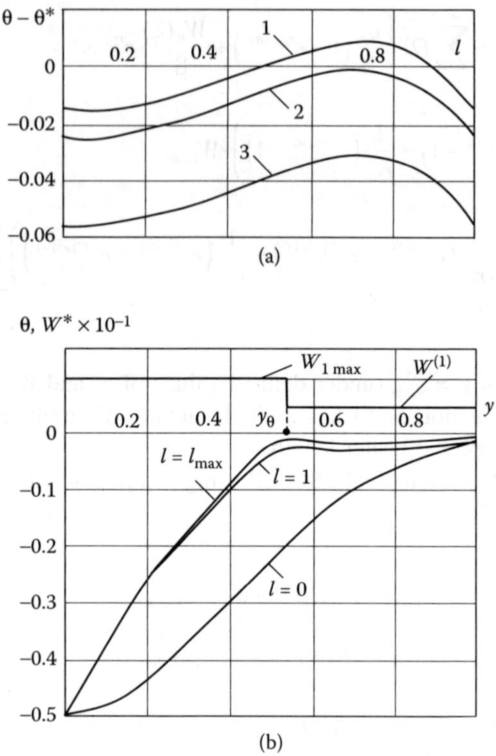

FIGURE 5.4 Optimal spatial control of continuous accelerated heating that provides a holding of temperature θ_{max} at the admissible level ($\xi = 4$; $\theta^0 = \theta_a = -0.5$; $Bi = 0.5$; $q(l) = -0.2$; $V = 100$; $\theta_{adm} = \theta^* = 0$; $\varepsilon = \varepsilon_{\min}^{(1)} = \varepsilon_{\inf}$). a: Temperature distribution along billet radius under control in the form of Equation (5.16) for $\chi = 1$ (1 – $W^{(1)}/W_{1\max} = 0.37$; $\beta^0 = 0.02$; $y_\theta = 0.52$; 2 – $W^{(1)}/W_{1\max} = 0.36$; $\beta^0 = 0.021$; $y_\theta = 0.54$; 3 – $W^{(1)}/W_{1\max} = 0.3$; $\beta^0 = 0.024$; $y_\theta = 0.61$). b: Distribution of heat power and temperature along the heater length ($W^{(1)}/W_{1\max} = 0.36$; $\beta^0 = 0.021$; $y_\theta = 0.54$; $l_{\max} = 0.93$).

Curves 1 through 3 in Figure 5.4 illustrate a process of searching for the value $W^{(1)}$ when minimizing deviation of θ_{max} from $\theta_{adm} = \theta^*$. Here, the optimal value of $W^{(1)}$ turns out to be equal to the value of 0.36 $W_{1\max}$ (curve 2). The obtained simple algorithm of spatial control can be relatively easily applied in engineering practice if each of the two sections of an induction heater has optimal length (depending on calculated value of y_θ), and applied power will be equal to the optimal value of $W^{(1)}$.

It is important to note that stepwise algorithms of optimal spatial control (Equation 5.6 or Equation 5.15, Equation 5.16) that have been found under idealized conditions of the relaxing constraint (Equation 5.4) cannot be accurately realized using practical design solutions of continuous induction heating installations. It is impossible to provide an absolutely uniform power distribution along each section with stepwise power change from one section to another due to a longitudinal electromagnetic end effect and gaps between sections.[2,6]

Keep in mind that obtained results can be extended onto other optimal control problems, subject to complicating real-life constraints, by using coil voltages applied to controlled sections as control inputs. Moreover, obtained results exhibit quantitative and qualitative properties of optimal spatial control and deal with estimation of a maximum of capability of an induction heater. This can be very helpful in order to have a rough idea with regard to number of sections, power levels, and required shop floor space.

5.1.3 Optimization of the Continuous Heating of Ferromagnetic Materials

The continuous induction machinery for heating ferromagnetic materials is widely used in industry. This process is governed by complex nonlinear equations. The nonlinear nature of magnetic permeability as a function of temperature is responsible for the presence of ferromagnetic and paramagnetic phases of heated material. The ferromagnetic phase takes place at temperatures below the Curie point. The paramagnetic phase takes place if temperature of a heated workpiece exceeds the Curie temperature.

Transition between these phases leads to significant changes in intensity and spatial distribution of internal heat power during the induction heating process. Also, magnetic permeability and other physical properties make a marked effect on temperature distribution of heated workpieces. It is imperative to keep in mind that electrical resistivity, thermal conductivity, and specific heat are also functions of temperature (Section 4.6.2).

The two-dimensional equation (Equation 5.20) properly describes steady-state temperature distribution of the heated cylindrical billet using a continuous induction heater[1]:

$$\frac{\partial^2\theta(l,y)}{\partial l^2}+\frac{1}{l}\frac{\partial\theta(l,y)}{\partial l}-\beta_1(y)\frac{\partial\theta(l,y)}{\partial y}+\alpha(y)W_d(l,y)W_1(y)=0; \qquad 0<l<1;\ 0<y<y^0; \tag{5.20}$$

with the boundary conditions:

$$\frac{\partial\theta(1,y)}{\partial l}=q(y)<0;\ \frac{\partial\theta(0,y)}{\partial l}=0;\ \theta(l,0)=\theta^0(l)\,. \tag{5.21}$$

Equation (5.20) and Equation (5.21) describe a temperature field during steady-state continuous heating under assumption that an effect of heat transfer in the axial direction (along the coordinate y) by heat conduction is negligibly small ($\partial^2\theta(l,y)/\partial y^2 \approx 0$).

Temperature dependencies of thermal conductivity, a, and heat conductivity, λ, and radial distribution of eddy current heat source density now will be taken into account by using the equivalent dependencies of heat source density, $W_d(l,y)$, and coefficients, β_1 and α, as functions of the axial coordinate y:

$$\beta_1(y) = \frac{VX^2}{Y^0 a(y)}; \quad \alpha(y) = \frac{\lambda_b}{\lambda(y)}. \tag{5.22}$$

Here, λ_b is a basic value of heat conductivity with respect to system of relative units (Equation 1.28 through Equation 1.35) and Y^0 is the inductor length in absolute units.

We shall limit our consideration to the analysis of stepwise constant approximation of functions $a(y)$, $\lambda(y)$, and $W_d(l,y)$ that take on constant values on appropriate intervals along the inductor length. These intervals represent "cold," "intermediate," and "hot" stages of the heating process.[7,8]

A distribution of internal heat power $W_1(y)$ along the heater length can be considered as a spatial control bounded by the following constraint:

$$0 \le W_1(y) \le W_{1\max}(y), \quad y \in [0, y^0]. \tag{5.23}$$

In contrast to Equation (5.4), the dependence $W_{1\max}(y)$ of admissible maximum power $W_1(y)$ should be considered as a step function of y that can be defined by well-known methods.[7,8]

Similarly to previous optimal control cases (Section 5.1.2), in the problem that minimizes heater length, the optimal control $W_1^*(y)$ algorithm consists of alternating piecewise intervals where power reaches its admissible limits (extreme values). Mathematically rigorous proof of this fact is provided in Rapoport.[1]

This means that the optimal design solution represents a heater consisting of alternating inline sections with heat ON and heat OFF (active and passive zones). While the billet moves through active zones, the heating takes place under maximum power. The passive zones provide temperature equalization without external power. It is important to mention that the value of maximum admissible heat power varies from each active section to another one, according to the dependence $W_{1\max}(y)$ in Equation (5.23). At the same time, in a linear case, the maximum heat power is constant along the whole coil length, according to Equation (5.6).

Obviously, the optimal power distribution along the heater length $W_1^*(y)$ can be considered as a scanning of the time-dependent optimal control algorithm (Equation 4.127) along the heater length.

Technological requirement demands that maximum value θ_{max} of the temperature within the heated workpiece during a heating process should be not above a certain admissible value θ_{adm}. In this case, one can obtain the following expression for optimal control algorithm $W_1^*(y)$ that is similar to Equation (4.127):

$$W_1^*(y) = \begin{cases} W_{1\max}^{(1)}, & 0 < y < y_{\max}^{(1)}; \\ W_{1\max}^{(2)}, & y_{\max}^{(1)} < y < y_{\max}^{(2)}; \\ W_{1\max}^{(3)}, & y_{\max}^{(2)} < y < y_\theta; \\ W^\theta(y), & y_\theta < y < \Delta_1^0 = y_1; \\ \dfrac{W_{1\max}^{(3)}}{2}[1+(-1)^{j+1}], & y_{j-1} < y < y_j, \quad j = \overline{2,N}. \end{cases} \tag{5.24}$$

Here,

$W_{1\max}^{(1)}$, $W_{1\max}^{(2)}$, and $W_{1\max}^{(3)}$ represent maximum admissible values of heating power density during the cold, intermediate, and hot stages of a heating process, respectively.

$y_{\max}^{(1)}$ and $y_{\max}^{(2)}$ are longitudinal coordinates of bound points of intermediate and hot stages.

All other denotations correspond to ones used in Expression (5.6) and Expression (5.15).

Both control algorithms (Equation 5.24 and Equation 4.127) provide heating above Curie temperature in the typical case when holding the maximum temperature $\theta_{\max}$ at its maximum permissible level θ_{adm}. The following stages of a heating cycle should be performed during the hot stage.

By analogy with Equation (5.15), one can obtain that $N = 1$ in Equation (5.24), if the required nonuniformity of heating $\varepsilon = \varepsilon_0$ exceeds the difference $\theta_{adm} - \theta^*$.

Assuming that the values $y_{\max}^{(1)}$ and $y_{\max}^{(2)}$ and y_θ are given, the problem can be reduced to determining the number, N, and lengths, Δ_i, $i = \overline{1,N}$, of alternating inline sections of an induction heater. This problem can be formulated in the form of Equation (5.7) and Equation (5.8).

The shape of radial temperature distribution at the inductor exit remains unchanged in all practicable cases when a nonlinear model (Equation 5.20 and Equation 5.21) is used. This means that the basic form of the previously described optimal control method could be applied for more general cases. Optimal control results discussed earlier for the linear IHP model (Equation 5.1 and Equation 5.2) can be transferred for more complex nonlinear cases as well.

It is important to underline at this point that the temperature field variations that take place during the heating cycle will be described by more complicated expressions. Thermal equations (Equation 5.20 and Equation 5.21) should be solved after substitution of a control input function (Equation 5.24) in order to obtain the function $\theta(l,y,\beta^*,\Delta^0, y^{(1)}_{max}, y^{(2)}_{max}, y_\theta)$ in its explicit form. This function defines the final temperature distribution at the inductor exit $y = y^0$ if it is assumed that $y^0 = 1$ with respect to relative (dimensionless) units.

Here, the variable β^* represents a sought parameter instead of y^0. The value $\beta^*_\cdot = \beta_1(a_b)$ can be computed by Expression (5.22) at the basic value $a(y) = a_b$. The lengths $\Delta^0_1, \Delta^0_2, \ldots, \Delta^0_{N-1}$ of separate inductor sections should be included in the set of sought unknowns as well. Similarly to Equation (5.17), the values $y^{(1)}_{max}$, $y^{(2)}_{max}$, and y_θ should be found from the following equalities that can be considered as their own definitions:

$$\theta\left(l, y, \beta^*, \Delta^0, y^{(1)}_{max}, y^{(2)}_{max}, y_\theta\right)\Big|_{y=y_\theta} = \theta_{adm}; \quad \frac{\partial\theta\left(l, y, \beta^*, \Delta^0, y^{(1)}_{max}, y^{(2)}_{max}, y_\theta\right)}{\partial l}\Bigg|_{y=y_\theta} = 0;$$

$$\theta\left(1, y, \beta^*, \Delta^0, y^{(1)}_{max}, y^{(2)}_{max}, y_\theta\right)\Big|_{y=y^{(1)}_{max}} = \theta_{Cp};$$

$$\theta\left(l_p, y, \beta^*, \Delta^0, y^{(1)}_{max}, y^{(2)}_{max}, y_\theta\right)\Big|_{y=y^{(2)}_{max}} = \theta_{Cp}. \tag{5.25}$$

Here,

θ_{Cp} is a temperature that would correspond to Curie point.

$l_p = 1 - \delta/X$,[7,8] where δ can be calculated from the formula (Equation 1.24) by substituting parameters of the hot stage.

Similarly to the set in Equation (5.17), the individual equations in Equation (5.25) cannot be solved separately; therefore, they should be included into a set of equations for determining parameters of the optimal control process.

The computations become sufficiently complicated in comparison with the case of neglected nonlinear properties due to a complexity of the heat source distribution resulting in a complex set of sought spatial coordinates, $y^{(1)}_{max}$, $y^{(2)}_{max}$, and y_θ, as a function of the heater length. Appropriate refining of optimal control parameters is performed using efficient iteration procedure. The electromagnetic computation of the inductor should be performed on each iteration step using the results of computation in previous step. After updating input data, the set of equations should be solved to obtain a temperature field.

Let us now consider as an example the design problem that minimizes length of a continuous inductor for heating cylindrical billets. Diameter of a billet is 0.08 m, applied frequency is 2500 Hz, initial temperature is 20°C, and heater production rate is 0.6 t/h ($V = 4.25 \cdot 10^{-3}$ m/s).[1]

It is possible to assume that heat losses flow, $q(y)$, in Equation (5.21) can be neglected during the cold and intermediate stages and is about 100 kW/m² during the hot stage.[8] Let us consider the problem of the billet heating up to required temperature $t^* = 1200$°C with given accuracy $\varepsilon = \varepsilon_0$. Maximum value θ_{max} of the

TABLE 5.1
Initial Data for Computation of Optimal Continuous Heating Process for Steel Cylindrical Billets

Parameter	Heating Stage		
	Cold	Intermediate	Hot
Specific electrical resistance, Om·m	$3.6 \cdot 10^{-7}$	$0.74 \cdot 10^{-6}$	$1.12 \cdot 10^{-6}$
Heat conductivity, W/(m · °C.	48	34	28
Thermal conductivity, m²/sec	$9.3 \cdot 10^{-6}$	$6.6 \cdot 10^{-6}$	$5.6 \cdot 10^{-6}$
Magnetic field intensity at the billet surface, A/m	$9 \cdot 10^{4}$	—	$10 \cdot 10^{4}$
Maximum heat power, W/m³	$75 \cdot 10^{6}$	$50 \cdot 10^{6}$	$25 \cdot 10^{6}$
Maximum heat power in relative units	$W_{1\max}^{(1)} = 1$	$W_{1\max}^{(2)} = 0.67$	$W_{1\max}^{(3)} = 0.33$
Parameter β_1 in Equation (5.22)	β^*	$1.4 \cdot \beta^*$	$1.67 \cdot \beta^*$
Coefficient α in Equation (5.22)	1	1.41	1.71

temperature within the heated billet should be equal or below the admissible value t_{adm} = 1250°C during the whole heating cycle.

All necessary input data and data resulting from electromagnetic computation of the inductor are presented in Table 5.1. The computation of an induction system has been conducted with respect to the given voltage of the power supply and certain coil design parameters.

In the system of relative units (Equation (1.28) it is possible to assume: $t_b = t^* = 1200°C$; $P_{\max} = P_{\max}^{(1)} = 75.0 \cdot 10^6$ W/(m³); $\lambda_b = 48$ W/(m² · °C); coordinates l and y in Expression (5.20) and Expression (5.21) represent a relation to inductor radius X and inductor length Y^0 and, therefore, $y^0 = 1$ in Equation (5.20).

According to the technique described in Section 4.1.3, the most typical case is $\varepsilon_{\min}^{(1)} = \theta_{adm} - \theta^* > \varepsilon_{\inf}$ at $\varepsilon_0 < \varepsilon_{\min}^{(1)}$. In this case, to obtain the heating accuracy $\varepsilon_0 = \varepsilon_{\min}^{(2)}$, the optimal control problem can be formulated by applying the control inputs in the form of Equation (5.24) at $N = 2$. In the case of $\chi = 1$ in Equation (5.16), the simplest approximation of $W^\theta(y)$ in Equation (5.24) can be considered as the constant function $W^\theta(y) = W^{(1)} = W^\theta = const$, $y \in (y_\theta, \Delta_1^0)$.

As a result, in the case of prescribed accuracy $\varepsilon_0 = \varepsilon_{\min}^{(2)}$, the optimal heat power distribution along the heater length can be written as:

$$W_1^*(y) = \begin{cases} W_{1\max}^{(1)}, & 0 < y < y_{\max}^{(1)}; \\ W_{1\max}^{(2)}, & y_{\max}^{(1)} < y < y_{\max}^{(2)}; \\ W_{1\max}^{(3)}, & y_{\max}^{(2)} < y < y_\theta; \\ W^\theta, & y_\theta < y < \Delta_1^0; \\ 0, & \Delta_1^0 < y < \Delta_1^0 + \Delta_2^0 = y^0 = 1. \end{cases} \tag{5.26}$$

General technique with respect to solving this problem leads to the set of equations of the type in Equation (3.16) with respective substitution of expressions for temperatures $\theta(l,y,\beta^*,\Delta_1^0,y_{max}^{(1)},y_{max}^{(2)},y_\theta)$ at appropriate points (l,y) where maximum admissible temperature deviation θ^* is reached:

$$\begin{cases} \theta\left(0,1,\beta^*,\Delta_1^0,y_{max}^{(1)},y_{max}^{(2)},y_\theta\right)-\theta^* = -\varepsilon_{min}^{(2)}; \\ \theta\left(1,1,\beta^*,\Delta_1^0,y_{max}^{(1)},y_{max}^{(2)},y_\theta\right)-\theta^* = -\varepsilon_{min}^{(2)}; \\ \theta\left(l_{e2},1,\beta^*,\Delta_1^0,y_{max}^{(1)},y_{max}^{(2)},y_\theta\right)-\theta^* = \varepsilon_{min}^{(2)}; \\ \dfrac{\partial\theta\left(l_{e2},1,\beta^*,\Delta_1^0,y_{max}^{(1)},y_{max}^{(2)},y_\theta\right)}{\partial l} = 0. \end{cases} \tag{5.27}$$

In the case of preset value W^θ in Expression (5.26), the system in Equation (5.27) combined with the equalities in Equation (5.25) can be solved with respect to the eight unknown variables: Δ_1^0, β^*, $\varepsilon_{min}^{(2)}$, l_{e2}, l_{max}, $y_{max}^{(1)}$, $y_{max}^{(2)}$, and y_θ. The value W^θ can be refined in an outer iteration loop when minimizing the deviation of the maximum temperature θ_{max} from θ_{adm}.

The following results have been obtained for input data presented in Table 5.1[1]:

$$y_{max}^{(1)} = 0.064 \ (0.045 \text{ m}); \ y_{max}^{(2)} = 0.189 \ (0.132 \text{ m}); \ y_\theta = 0.461 \ (0.320 \text{ m});$$

$$l_{max} = 0.984; \ l_{e2} = 0.670; \ \beta^* = 1.042 \ \ (Y^0 = 0.7 \text{ m});$$

$$\Delta_1^0 = 0.872 \ \ (0.61 \text{ m}); \ \Delta_2^0 = 1 - \Delta_1^0 = 0.128 \ \ (0.09 \text{ m});$$

$$\varepsilon_{min}^{(2)} = 0.42 \cdot 10^{-2} \ (10°\text{C}); \ W^\theta = 0.085 \ (6.48 \cdot 10^6 \ \text{W/m}^3).$$

The minimum inductor length is equal to 0.7 m. An optimal design solution leads to three inline sections. An accelerated heating takes place inside the first section of the length 0.32 m (this would correspond to calculated value y_θ = 0.461). Maximum power that is equal to $P_{max}^{(1)}$ is applied on the first interval of the length 0.045 m. $P_{max}^{(2)}$ is power of the second interval of the length 0.132 – 0.045 = 0.087 m and $P_{max}^{(3)}$ on the last interval of the length 0.320 – 0.132 = 0.188 m.

At the end of the first section, temperature at internal point of the billet (l_{max} = 0.984, y_θ = 0.461) reaches maximum admissible value 1250°. Inside the second section of the length 0.61 – 0.32 = 0.29 m, the maximum temperature within the billet cross-section is held at the level of 1250°C under the reduced heating power W^{θ}. The third section represents a holding zone that consists of refractory only (its length is about 0.09 m), where the temperature equalization within the billet cross-section takes place. It is assumed that heat losses from the billet's surface are relatively small thanks to thermal insulation. As can be seen, temperature differential of 10° represents the utmost heating accuracy/uniformity that can be achieved in the class of control inputs of the type in Equation (5.26). Total required coil active power is approximately 182 kW.

Results of computations show that, under condition $\theta_{adm} - \theta^* = 0$, the heater that provides the same production rate and heating accuracy as one described previously should have a length in the range 1.1 to 1.2 m.

Figure 5.5 presents some optimization results. Figure 5.5a shows temperature distribution along the minimum length heater. It is clear that, under $W^{\theta} = const$, a temperature maximum is held at the required level. Figure 5.5b shows a radial temperature profile after exiting an induction coil that is identical to the temperature profile for the similar case of time-optimal process (see Figure 3.8).

Figure 5.5c shows the optimal control algorithm (Equation 5.26) with respect to absolute units for the surface density $P_0 = PX$ according to Equation (1.25). The optimization was performed for input data presented in Table 5.1. Optimal modes of continuous heating have been calculated for different values of magnetic field intensity at the surface of the billet.[1]

Improving technical and economical parameters in the course of an optimization procedure allows using them as optimal design solutions when it is necessary to minimize the heater length. It can be shown that the following spatial control, of the conventional type

$$W_1^*(y) = \begin{cases} W_{1\max}^{(1)}, & 0 < y < y_{\max}^{(1)}; \\ W_{1\max}^{(2)}, & y_{\max}^{(1)} < y < y_{\max}^{(2)}; \\ W_{1\max}^{(3)}, & y_{\max}^{(2)} < y \le \Delta_1^0 = y^0 = 1, \end{cases} \tag{5.28}$$

cannot provide the required heating nonuniformity $\varepsilon_0 < 25$°C. Higher accuracy can be obtained after utilizing the control algorithm of Equation (5.28) by decreasing specific power $W_{1\max}$ of internal heat sources. As a result, an optimal heater length should be increased. At the same time, electrical and thermal efficiencies will be reduced.

In contrast to Equation (5.28), it is advantageous to use the optimal power control algorithm (Equation 5.26) in the case of $\Delta_1^0 = y^0$, $\Delta_2^0 = 0$, when a heating cycle includes a stage of holding maximum temperature at the admissible level. At the same time, the soaking stage is not envisaged in this case. It is possible to show that, for given heating accuracy ε_0, which can be attainable under control input in the form of Equation (5.26) at $\Delta_2 = 0$, the minimum heater length as a

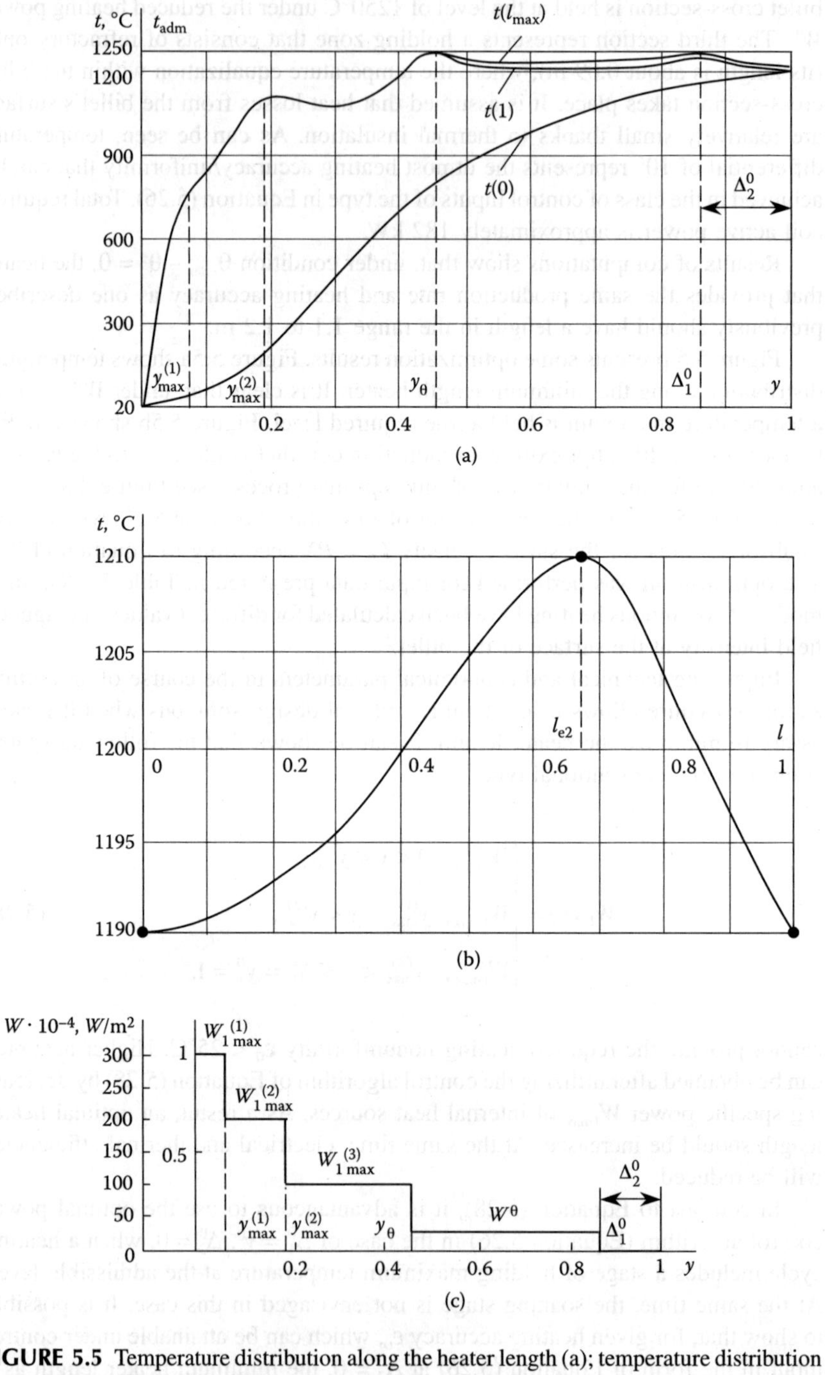

FIGURE 5.5 Temperature distribution along the heater length (a); temperature distribution along billet radius at the inductor exit (b) and optimal spatial control (c) for continuous heating of steel cylindrical billets.

function of specific heating power has a well-defined minimum that depends on ε_0 and θ_{max}.

For example, to satisfy condition $\theta_{max} - \theta_{adm} = 0$, the optimal process of accelerated heating provides the utmost admissible heating accuracy $\varepsilon_{inf} = \varepsilon_{min}^{(1)}$, which is equal to 13°. The value of ε_{inf} can be varied up to ±3.5° when heat power P_{max1} increases from 36 to 73 W/m³.

In a number of cases, the optimal controls of the type in Equation (5.26) under $\Delta_2^0 > 0$ provide appreciable advantages over conventional heating technology. In the case of required heating accuracy $\varepsilon_0 = 10 \pm 25°$, these algorithms provide decrease of the heater length by 2 ÷ 2.5 times and increase of electrical efficiency by 5 to 7%.[1] The utmost heating accuracy ε_0, which coincides with the minimax value $\varepsilon_{min}^{(2)}$ with respect to the class of control inputs (Equation 5.26), will be equal to 8 to 10°C and weakly depends on the maximum level of heating power.

As was mentioned in Section 5.1.2, the heater section of the length Δ_2^0 that has no power ($W_1^*(y) = 0$) can be excluded because the temperature will be equalized during the transfer stage to hot working equipment. In this case, the value of $\tilde{\varepsilon}_{min}^{(2)}$ increases up to 20 to 25° for time of transfer stage and level of heat losses in the open air that would correspond to real-life conditions.

5.1.4 Optimization of the Continuous Heating Process Controlled by a Power Supply Voltage

Section 5.1.2 and Section 5.1.3 provided several examples of solution of a design problem that minimizes the heater length. *Optimal design solution* represents a heater consisting of alternating inline sections with heat ON and heat OFF (active and passive zones). The problem can be further reduced to determining the number of inline coils with different geometry and power. If the number of sections, N, and their sizes are defined by the given IHI design, the chosen voltages $U_1, U_2, \ldots, U_N$ (or appropriate heat powers $u_1, u_2, \ldots, u_N$) for all N coils of an induction heater can be considered as spatial control inputs. In this case, a problem of determining optimal coil voltages can be formulated as the problem in Equation (2.38) of obtaining the maximum heating accuracy (Section 5.1.1).

Let us consider one industrial continuous heater with ten coils designed for heating of steel billets up to 1280°. All the computations have been done using a complex electrothermal model of temperature distribution within the ferromagnetic billets. This model has been developed at the Institute for Electrothermal Processes of Hannover University.[4,5] The specialized software package performs electrical calculation of induction heating installation and represents temperature profiles within billet cross-sections that will depend upon coil voltages. Design specifics and input data are presented in Table 5.2.

The optimal values of coil voltages $U_i^0,\ i = \overline{1, N}$ can be found according to a universal technique of alternance method in the typical cases of single-section and two-section heater design ($S = 1$ and $S = 2$) that allows one to obtain heating accuracy at the level of $\varepsilon_{min}^{(1)}$ and $\varepsilon_{min}^{(2)}$ (Figure 5.6) (see Section 5.1.1).

TABLE 5.2
Initial Data for Numerical Simulation

Inductor	
Length, m	8.43
Diameter of coils, m	0.273
Nominal throughput, kg/h	7000
Billet	
Length, m	0.446
Square cross-section, mxm	0.17 × 0.17
Initial temperature, °C	20
Required temperature, °C	1280
Ambient temperature, °C	20
Material	Steel

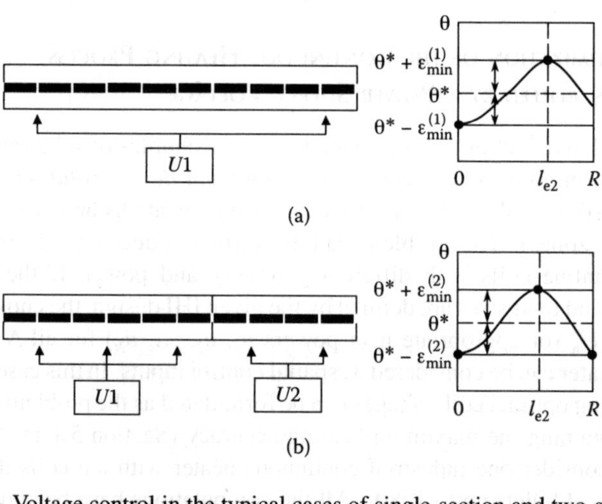

FIGURE 5.6 Voltage control in the typical cases of single-section and two-section heater design (a: $S = 1$, $\varepsilon = \varepsilon^{(1)}_{min}$. b: $S = 2$, $\varepsilon = \varepsilon^{(2)}_{min}$).

The computational results confirm that the optimal temperature profiles will be identical to the temperature profiles for the similar cases of the general time-optimal problem. Computational sets of required equations are similar to the sets in Equation (3.12) and Equation (3.16) and can be written for $S = 1$ and $S = 2$, respectively, as:

$$\begin{cases} \theta(0, U_1^0) - \theta^* = -\varepsilon_{\min}^{(1)}; \\ \theta(l_{e2}, U_1^0) - \theta^* = \varepsilon_{\min}^{(1)}; \\ \dfrac{\partial\theta(l_{e2}, U_1^0)}{\partial l} = 0; \end{cases} \quad (5.29)$$

$$\begin{cases} \theta(0, U_1^0, U_2^0) - \theta^* = -\varepsilon_{\min}^{(2)}; \\ \theta(l_{e2}, U_1^0, U_2^0) - \theta^* = \varepsilon_{\min}^{(2)}; \\ \theta(1, U_1^0, U_2^0) - \theta^* = -\varepsilon_{\min}^{(2)}; \\ \dfrac{\partial\theta(l_{e2}, U_1^0, U_2^0)}{\partial l} = 0. \end{cases} \quad (5.30)$$

These sets can be solved by conventional iteration methods with respect to all unknown variables including minimax values $\varepsilon_{\min}^{(1)}$ and $\varepsilon_{\min}^{(2)}$. The numerical model allows computing the final maximum and minimum temperature deviations from the required temperature θ^* at nodes of the spatial mesh. All induction heating coils should be grouped and it should be decided how many sections are necessary and how many coils would be located in those sections.

Figure 5.7 shows how a number of inductor coils in the second controlled section ($S = 2$) affects the minimax value $\varepsilon_{\min}^{(2)}$ and maximum temperature that took place during a heating cycle.[4,5] Computations have been conducted for heating of slabs with cross-section 170 mm × 170 mm at a nominal throughput

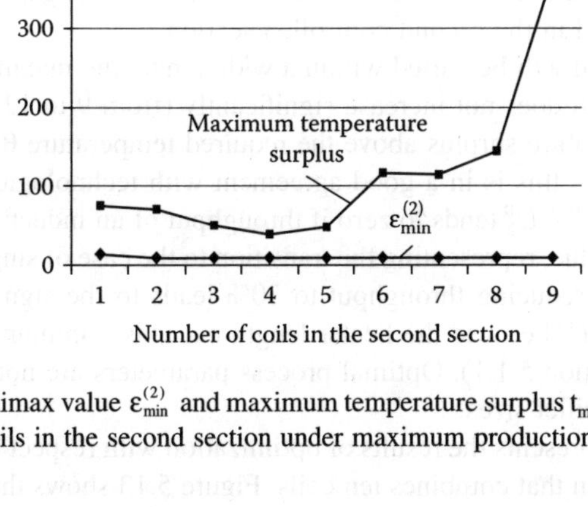

FIGURE 5.7 Minimax value $\varepsilon_{\min}^{(2)}$ and maximum temperature surplus $t_{max} - t^*$ as a function of a number of coils in the second section under maximum production rate.

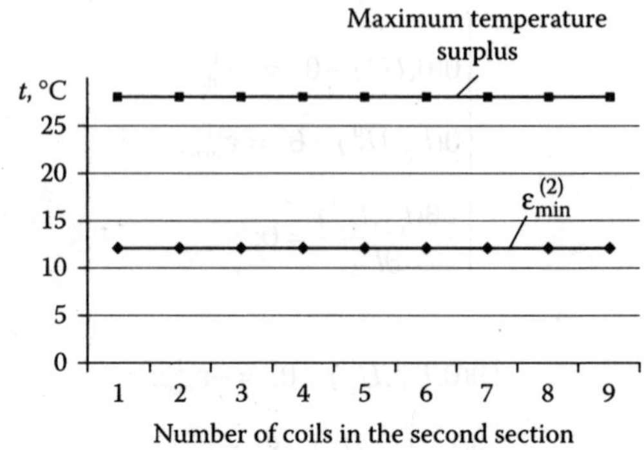

FIGURE 5.8 Minimax value $\varepsilon^{(2)}_{min}$ and maximum temperature surplus $t_{max} - t^*$ as a function of a number of coils in the second section under production rate of 30%.

of 7000 kg/h. As one can see, the attainable accuracy of heating does not depend on a number of coils in the second section. At the same time, the maximum temperature surplus as a function of a number of coils in the second section has a well-pronounced minimum. Therefore, the minimum overheat can be reached for the case when six inductors are included in the first controlled section and four inductors in the second section. Based on this conclusion, this scenario can be recommended as an optimal design solution for multicoil heater under maximum temperature technological constraint (Equation 5.14).

The optimally controlled induction heater of the given design has been investigated for different throughputs. Even when throughput decreases from 100 to 30%, the maximum temperature deviation $\varepsilon^{(2)}_{min}$ does not increase significantly (from 9 to 12°). In this case, the overheat value $\theta_{max} - \theta^*$ and the value of $\varepsilon^{(2)}_{min}$ do not depend on the number of coils in the second section (Figure 5.8).

Figure 5.9 through Figure 5.11 show the optimal coil voltages $U_1{}^0$ and $U_2{}^0$, minimax $\varepsilon^{(2)}_{min}$, and overheat $\theta_{max} - \theta^*$ as functions of throughput when there are four coils located in the second controlled section.

If billet speed will be varied within a wide range, the maximum temperature deviation $\varepsilon^{(2)}_{min}$ still does not increase significantly (from 9 to 12°C). At the same time, the temperature surplus above the required temperature $\theta_{max} - \theta^*$ does not exceed 15 to 40°; this is in a good agreement with technological requirements. The difference $U_1^0 - U_2^0$ tends to zero if throughput of an induction heater is 50% of its nominal value, representing the transition to the case of single section under control. Further reducing throughput to 30% leads to the sign inversion of the difference $U_1^0 - U_2^0$ because the heater length exceeds minimum length required for heating (Section 5.1.1). Optimal process parameters are not greatly affected by a variety of billet sizes.

Figure 5.12 presents the results of optimization with respect to using a single-controlled section that combines ten coils. Figure 5.13 shows the same results of optimization in the case of two individually controlled sections (six and four

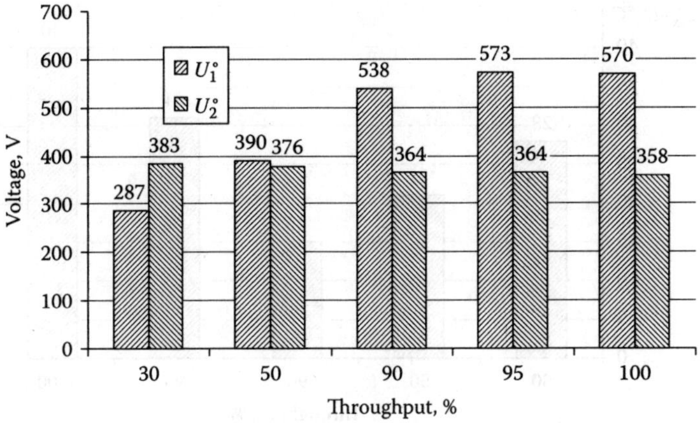

FIGURE 5.9 Optimal voltages U_1^0 and U_2^0 as function of throughput in the case of four coils located in the second controlled section.

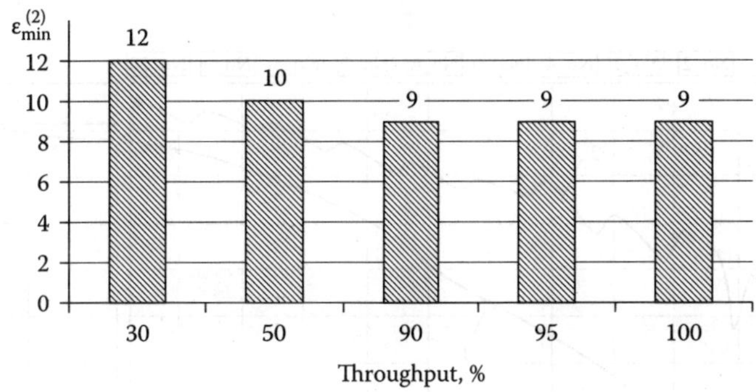

FIGURE 5.10 Minimax value $\varepsilon_{min}^{(2)}$ as function of throughput in the case of four coils located in the second controlled section.

coils, respectively). Figure 5.12 and Figure 5.13 show the temperature distributions at optimal operating mode at the nominal throughput of 100%. Nonmonotonous character of the surface temperature variation can be explained by intersection gaps and nonlinear effects due to the transition between ferromagnetic and paramagnetic phases of heated material at the Curie point.

When throughput decreases from nominal value to 30%, the temperature drops within the billet cross-section reduce significantly (Figure 5.14) because of sign inversion of the difference $U_1^0 - U_2^0$ (Figure 5.9).

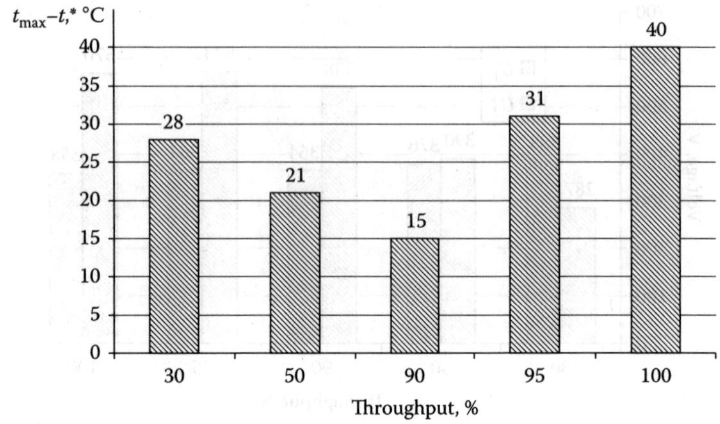

FIGURE 5.11 Maximum temperature surplus $t_{max} - t^*$ as function of throughput in the case of four coils located in the second controlled section.

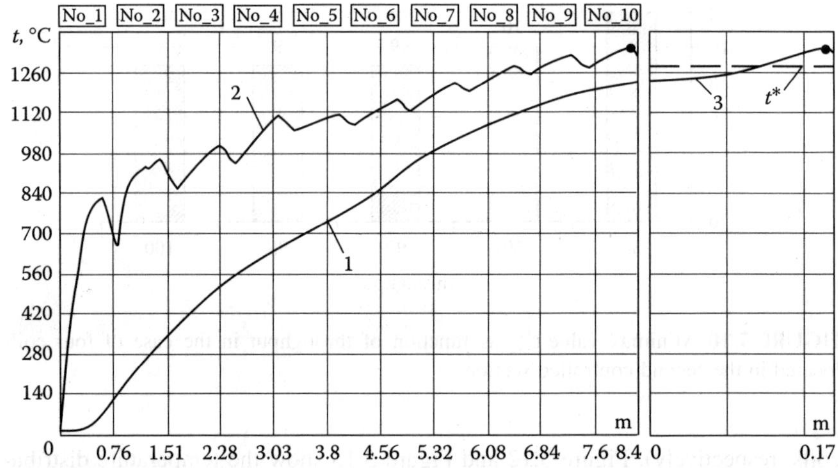

FIGURE 5.12 Temperature distribution along the heater length in the billet center (1) and at the billet surface (2) and temperature distribution along billet radius at the inductor exit (3) in the case of single-controlled section for maximum throughput at heating accuracy $\varepsilon = \varepsilon_{min}^{(1)}$.

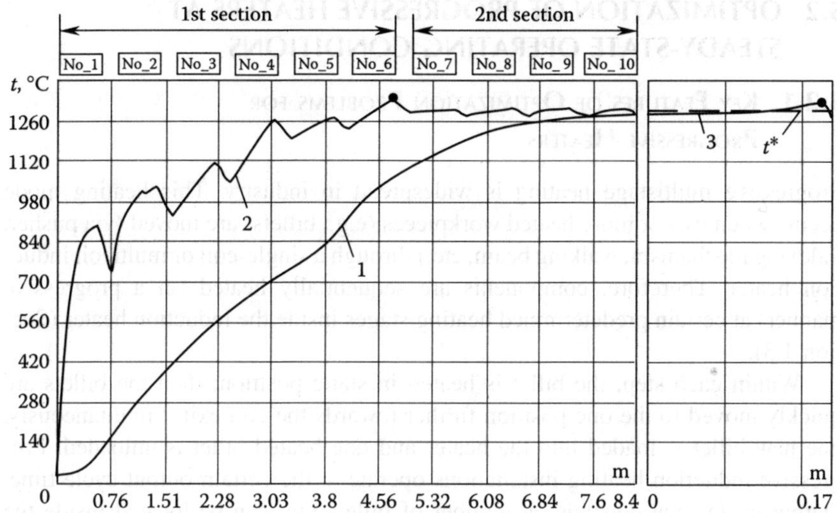

FIGURE 5.13 Temperature distribution along the heater length in the billet center (1) and at the billet surface (2) and temperature distribution along billet radius at the inductor exit (3) in the case of two individually-controlled sections for maximum throughput at heating accuracy $\varepsilon = \varepsilon_{\min}^{(2)}$.

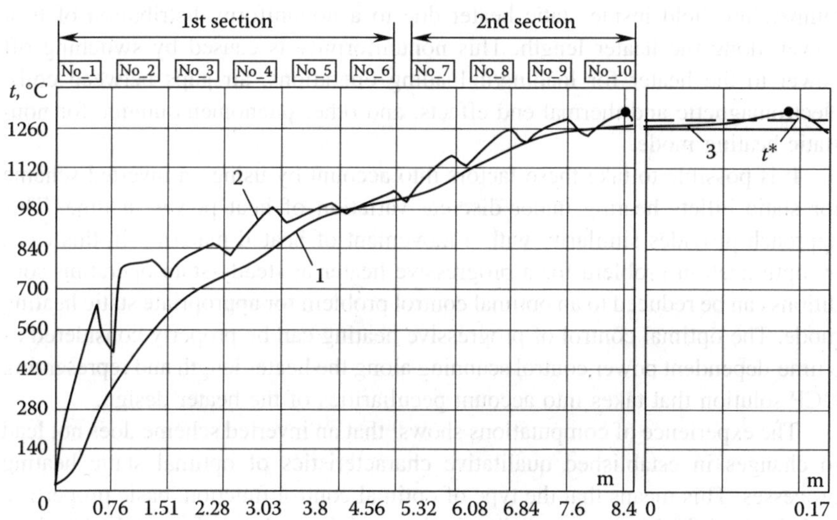

FIGURE 5.14 Temperature distribution along the heater length in the billet center (1) and at the billet surface (2) and temperature distribution along billet radius at the inductor exit (3) in the case of two individually-controlled sections for 30% throughput at heating accuracy $\varepsilon = = \varepsilon_{\min}^{(2)}$.

5.2 OPTIMIZATION OF PROGRESSIVE HEATERS AT STEADY-STATE OPERATING CONDITIONS

5.2.1 Key Features of Optimization Problems for Progressive Heaters

Progressive multistage heating is widespread in industry. This heating mode occurs when two or more heated workpieces (e.g., billets) are moved (via pusher, indexing mechanism, walking beam, etc.) through a single-coil or multicoil induction heater. Therefore, components are sequentially heated (in a progressive manner) at certain predetermined heating stages inside the induction heater (Section 1.3).

Within each step, the billet is heated in static position; then the billets are quickly moved to the one position further towards the coil exit. Simultaneously, one new billet is loaded into the heater and one heated billet is unloaded. Progressive induction heating installations operate at the certain output cycle time. Output cycle time depends on number of billets that can be located inside the heater. The value of output cycle time can be changed during transient operating modes.

The heating of each billet in a progressive manner differs from static heating by the fact that the billet is heated step by step in a number of static positions inside a heater instead of heating in a single constant position. Therefore, a temperature distribution within the inductor load is distorted in comparison with temperature field inside static heater due to a nonuniform distribution of heat power along the heater length. This nonuniformity is caused by switching off power to the heater for loading/unloading operations, air gaps between coils, electromagnetic and thermal end effects, and other phenomena unique for non-static heating mode.

It is possible to take these factors into account by using an inverted scheme for static billets heating under discrete variation of heat power in time. This approach provides similarity with a movement of heated material. In this case, an optimization problem for a progressive heater at steady-state operating conditions can be reduced to an optimal control problem for appropriate static heating mode. The optimal control of progressive heating can be properly considered as a time-dependent power control scanning along the heater length and representing OCP solution that takes into account peculiarities of the heater design.

The experience of computations shows[1] that an inverted scheme does not lead to changes in established qualitative characteristics of optimal static heating processes. This means that the type of optimal control function, basic properties, and shape of final temperature distribution remain unchanged as applied to optimization problems for static heating.

Based on these conclusions, the alternance method can be extended onto a wide range of IHP and IHI optimization problems for progressive heaters. However, essential features arise when applying the alternance method. They deal with unique features of the progressive heating conditions.

First of all, the more complicated electrothermal models are needed for the accurate simulation of progressive heating modes. Moreover, the specific constraints on the coil lengths appear when scanning the optimal time-dependent program of heating power variation along the heater length. It is necessary that lengths of inductor coils should be divisible by the billets' length for each production run. If this requirement is not satisfied, then at least one of the billets will not be heated optimally because, during the heating in static position, it is located inside two coils with different powers and/or frequencies. This can also lead to an appearance of unacceptable temperature gradients along the billets' length.

Finally, one of the most typical implementations of a space–time control approach deals with a multicoil design with independently controlled power of each coil. This approach provides desired change of heating power in time and along the heater length. Therefore, the part of the optimal program for control input variation (including, for example, intervals of temperature holding under control $u^{\theta}(\varphi)$ in the form of Equation 4.3 and Equation 4.13) can be performed for each billet during an appropriate step between unloading of the neighboring billets. This is similar to the static heating mode. Therefore, in contrast to continuous heating mode with constant speed, V, of workpiece movement through an inductor, optimization of IHI with progressive movement can generate a need for consideration of space–time control functions (Section 2.4).

The required number of inline independently controlled coils depends upon the scanning version of time-dependent power control for static heating similar to continuous heating at a constant speed (Section 5.1). For the majority of practical problems, a two-section inductor with one independently controlled coil represents optimal design solution. Only the one billet should be located inside the coil under autonomous control (Section 5.2.2).

Multicoil inductor design exhibits the greatest potential for control if all inline coils may be controlled individually and only one billet is located inside each coil. The induction heating installations of similar designs are complex and expensive. They are only used in critical applications for heating large billets or materials made from expensive alloys.

In contrast to this case, the simple operational modes of induction installations are utilized in many practical applications. Stabilizing a common optimal inductor voltage represents one of the simplest means to control the process of induction heating. The single-section heater represents the simplest inductor design that does not provide independent control of individual coils. A single-section heater can be used in practice as a single coil of required length or as inline coils that may have different lengths and windings, at the same time. Common voltage of power supply should be applied to all coils (Chapter 1). A multicoil induction heater design provides symmetrical loading of three-phase power supply. Obviously, in this case, it is impossible to provide optimal time-dependent variation of heat power with respect to each separate billet.

Based on these general conclusions, some particular optimization problems will be considered with regard to different induction heaters that operate in progressive mode.

5.2.2 Optimization of Induction Heater Design and Operating Modes

Similarly to Chapter 3 and Chapter 4, let us consider a time-optimal control problem for progressive induction heating with prescribed accuracy ε at the exit of an inductor.

As was shown in Section 5.2.1, the optimal control problem for progressive heating mode can be reduced to a time-optimal control problem for inverted scheme of static heating mode. This means that optimal control of progressive heating can be properly considered as a time-dependent power control scanning along the heater length. It is important to underline that all essential features and appropriate relationships between time-dependent control and spatial scanning along the heater length should be taken into account.

The shape of the optimal control algorithm remains similar to the case of static induction heating. The typical algorithm of time-optimal power control with consideration of the "real-life" technological constraints can be represented by the expressions in Equation (4.1) through Equation (4.5). The difference only deals with the expressions that describe temperature distribution. These expressions will be modified and become more complicated to simulate the movement of billets.

Results of computation show that, in a majority of practical cases, the first stage of time-optimal control for static heaters represents heating with maximum power $u \equiv U_{\max}$ or using power close to its maximum value (in case it is necessary to hold tensile thermal stress at admissible levels according to expressions in Equation 4.1 through Equation 4.5, Figure 4.2 through Figure 4.5, and Figure 4.12). The value of φ_h represents duration of this control stage.

Let us denote the time required for heating of one billet as φ^0_{opt} and the duration of one step of progressive heating (output cycle time) as φ^*. If the following condition:

$$\varphi^0_{opt} - \varphi_h \le \varphi^* , \tag{5.31}$$

is satisfied, then the optimal design solution represents a heater that consists of two sections of different lengths.

In the first section, where $B \ge 1$ billets are located simultaneously, the maximum constant voltage or voltage close to its maximum admissible value is applied for duration $B\varphi^*$. This stage provides intensive heating under the condition that the maximum value of tensile thermal stresses would not exceed prescribed admissible value σ_{adm}.

In the second independently controlled section, only one billet of maximum length can be located. The second control stage of duration φ^* takes place in this section according to the optimal control algorithm. Each section can be used in practice as a single coil of required length or as inline coils that may have different lengths and windings but should be fed from a common power supply. The length

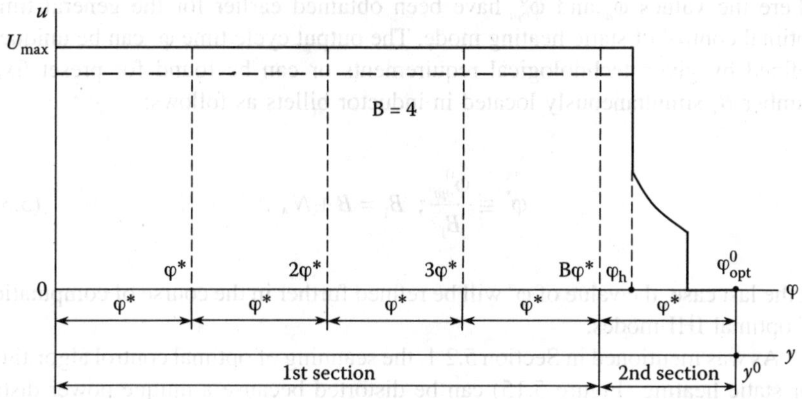

FIGURE 5.15 Optimal control algorithm for the heater with the second independently controlled section.

of each section depends upon the length and number of billets that can be located inside this section simultaneously.

Figure 5.15 presents an optimal control mode when it is necessary to provide the maximum heating accuracy $\varepsilon = \varepsilon^{(2)}_{\min}$ using a two-stage, time-optimal control ($N = 2$) under the condition in Equation (5.31). During the whole heating process, the maximum temperature was held at the admissible level θ_{adm}. Figure 5.15 shows how power should be changed during a heating cycle according to scanning of the optimal control algorithm (Equation 4.3) along the heater length, if the equality takes place in Equation (5.31).

If, during the optimal process, the value of tensile thermal stress reaches its admissible maximum value $\sigma_{\max}$, then solution of the problem should be found among optimal control functions in the form of Equation (4.1) or Equation (4.2). In this case, the heating in the first section takes place under power that should be less than $U_{\max}$ in order to hold tensile thermal stress at admissible level σ_{adm} (Figure 5.15, dashed line). Obviously, in this case there is a certain fallibility that would lead to quasioptimal control.

If the output cycle time exceeds time required for the last control stage, i.e., $\varphi^0_{opt} - \varphi_h < \varphi^*$, then the optimal heating cycle is over before billet unloading (Figure 5.15). In this case, the heating cycle could be longer than minimal possible time required for heating of a separate billet. When a billet is located inside the inductor outlet section, it is necessary to stabilize its temperature during the remaining time of this cycle.

If the condition $\varphi^0_{opt} - \varphi_h > \varphi^*$ is met, then optimal design solution becomes more complicated because of the necessity to increase the number, N_A, of independently controlled sections where only one billet of maximum length can be located.

Minimum integer number N_A can be found using the expression:

$$N_A - 1 < (\varphi^0_{opt} - \varphi_h) / \varphi^* \le N_A, \tag{5.32}$$

where the values φ_h and φ^0_{opt} have been obtained earlier for the general time-optimal control of static heating mode. The output cycle time φ^* can be uniquely defined by given technological requirements or can be found for preset fixed number B_1 simultaneously located in inductor billets as follows:

$$\varphi^* \cong \frac{\varphi^0_{opt}}{B_1};\ B_1 = B + N_A\ . \tag{5.33}$$

In the last case, the value of φ^* will be refined further in the course of computation of optimal IHI modes.

As was mentioned in Section 5.2.1, the scanning of optimal control algorithms for static heating (Figure 5.15) can be distorted because a unique power distribution along the billet length takes place during each step. This can be simply explained by nonuniform distribution of magnetic field along the heater length due to electromagnetic longitudinal end effects, air gaps between coils, and interaction between fields of different phases of power supply. These factors affect the final temperature distribution within the heated workpiece, and proper mathematical models should describe such an influence.

As a result, the inverted computation scheme of the heating process should take into account the fact that the spatial distribution of internal heat power within the billet will be changed in time from step to step. For example, the function $W_d(\xi,l,y)$ for a two-dimensional heating process, according to Equation (4.72) through Equation (4.75), can be written with regard to the optimal control algorithm shown in Figure 5.15 as a stepwise function for each cycle of the cylindrical billet heating in the form:

$$W_d(\xi,l,y,\varphi) = W_d^{(k)}(\xi,l,y),\ \varphi \in (\bar{\varphi}_{k-1},\bar{\varphi}_k),\ \bar{\varphi}_k = k\varphi^*,\ k = \overline{1, B+1},\ \varphi_0 = 0\ . \tag{5.34}$$

The function $W_d(\xi,l,y,\varphi)$ describes step-by-step movement of the workpiece through the inductor. The function $W_d^{(k)}(\xi,l,y)$, which describes the distribution of internal heat sources during the kth step, can be defined a priori by any well-known method based on electromagnetic coil computation.[2,6,7]

The part of the first stage of optimal control algorithms in Equation (4.1) through Equation (4.5) would be performed during the interval $[0,\varphi_h]$, where $\varphi_h \geq B\varphi^*$. The function $W_d(\xi,l,y,\varphi)$ in this case takes on all values from $W_d^{(1)}$ through $W_d^{(B)}$ sequentially (Figure 5.15). During the last stage $(\varphi_h,\varphi^0_{opt})$, the expression $W_d = W_d^{(B+1)}$ is valid as the remaining part of the optimal program. The expressions for final temperature distribution will look quite different in comparison with temperature distribution at the end of optimal static heating due to the difference between stepwise function $W_d^{(k)}$ and constant function $W_d(\xi,l,y)$.

As an example, let us consider the optimal control algorithm in the form of Equation (2.27) with neglected technological constraints on σ_{max} and θ_{max} during the heating process. In this case, instead of Equation (4.95), the following expression for temperature distribution can be used under boundary conditions of the third kind[1]:

$$\theta(l,y,\Delta)=\Phi_3\left(l,y,\sum_{i=1}^{N}\Delta_i\right)+U_{\max}\sum_{r=1}^{\infty}\sum_{n=1}^{\infty}\frac{D_{rn}}{\eta_{rn}}\left(\cos\lambda_n y+\frac{Bi_0}{\lambda_n}\sin\lambda_n y\right)\cdot J_0(\mu_r l)\times$$

$$\times\left\{\sum_{k=1}^{B}W_{rn}^{(k)}\left[\exp\left(-\eta_{rn}\left(\sum_{i=1}^{N}\Delta_i-k\varphi^*\right)\right)-\exp\left(-\eta_{rn}\left(\sum_{i=1}^{N}\Delta_i-(k-1)\varphi^*\right)\right)\right]+\right.$$

$$\left.+W_{rn}^{(B+1)}\left[\left(1-\exp\left(-\eta_{rn}\left(\sum_{i=1}^{N}\Delta_i-B\varphi^*\right)\right)\right)+\sum_{j=2}^{N}(-1)^{j+1}\left(1-\exp\left(-\eta_{rn}\sum_{i=j}^{N}\Delta_i\right)\right)\right]\right\}, \tag{5.35}$$

where

$$\eta_{rn}=\mu_r^2+\beta^2\lambda_n^2. \tag{5.36}$$

The variables $W_{rn}^{(k)}$, $k=1,B+1$ can be calculated similarly to $W_{rn}(\xi)$ in Equation (4.83). They represent the coefficients of expansion of functions $W_d^{(k)}(\xi,l,y)$ in infinite multiple series in terms of fundamental functions.

In the general case of $N_A = 1$, the following expression directly links the output cycle time φ^* in Equation (5.35) with a total process time φ^0:

$$(B+1)\varphi^*=\sum_{i=1}^{N}\Delta_i=\varphi^0. \tag{5.37}$$

In the general case of $N_A \geq 1$, the following equation can be obtained instead of Equation (5.37):

$$(B+N_A)\varphi^*=\sum_{i=1}^{N}\Delta_i. \tag{5.38}$$

Similarly to the one-dimensional case (Section 4.1 and Section 4.2), the technological constraints on temperature $\theta_{\max}$ and thermal stress $\sigma_{\max}$ can be taken into account if they are violated during the heating cycle under optimal control (Equation 2.27). The time of billet transfer in the open air can be considered as well.

The temperature distribution within the billet is distorted due to nonuniform distribution of heat power along the heater length. This nonuniformity is caused by switching off power to the heater for loading/unloading operations, air gaps between coils, end effects, and other process features that are unique for nonstatic heating mode. These effects can be described if the optimal control program $u^*(\varphi)$ would be supplemented by powerless intervals where applied power $u(\varphi) \equiv 0$. The influence of contacts between butt-ends of billets usually can be neglected.

When nonlinear mathematical models of heating process are applied, scanning of time-dependent power control for static heating should be modified because the heating power depends upon temperature distribution within the workpiece. In this case, the temperature distribution can be computed by using the method of successive approximations or by applying the model that describes all stages of the heating process: from the nonstationary stage under given initial conditions until the steady-state stage.[1,2]

As a result, the expressions for temperature distribution within each billet at the inductor exit can be obtained in the explicit form as functions of the set of optimal process parameters Δ_i, $i = \overline{1, N}$. These functions can be calculated analytically or numerically for any given inductor design. Similarly to the previously described solution, the values Δ_i, $i = \overline{1, N}$ represent durations of stages of power control functions that can be described in the form of Equation (4.1) through Equation (4.5). These functions will be used in sets of the equation that ought to be solved with respect to unknown process parameters Δ_i, $i = \overline{1, N}$. The computational technique remains unchanged in comparison with that described in Chapter 3 and Chapter 4 with respect to static heating.

Two different statements of the time-optimal problem will arise with regard to progressive heating that describe various aspects of requirements to output cycle time φ^* for heated billets. Based on Expression (5.33), Expression (5.35), and Expression (5.37), one can conclude that optimal control algorithms can be defined uniquely by using the set of parameters $\Delta = (\Delta_i)$, $i = \overline{1, N}$ in the case when one of the values φ^* and B_1 is known a priori. Otherwise, the problem has no single solution.

Let us assume that the number B_1 of billets, which can be located simultaneously inside the inductor of given length, is fixed. In this case, a number of independently controlled sections, N_A, can be found by Expression (5.32) and Expression (5.33) using previously obtained results for the static heating process. A number of billets that can be located simultaneously inside the first section should be calculated by the formula $B = B_1 - N_A$.

The problem can be reduced to the problem of minimizing the value $\varphi^*_{\min}$ that provides a minimum output cycle time under the condition that required temperature will be reached with desired accuracy at the inductor exit. The value of $\varphi^*_{\min}$ can be calculated by Equation (5.38) after substitution of values Δ_i^0, $i = \overline{1, N}$, under given numbers B and B_1.

If the computations reveal that the condition in Equation (5.32) is not violated under obtained values φ^0_{opt} and φ_h at given number N_A, then the obtained design solution should be considered as an optimal. Otherwise, the design solution should be changed and optimal control parameters should be refined by the method of successive approximations. It is assumed here that a heater operates under condition $N_A - 1 < (\varphi^0_{\text{opt}} - \varphi_h)/\varphi^* = N_A$ (in comparison with Equation 5.32) — i.e., the billet will be not delayed in the last section (Figure 5.15).

At this point, it is necessary to compare obtained numerical values N_A for the whole variety of billets available for heating and to select a multicoil design solution that would maximize its value. This design solution should be considered as optimal for any production run of the induction heater that operates in a steady-state mode.

If output cycle time is fixed by technological charts, then the integer numbers B, N_A, and $B_1 = B + N_A$ can be found as initial approximations using Expression (5.32) and Expression (5.33). These numbers should be refined by an iterative method when solving sets of equations with respect to vector Δ^0 under given values φ^*, B, and N_A. If, after completion of the previous step, it is found that the expression $\varphi^0_{opt} - \varphi_h > \varphi^* N_A$ is valid, then it is possible to conclude that billets' unloading will take place with certain delays in comparison with an optimal rate.

The problem can be reduced to the problem that minimizes heater length, assuming that only integer number of billets of one production run can be located simultaneously inside the inductor. This means that a minimum possible number of billets should be heated simultaneously.

Optimization of the progressive induction heating process using an inductor with independently controlled sections leads to the complex design solutions. In the case of a multicoil inductor design, an appropriate number of inline coil sections will be used. Each section may have coils of different length, and number of turns, meaning that each section can have different applied power. This allows providing a stepwise power distribution along the heating line. Such a distribution of power can be interpreted as a scanning of a time-dependent heat power program along the heating line that is similar to the case of continuous heating at constant speed (Section 5.1). It is important to underline that heat power can be varied from section to section but does not change in time.

Therefore, the number of coil sections, their lengths, and their powers should be found assuming that only an integer number of billets of one production run can be located simultaneously inside each section. It is assumed that output cycle time φ^* will be constant during the whole heating cycle. Each section can consist of one or more separate coils. The number of sections should correspond to the number of intervals of the optimal time-dependent heat power program for static heating that has been "scanned" along the heater length.

To simplify inductor design, the problem can often be solved in the simple case of providing heating accuracy $\varepsilon \geq \tilde{\varepsilon}^{(2)}_{\min}$ under condition $\alpha_T = 1$ that allows performing temperature equalization during the billet transportation to the metal forming operation (Section 4.2). In this case, as was shown in Section 4.2, the time-optimal control process will be completed during the first stage of the control algorithm (Equation 4.1) that would be performed under maximum permissible level of power. It is important to notice that this control process will be time-optimal and at the same time will provide the best accuracy of heating as well as it will require the minimum of energy consumption. If we limit our consideration to the simplest practical algorithm (Equation 4.2), then the three-section inductor design (Figure 4.3) will be required.

In the first section, an accelerated heating takes place under maximum power restrained only by maximum permissible power level of the power supply. In the second and third sections, the maximum thermal stress and maximum temperature within billets should be held at maximum admissible levels.

If, during the first heating stage under applying the maximum power, the temperature maximum reaches admissible value θ_{adm} and, at the same time, the value $\sigma_{\max}$ still remains less than σ_{adm}, then holding interval for $\sigma_{\max}$ should be excluded from the optimal control algorithm. If the first heating stage is completed before $\theta_{\max}$ reaches θ_{adm}, then the interval of holding of $\theta_{\max}$ should be excluded from the optimal heat power control algorithm.

In both cases, the two-section heater can be considered as an optimal design solution (see algorithms in Equation 4.3 through Equation 4.5) as well as in the case when powers of the first and second sections are slightly different. On the other hand, if the duration of interval of holding the maximum temperature is relatively long, then there might be unacceptable accuracy of temperature stabilization under constant inductor power. In this case, to conduct the required control algorithm, at least two inductor sections with different powers should be used. The final design solution can be selected by method of successive approximation.

Approximate values of heat powers for separate sections can be found similar to Section 5.1.2 by minimizing deviations of maximum temperature or thermal stresses from their admissible values. A simplified approach can be introduced to determine an optimal value of heat power by utilizing the conditions $\sigma_{\max} = \sigma_{adm}$ and $\theta_{\max} = \theta_{adm}$. These conditions should be satisfied at the end of holding intervals. Our experience shows[1] that, during the whole holding stage, this approach guarantees that maximum temperature $\theta_{\max}$ and/or thermal stress $\sigma_{\max}$ will not exceed admissible values θ_{adm} and/or σ_{adm}. Moreover, the accuracy of approximation for $\sigma_{\max}$ to σ_{adm} and/or $\theta_{\max}$ to θ_{adm} will also be sufficient.

If it is necessary to provide the maximum production rate for the given number of billets located simultaneously inside the inductor, then an optimal design solution can also be found according to the previously described technique. After computation of approximate value of φ^* using Expression (5.33), the appropriate initial values of number of sections and their lengths as well as numbers of billets that will be simultaneously located in each section can be determined.

Obtained design parameters should be in good agreement with durations of appropriate holding intervals. Optimal design parameters can be refined by the method of successive approximations. In each iteration step, the solutions of appropriate sets of equations should be used in the form of Expression (5.35) for final temperature distribution.

Similarly to the technique described in Section 4.1, optimal values of heat powers for separate sections can be found by using an inverted scheme for static heating. Therefore, these values can be considered as given a priori in the course of solving sets of equations.

In the case of fixed value φ^*, which depends upon required heater production rate, the time of heating inside each vth section proves to be fixed and equal to $B^{(v)}\varphi^*$, where $B^{(v)}, v=\overline{1,r}$ is a number of billets that can be located simultaneously in the section with number v, and r is a number of sections.

Under given values $B^{(v)}$, the problem can be reduced to searching for number of sections, r, and optimal heat powers, P_v, for each separate section.

The heat power of the first section, P_1, can be defined from the following condition that is similar to Expression (4.8):

$$\sigma_{\max}\left(B^{(1)}\varphi^*\right)=\sigma_{adm}\,. \tag{5.39}$$

According to Equation (5.39), the maximum thermal tensile stress $\sigma_{\max}$ reaches its admissible value at the end of optimal heating stage $B^{(1)}\varphi^*$ in the first section.

Expression (4.8) can be considered as an equation with respect to time φ_1 when thermal tensile stress reaches its admissible value $\sigma_{\max}$ under given maximum heat power $U_{\max}$. In contrast to that, Expression (5.39) can be considered as an equation with respect to heat power P_1 under given time $\varphi_1 = B^{(1)}\varphi^*$.

Computation of the value P_2 according to the suggested method can be performed similarly to the case of P_1 by using the following condition:

$$\sigma_{\max}\left(\left(B^{(1)}+B^{(2)}\right)\varphi^*\right)=\sigma_{adm}\,. \tag{5.40}$$

The condition in Equation (5.40) takes place at the end of an interval when thermal stress is held at permissible level. Unknown heat power of the second section can be equal to the value P_2 obtained from Equation (5.40), assuming that constraint on $\theta_{\max}$ is not violated. This would be the case if the following condition holds true:

$$\theta_{\max}\left(\left(B^{(1)}+B^{(2)}\right)\varphi^*\right)\le\theta_{adm}\,. \tag{5.41}$$

Otherwise, unknown value P_2 can be obtained from the following expression:

$$\theta_{\max}\left(\left(B^{(1)}+B^{(2)}\right)\varphi^*\right)=\theta_{adm}\,. \tag{5.42}$$

Resulting heat power will be less than the value of P_2 that is obtained from Equation (5.40). Therefore, the value of $\sigma_{\max}$ should be reduced in comparison with permissible level.

Heat powers P_ν, $\nu = \overline{3, r}$ of the third and other sections can be found under condition $\varepsilon \geq \tilde{\varepsilon}^{(2)}_{\min}$ from equations similar to Equation (5.42):

$$\theta_{\max}\left(\varphi^*\sum_{k=1}^{\nu} B^{(k)}\right) = \theta_{adm}\ . \tag{5.43}$$

Equation (5.43) is obtained with regard to the end of the cycle when the billet is located inside the section of number ν.

Particular dependencies of $\sigma_{\max}$ and $\theta_{\max}$ on their arguments in Equation (5.39), Equation (5.40), Equation (5.42), and Equation (5.43) can be defined using appropriate mathematical models. Under fixed output cycle time, the obtained heating accuracy at the heater exit can be uniquely defined by chosen values of P_ν.

If number of billets $B^{(\nu)}$ that can be simultaneously located in each of the sections is not given a priori, then it leads to a different design solution with different correlation between P_ν and $B^{(\nu)}$.

Figure 5.16 shows some results[1] of computation that can serve as a good example of the proposed method in application to time-optimal induction heating of aluminum alloy slabs 7000 × 700 × 1580 mm^3. The required temperature is 470°. Induction heating installation consists of three sections ($r = 3$; $B^{(\nu)} = 1$; $\nu = 1, 2, 3$); production rate is 600 t/h; $\theta_{adm} = 520°$; and $\sigma_{adm} = 80$ MPa. Time of transportation from one section to another is assumed to be equal to 120 s. Time required for heating in each section corresponds to preset output cycle time being equal to 700 s.

5.2.3 Optimal Control of a Single-Section Heater

In the simplest cases, an optimal control algorithm for each separate billet can be reduced to holding control input at maximum permissible level during the whole heating cycle. It is important to notice that the simplest possible cases are, at the same time, the most practical ones. As was shown in Chapter 3 and Chapter 4, such simplified control strategy can be used under the following conditions:

- The values of maximum temperature $\theta_{\max}$ and maximum thermal stress $\sigma_{\max}$ do not reach their maximum admissible values during the optimal control process.
- The required heating accuracy $\varepsilon = \varepsilon_0$ coincides with minimax value $\varepsilon^{(1)}_{\min}$ (when time of transportation is neglected: $\Delta_T \approx 0$) or the expression $\varepsilon_0 \geq \tilde{\varepsilon}^{(2)}_{\min}$ holds true when the temperature equalization takes place during the transfer stage at $\alpha_T = 1$ (Section 4.2).

Final temperature distribution within the billet at the exit of the progressive induction heater can be described by expression of the type in Equation (5.35) at $N = 1$.

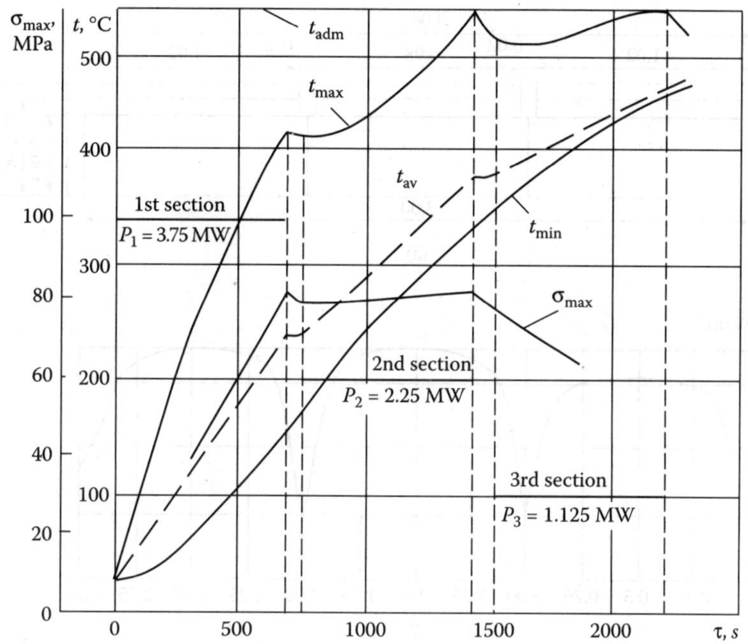

FIGURE 5.16 Time-optimal induction heating of aluminum alloy slabs with constraints on t_{max} and σ_{max} in three-section heater.

Optimal process time Δ_1^0, being equal to $\varphi^* B = \varphi^* B_1$ according to Equation (5.33), can be defined by solving the appropriate set of equations. Optimal output cycle time φ^*_{min} can be defined as a result of solution under preset number B_1. At the same time, the number of billets that can be located inside the heater can be defined under preset output cycle time.

Considered optimal heating modes can be relatively easily applied in engineering practice by using single-section heaters when common voltage of power supply should be applied to all coils. In this case, design solution leads to one or more inline coils without gaps between coils (standard modules) that would be electrically connected with common power supply.

Figure 5.17 and Figure 5.18 show some results of computation for steady-state induction heating of large aluminum alloy billets prior to extrusion. The time-optimal transportation problem has been solved under conditions $\varepsilon = \tilde{\varepsilon}^{(2)}_{min}$, $\Delta_T = \Delta^*_{T2}$ (Section 4.2), and $B_1 = 3$. In this case, the set of equations can be reduced according to the alternance method to a two-dimensional analogue of set of the type in Equation (4.40). Solution of this set with respect to required heating time Δ_1 allows one to define optimal output cycle time according to expression $\varphi^*_{min} = \Delta_1^0/3$.

Initial data are presented in Table 4.3 (Section 4.6.3). The computation has been conducted with respect to progressive multistage heating mode using a technique described in Section 4.6.3. Electromagnetic and temperature fields of

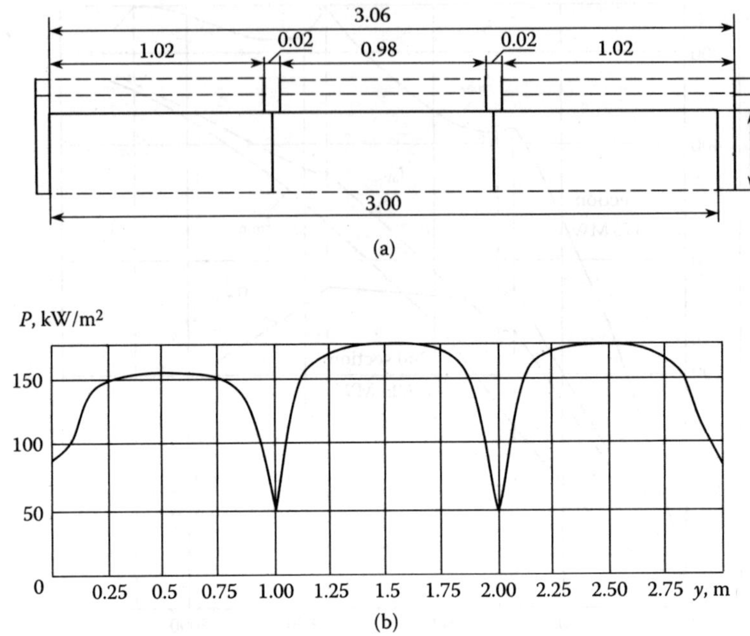

FIGURE 5.17 (a) Induction heating system and (b) heat power distribution along the heater length for heating of aluminum alloy billets before hot forming.

induction system "heater–billet" has been simulated based on a complex two-dimensional numerical model using external iteration cycle to define the output cycle time.* The mathematical model allows taking into account switching off the power supply for 60 s.

The induction heater consists of three inline coils without gaps between them, and common voltage of power supply is applied to all coils. Three billets can be located simultaneously inside the inductor (Figure 5.17). Figure 5.17b shows a distribution of heat power along the heater length.

As can be seen in Figure 5.18b, the shape of final temperature distribution and number of points, where maximum temperature deviations are reached, are identical to the temperature profile for the case of the static heating process (Section 4.5.3, Figure 4.39).

A special situation takes place in the cases of single-section heater optimization, when optimal control algorithms cannot be reduced to holding voltage at constant level. Such algorithms cannot be applied to each separate billet. Taking into account single-section heater design, it is necessary to define the optimal control algorithm that could be applied simultaneously to all billets located inside the heater.

* The computations have been performed by V.B. Demidovich.

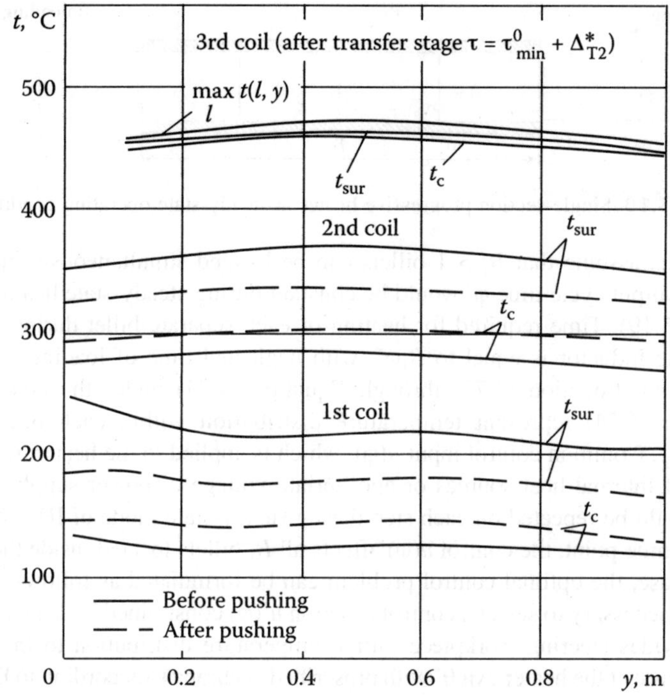

(a)

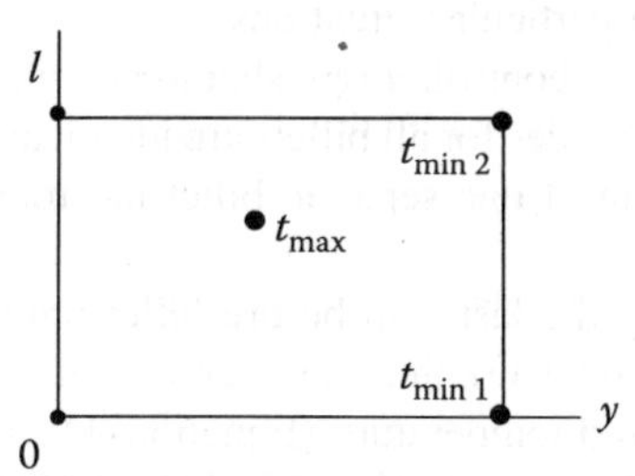

(b)

FIGURE 5.18 Results of optimization of steady-state induction heating of aluminum alloy billets in single-section heater. a: Temperature distribution along the billet length (t_{sur}, t_c — temperatures at the surface and in the center). b: Location of points within billet volume where maximum deviations of final temperature from required temperature are reached ($\varepsilon = \tilde{\varepsilon}^{(2)}_{min} = 10\,°C$; $t_{max} = 469°C$; $t_{min1} = t_{min2} = 449°C$; $\Delta_1^0 = 1248$ s; $\Delta_2^0 = 0$; $\Delta_T^0 = \Delta^*_{T2} = 134$ s; $\varphi^* = 376$ s).

Let us consider the problem of providing maximum production rate with respect to time-optimal heating of cylindrical billets. The two-dimensional linear equation of heat transfer (Equation 4.72 through Equation 4.75) under $\Gamma = 1$ will be considered further as a basic mathematical model of the induction heating process.

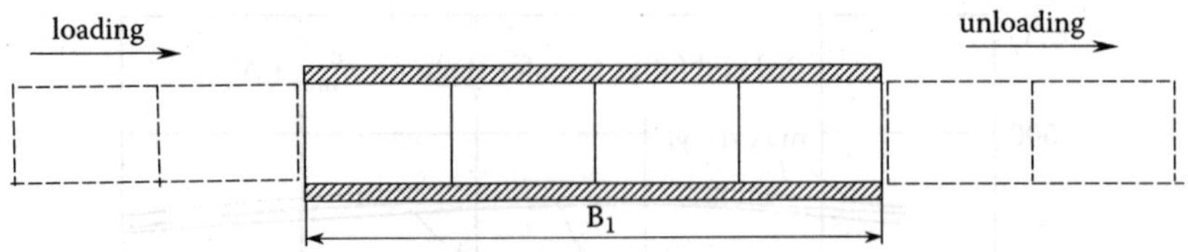

FIGURE 5.19 Single-section progressive heater at steady-state operating conditions.

Let us assume that $B_1 > 1$ billets can be located simultaneously inside the heater. Output cycle time φ^* would be constant during steady-state heating mode (Figure 5.19). Time required for heating of each separate billet during B_1 steps inside the inductor is equal to $B_1\varphi^*$ (with neglected time of loading/unloading operations). Equation (4.72) through Equation (4.74) under the condition in Equation (5.34) represent temperature distribution within each billet under $k = \overline{1, B_1}$. Common control input $u(\varphi)$, which is applied to the heater, represents power of internal heat sources or appropriate voltage of power supply. Control $u(\varphi)$ should be repeated on each step during steady-state mode of IHI operation. At each time point, the control $u(\varphi)$ affects all B_1 billets located inside the heater. In this case, the optimal control problem can be formulated as follows.

It is necessary to select a control function $u^*(\varphi)$ constrained by Equation (2.7) that provides steering workpiece initial temperature distribution to the desired temperature at the heater exit θ^* with prescribed accuracy ε (according to Equation 4.92) under maximum output rate (i.e., under minimum value of φ^*). Similarly to problems described in Chapter 4, different problem statements can be formulated depending upon particular situations.

Because the optimal control, $u^*(\varphi)$, should be repeated on all steps of steady-state process, heating modes for all billets are identical. Therefore, it is reasonable to consider the heating of one separate billet in order to define optimal control algorithms.

Obviously, during the last step before billet unloading, the optimal control should provide steering workpiece temperature distribution at the end of the previous step to desired temperature (Equation 4.92) in minimal possible time. It has been proved earlier (see Section 2.8) that, within the limits of the considered process step, the time-optimal control algorithm consists of alternating stages of heating with maximum power $u \equiv U_{max}$ (heat ON) and subsequent soaking/cooling under $u \equiv 0$ (heat OFF) cycles. Consequently, optimal control $u^*(\varphi)$ takes the form of Equation (2.27).

Because the optimal control function $u^*(\varphi)$ is identical for all steps of the steady-state heating process, the consequence of optimal control programs in the form of Equation (2.7) can be considered as the optimal control algorithm in this case. This algorithm could be improved by supplementing the process with the further stage of duration Δ_c for performing loading/unloading operations under power switched off.

As a result, it is possible to perform preliminary parameterization of sought optimal control inputs. The initial problem is reduced to searching for number,

N, and optimal values of parameters Δ_i^0, $i = \overline{1,N}$ that uniquely specify durations of optimal control $u^*(\varphi)$ stages.

The optimal process of heating each separate billet represents consequence of repetitive identical stages, the number of which is equal to B_1. Final temperature distribution can be defined by the integral expression in Equation (4.77) after substitution of cyclically repeated control function $u^*(\varphi)$:

$$u^*(\varphi) = \begin{cases} \frac{U_{max}}{2}\left[1+(-1)^{j+1}\right], & \sum_{i=1}^{j-1}\Delta_i^0 < \varphi < \sum_{i=1}^{j}\Delta_i^0, \; j=\overline{1,N}; \\ 0, & \sum_{i=1}^{N}\Delta_i^0 < \varphi < \sum_{i=1}^{N}\Delta_i^0 + \Delta_c. \end{cases} \tag{5.44}$$

$$u^*\left(\varphi + \nu\left(\sum_{i=1}^{N}\Delta_i^0 + \Delta_c\right)\right) = u^*(\varphi), \; \nu = \overline{1, B_1 - 1}, \quad 0 < \varphi < B_1\left(\sum_{i=1}^{N}\Delta_i^0 + \Delta_c\right). \tag{5.45}$$

Here, Δ_i^0, $i = \overline{1,N}$ is duration of stage with number i of N-stage control input that is applied *during each step* of steady-state heating mode.

Instead of Equation (5.35), the final temperature distribution $\theta(l,y,\Delta^0)$ can be described under the condition in Equation (5.44) and Equation (5.45) by the following expression[1]:

$$\begin{aligned} \theta\left(l,y,\Delta^0\right) &= \Phi_3\left(l,y,\sum_{i=1}^{N}\Delta_i^0\right) + U_{max}\sum_{r=1}^{\infty}\sum_{n=1}^{\infty}\frac{D_{rn}}{\eta_{rn}}J_0\left(\mu_r l\right)\times \\ &\times\left(\cos\lambda_n y + \frac{Bi_0}{\lambda_n}\sin\lambda_n y\right)\left\{\sum_{k=1}^{B_1}W_{rn}^{(k)}\sum_{j=1}^{N}(-1)^{j+1}\times\right. \\ &\left.\times\left[\exp\left(\eta_{rn}\left(k\varphi^* - \Delta_c\right)\right) - \exp\left(\eta_{rn}\left(\varphi^*(k-1) + \sum_{i=1}^{j-1}\Delta_i^0\right)\right)\right]\right\}\times \\ &\times\exp\left(-\eta_{rn}\sum_{i=1}^{N}\Delta_i^0\right). \end{aligned} \tag{5.46}$$

Here, optimal output cycle time, φ^*_{min}, and total time, φ^0_{min}, required for heating of each billet can be obtained as the following functions of durations Δ_i^0 of $u^*(\varphi)$ constancy intervals:

$$\varphi^*_{\min} = \sum_{i=1}^{N} \Delta_i^0 + \Delta_c; \; \varphi^0_{\min} = B_1 \varphi^*_{\min} \,. \tag{5.47}$$

If the technological restrictions should be taken into account, then the optimal control algorithm (Equation 5.44) within each process step should be complicated and can be obtained in the form of Equation (4.1).

Problems of searching for optimal control functions $u^\sigma(\varphi)$ and $u^\theta(\varphi)$, which provide holding of maximum temperature and thermal stress at their maximum permissible levels, can be noticeably different in comparison with static heating processes (Section 4.1). Under cyclically repeated control function $u^*(\varphi)$, the initial temperature states at times φ_σ and φ_θ (when thermal stress and maximum temperature reach their maximum permissible levels $\sigma_{\max}$ and $\theta_{\max}$) prove to be dependent on control functions $u^\sigma(\varphi)$ and $u^\theta(\varphi)$. The following iteration procedure can be used to solve the problem.[1]

Each kth approximation $u_k^\sigma(\varphi)$ and $u_k^\theta(\varphi)$ of $u^\sigma(\varphi)$ and $u^\theta(\varphi)$ can be found according to the technique described in Section 4.1 during the step of number B_1 of the heating process with respect to each separate billet. Optimal control during all previous $B_1 - 1$ steps is calculated for previous iterations $u_{k-1}^\sigma(\varphi)$ and $u_{k-1}^\theta(\varphi)$. The control algorithm during $B_1 - 1$ stages with neglected technological constraints can be considered as the first approximation. Iterative procedure is over when the functions $u_k^\sigma(\varphi)$ and $u_k^\theta(\varphi)$ coincide with $u_{k+1}^\sigma(\varphi)$ and $u_{k+1}^\theta(\varphi)$ with required accuracy.

In typical cases of a transportation problem that have been considered earlier (Section 4.2) when it is necessary to provide heating accuracy $\varepsilon \geq \tilde{\varepsilon}^{(2)}_{\min}$ at $\alpha_T = 1$ under technological constraint on $\theta_{\max}$, the optimal control (Equation 4.1) can be reduced to the particular form of Equation (4.3). In similar cases, during the first heating stage under applying the maximum power, the temperature maximum reaches admissible value θ_{adm}; however, at the same time the value $\sigma_{\max}$ still remains less than σ_{adm}. Therefore, the interval of $\sigma_{\max}$ holding should be excluded (Section 4.1 and Section 4.2). In contrast to Equation (4.3), the optimal control function $u^*(\varphi)$ should be supplemented by the stage of duration Δ_c with heat power switched on.

The following expression can be obtained for $u^*(\varphi)$ instead of Equation (5.44) within the limits of each step:

$$u^*(\varphi) = \begin{cases} U_{\max}, \; 0 < \varphi < \varphi_\theta; \\ u^\theta(\varphi), \varphi_\theta < \varphi < \Delta_1^0; \\ 0, \quad \Delta_1^0 < \varphi < \Delta_1^0 + \Delta_c. \end{cases} \tag{5.48}$$

Here, φ_θ and Δ_1^0 are appropriate times within the limits of one step; the expression in Equation (5.48) should be considered together with Equation (5.45).

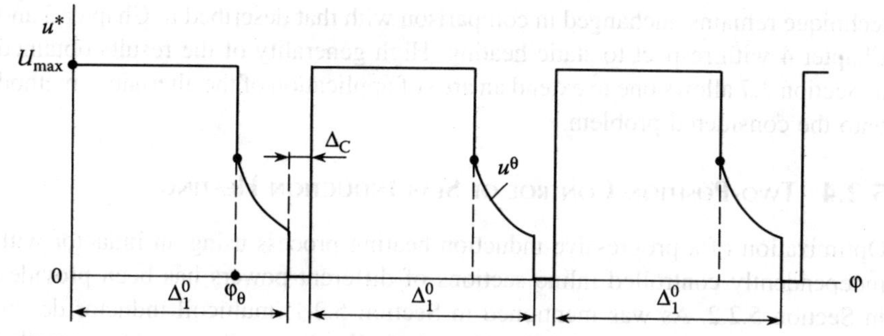

FIGURE 5.20 Optimal control algorithm for single-section progressive heater at steady-state operating conditions.

Figure 5.20 shows a general view of the optimal control algorithm $u^*(\varphi)$ in the form of Equation (5.48) and Equation (5.45) with respect to heating of each separate billet.

If the control input function $u^{\theta}(\varphi)$ is written in the form of Equation (4.13), then final temperature distribution at the heater exit can be represented instead of Equation (5.46) by the following expression, which is similar to Equation (4.15) and Equation (4.16)[1]:

$$
\begin{aligned}
\theta(l, y, \Delta^0) = {} & \Phi_3\left(l, y, \sum_{i=1}^{N} \Delta_i^0\right) + U_{\max} \sum_{r=1}^{\infty} \sum_{n=1}^{\infty} \frac{D_{rn}}{\eta_{rn}} J_0(\mu_r l)\left(\cos\lambda_n y + \frac{Bi_0}{\lambda_n}\sin\lambda_n y\right) \times \\
& \times \left\{ \sum_{k=1}^{B_1} W_{rn}^{(k)} \left[\exp\left(\eta_{rn}\left(\varphi_\theta + (k-1)(\Delta_1^0 + \Delta_c)\right)\right) - \exp\left(\eta_{rn}(k-1)(\Delta_1^0 + \Delta_c)\right) + \right.\right. \\
& + a_\theta \left(\exp\left(\eta_{rn}(k\Delta_1^0 + (k-1)\Delta_c)\right) - \exp\left(\eta_{rn}\left(\varphi_\theta + (k-1)(\Delta_1^0 + \Delta_c)\right)\right) \right) + \frac{b_\theta \eta_{rn}}{\eta_{rn} - \beta_\theta} \times \\
& \times \left(\exp\left((-\beta_\theta + \eta_{rn})(k\Delta_1^0 + (k-1)\Delta_c)\right) - \exp\left((-\beta_\theta + \eta_{rn})(\varphi_\theta + (k-1)(\Delta_1^0 + \Delta_c))\right) \right) \times \\
& \left.\left. \times \exp\left(\beta_\theta(\varphi_\theta + (k-1)(\Delta_1^0 + \Delta_c))\right) \right] \exp\left(-\eta_{rn} \sum_{i=1}^{N} \Delta_i^0 \right) \right\}.
\end{aligned}
\tag{5.49}
$$

Using an analogy to the previously described transportation problem (Section 4.2), a time, Δ_T, of billet transfer to hot forming equipment can be taken into account.

Our experience shows that properties of final spatial temperature distribution remain unchanged in comparison with properties established in Chapter 3 and Chapter 4 in the case of static heating processes. Thus, the computational technique remains unchanged in comparison with that described in Chapter 3 and Chapter 4 with respect to static heating. High generality of the results obtained in Section 3.7 allows one to extend an area of application of the alternance method onto the considered problem.

5.2.4 Two-Position Control of Slab Induction Heating

Optimization of a progressive induction heating process using an inductor with independently controlled inline sections of different powers has been provided in Section 5.2.2. As was mentioned in Section 5.2.1, multicoil inductor design exhibits the greatest potential for control if all inline coils may be controlled individually and only one billet is located inside each coil. Within each step, the billet is heated in static position inside one coil, and then billets are quickly moved to the next coil further towards the inductor exit. The required heating mode can be provided by applying autonomous voltage control to each coil separately. The difference with optimal static heating only deals with effects of breaks needed for billets' movement from one coil to another.

If gaps between coils are large enough, then loading/unloading operations can be performed separately, thus allowing one to use workpiece positioning inside the coil as an additional means of spatial control of induction heating.

Two-position slab heating can serve as a good example of effective use of this type of spatial control.[9] According to this approach, a slab is heated inside two separate coils, and the angle between slab axis inside the second coil and vector of magnetic intensity is equal to 90°. The temperature profiles along the slab length and width are affected by, among other factors, a distortion of electromagnetic field at its end and edge areas. These effects and the field distortion caused by them are primarily responsible for nonuniform temperature profiles within the slab. In the case of two-position heating, the temperature gradients due to end and edge effects can be compensated (Figure 5.21a). As a result, the heating accuracy becomes higher than in continuous two-section heater under constant in time maximum voltages applied to both coils.

Figure 5.21b shows some results of computation in application to two-position heating of A6 alloy slabs 1250 × 1250 × 200 mm^3; frequency is 50Hz and active power of each coil is 200 kW.[1]

A special piecewise linear three-dimensional model that takes into account surface heat-generating sources has been used to describe temperature distribution within the slab. As can be seen, heating nonuniformity reduces at the slab's surface due to reorientation of slab in space, which means the use of the longitudinal end effect as a means of control.

Durations of stages of applied voltage constancy can be considered in this case as control parameters as well as one additional parameter, K_p. This parameter characterizes optimal heating mode and shows how the longitudinal end effect acts.[1,9] Factor K_p can be considered as a relation of average values of surface heat power densities in central part and longitudinal end effect area of heated slab, and can be calculated according to limiting values of these densities P_A and P_B

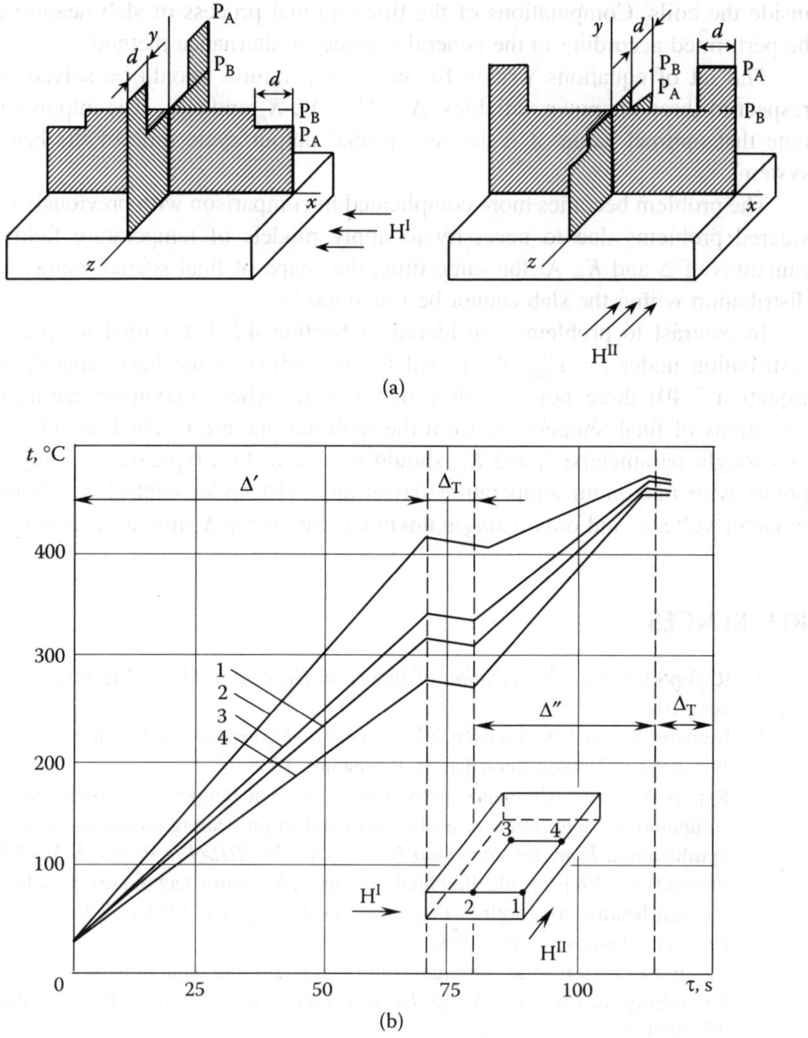

FIGURE 5.21 Two-position heating of slabs. a: Heat power distribution at slab surface for longitudinal and transversal orientation of vector of magnetic intensity. b: Time–temperature history for different points.

(Figures 5.21a). The value of K_p can be changed by different means of spatial control, including but not limited to different electrical connections, coil overhang, etc.

In the simplest case of single-stage voltage control, it is necessary to provide heating accuracy $\varepsilon = \varepsilon^{(1)}_{\min}$ under the condition that time, Δ', required for heating inside the first coil is equal to appropriate time, Δ'', that is required for heating inside the second coil. Values of K_p are the same for both cases of slab positioning inside the coils. Computations of the time-optimal process of slab heating can be performed according to the general scheme of alternance method.

The set of equations written for final temperatures should be solved with respect to three unknown variables: $\Delta = \Delta' = \Delta''$, K_p and $\varepsilon^{(1)}_{\min}$. It is important to note that optimal value of Δ defines optimal output cycle time of the heating system.

The problem becomes more complicated in comparison with previously considered problems due to necessity to apply models of temperature fields as functions of Δ and K_p. At the same time, the shape of final spatial temperature distribution within the slab cannot be found easily.

In contrast to problems considered in Section 4.5.4, for final temperature distribution under $\varepsilon = \varepsilon^{(1)}_{\min}$, there will be (according to the basic condition in Equation 3.40) three points within slab volume where maximum admissible deviations of final temperature from the required one are reached. In this case, *two* sought parameters, Δ and K_p, should be found. In a typical case, only two points with maximum temperature deviations exist under control by common inductor voltage, and only a *single* unknown parameter Δ should be found.

REFERENCES

1. Rapoport, E.Ya., *Optimization of Induction Heating of Metal,* Metallurgy, Moscow, 1993.
2. Nemkov, V.S. and Demidovich, V.B., *Theory and Computation of Induction Heating Installations,* Energoatomizdat, St. Petersburg, 1988.
3. Rapoport, E.Ya., Alternance properties of optimal temperature distribution and computational algorithms in problems of induction heating processes parametric optimization, *Proc. Int. Induction Heating Semin. (IHS-98)*, Padua, 443, 1998.
4. Pleshivtseva, Yu.E. et al., Potentials of optimal control techniques in induction through heating for forging, *Proc. Int. Sci. Colloquium Modeling Electromagn. Process.,* Hannover, 145, 2003.
5. Pleshivtseva, Yu.E. et al., Optimal control techniques in induction through heating for forging, in *Proc. Int. Symp. Heating Electromagn. Sources HES-04,* Padua, 97, 2004.
6. Rudnev, V.I. et al., *Handbook of Induction Heating,* Marcel Dekker, New York, 2003.
7. Sluhotskii, A.E., Nemkov, V.S., Pavlov, N.A., and Bamunauer, A.V., *Induction Heating Installations,* Energoatomizdat, St. Petersburg, 1981.
8. Pavlov, N.A., *Engineering Thermal Computation of Induction Heaters,* Energia, Moscow, 1978.
9. Nemkov, V.S. et al. End effects as a means for control of induction heating processes of rectangular shape workpieces, in *Algoritmy i systemi upravlenija tehnologicheskimi processami v mashinostroenii*, Kujbyshev, 141, 1986.

6 Combined Optimization of Production Complex for Induction Billet Heating and Subsequent Metal Hot Forming Operations

This section discusses a system approach for optimization of induction heaters. This system approach takes into consideration the whole sequence of the interrelated technological operations including billet heating by induction, its cooling during transportation, and subsequent plastic deformation during hot working. All these subsequent processes will be considered as a technological complex and combined optimization criteria will be used for its optimization. Depending upon application, metal hot working operations may be represented by forging, rolling, stamping, upsetting, extrusion, and other operations where metal undergoes plastic deformation.

For a particular technological complex, it is important to reveal the cost function for each operation or stage of the technological process. As was mentioned in Section 2.2, a proper selection of local cost function for IHI optimization is based upon valid decomposition of a combined optimization problem into separate optimization problems for processes of heating and hot working.

A local problem of IHI optimization could be solved only within the rigid frame of a priori given technological instructions. This means that optimal control inputs can be found only in the case when required final temperature distribution of heated billets is a priori known.

Initial temperature distribution at the beginning of the billet transportation stage complies with final temperature distribution at the end of the heating stage. The temperature distribution before the hot forming complies with final temperature distribution at the end of the billet transportation stage. These facts are imperative to bind workpiece temperature states during all stages of the "induction heating–metal hot forming" technological cycle.

As most metal forming processes, hot working operations impose certain requirements for initial temperature distribution. For example, there is the maximum range of admissible temperature variations within the heated workpiece prior to its hot forming. The temperature distribution within a heated billet before the forming stage affects essentially the overall performance index of a technological complex. Required temperature distribution at the inductor exit is constrained by preset technological requirements. As a result, the value of overall cost function can deviate from the optimum one.

New possibilities appear when solving a general optimization problem with respect to combined technical and economic indexes of complex operation as a whole. A technical index can be represented by such criteria as maximum production, minimum required floor space, utilization of existing power supply sources, minimum power consumption, and others. An economical index can be represented by cost criteria, metal loss minimization criteria, etc.

A general optimization problem can be solved under condition of having a maximum freedom in choosing parameters of heating and hot forming operations. In this case, it might be possible to find optimal parameters (according to certain optimization criteria) of each operation as standalone process as well.

If a designer has unlimited freedom in choosing the final thermal condition of a heated billet, then it is possible to:

- Reveal local cost functions for optimization of IHI and deforming equipment separately, utilizing means of valid system decomposition
- Find the final temperature distribution within a heated billet that is optimal with respect to overall cost function
- Define control inputs during the stages of heating and hot forming that provide desired temperature distribution and hold overall cost function at extremum value

The described approach requires determining capabilities of a control algorithm out of the framework of traditional "servicing" of technology and ensures active participation in technology formation. This system approach deals with finding optimal technological parameters within process flow sheets and optimal design of a technological site to provide optimal operational modes for each separately considered technological operation.[1–6]

Optimization of a technological complex of induction heating of cylindrical billets and subsequent hot forming operation using hydraulic presses will be used as an example of application of alternance method.[5,6]

6.1 MATHEMATICAL MODELS OF CONTROLLED PROCESSES

The ultimate goal of optimal control strategy is to provide required thermal conditions of workpieces just before hot working operations.[7] This is the reason why the temperature distribution within the billet can be regarded as an output-controlled function. Therefore, as a subject of optimal control, a technological complex “induction heater–hot forming” can be described by a system of heat transfer equations that represent heating, cooling (billet transfer), and hot forming stages (see Figure 2.5).

Required accuracy of temperature distribution within the billet is usually provided by an acceptable level of three-dimensional temperature variations. For the majority of practical applications, the temperature distribution within the heated workpiece can be described by two- and three-dimensional mathematical models. The majority of induction billet heaters utilize multiturn solenoid-type induction coils. Because they are electromagnetically long systems, such inductors typically have unappreciative temperature variations along the billet perimeter due to symmetrical positioning of billets inside inductors. Therefore, a two-dimensional mathematical model can be effectively applied for mathematical modeling of cylindrical systems. Two-dimensional models provide sufficient information regarding temperature profiles in axial and radial directions.

Temperature of cylindrical billet $\theta_1(l,y,\varphi)$ during a heating process can be described by the following system of two-dimensional heat transfer equations (Equation 4.72 through Equation 4.75) for $\Gamma = 1$ (Figure 4.37):

$$\frac{\partial\theta_1(l,y,\varphi)}{\partial\varphi} = \frac{\partial^2\theta_1(l,y,\varphi)}{\partial l^2} + \frac{1}{l}\frac{\partial\theta_1(l,y,\varphi)}{\partial l} + \\ + \beta^2\frac{\partial^2\theta_1(l,y,\varphi)}{\partial y^2} + W_d(\xi,l,y)u(\varphi); \tag{6.1}$$

$$l,y\in(0,1);\ 0<\varphi\le\varphi_1^0;$$

$$\begin{cases} \dfrac{\partial\theta_1(1,y,\varphi)}{\partial l} = Bi_h\left(\theta_a-\theta_1(1,y,\varphi)\right); \quad \dfrac{\partial\theta_1(0,y,\varphi)}{\partial l} = 0; \\ \dfrac{\partial\theta_1(l,1,\varphi)}{\partial y} = Bi_h\left(\theta_a-\theta_1(l,1,\varphi)\right); \\ -\dfrac{\partial\theta_1(l,0,\varphi)}{\partial y} = Bi_h\left(\theta_a-\theta_1(l,\ 0,\varphi)\right); \end{cases} \tag{6.2}$$

$$\theta_1(l,y,0) = \theta_{10}(l,y). \tag{6.3}$$

Here, φ_1^0 is required total heating time; Biot number Bi_h corresponds to heat losses from the surface of the billet during the whole heating process inside the induction heater.

Temperature distribution during billet transportation stage $\theta_2(l,y,\varphi)$ can also be described by Equation (6.1) except that there are no heat sources:

$$\frac{\partial\theta_2(l,y,\varphi)}{\partial\varphi} = \frac{\partial^2\theta_2(l,y,\varphi)}{\partial l^2} + \frac{1}{l}\frac{\partial\theta_2(l,y,\varphi)}{\partial l} + \beta^2\frac{\partial^2\theta_2(l,y,\varphi)}{\partial y^2}; \tag{6.4}$$

$$l, y \in (0,1);\ \varphi_1^0 \le \varphi \le \varphi_1^0 + \Delta_T;$$

$$\begin{cases} \dfrac{\partial\theta_2(1,y,\varphi)}{\partial l} = Bi_T\left(\theta_a - \theta_2(1,y,\varphi)\right); \quad \dfrac{\partial\theta_2(0,y,\varphi)}{\partial l} = 0; \\ \dfrac{\partial\theta_2(l,1,\varphi)}{\partial y} = Bi_T\left(\theta_a - \theta_2(l,1,\varphi)\right); \\ -\dfrac{\partial\theta_2(l,0,\varphi)}{\partial y} = Bi_T\left(\theta_a - \theta_2(l,0,\varphi)\right); \end{cases} \tag{6.5}$$

$$\theta_2(l,y,\varphi_1^0) = \theta_1(l,y,\varphi_1^0). \tag{6.6}$$

Time Δ_T should be considered here as a total transportation time. Heat losses during the billet transfer stage can be estimated by value Bi_T that vastly increases in comparison with heat losses inside the induction coil ($Bi_T > Bi_h$). The thermal refractory located inside the induction coil noticeably reduces surface heat losses. In contrast, there will be intensive surface cooling during billet transportation on air.

Both heat transfer equations (Equation 6.1 and Equation 6.4) are complemented by boundary conditions of the third kind (Equation 6.2 and Equation 6.5). Appropriate value of Biot criterion $Bi_T > Bi_h$ takes into consideration an increase of heat losses from the billet surface during billet transportation from heating operation to metal hot forming stage.

To simplify the study, the ambient temperature in Equation (6.2) and Equation (6.5) is considered as uniform within billet surface and equal to constant value θ_a.

The linear axis-symmetric model of temperature distribution $\theta_3(l,y,\varphi)$ within the cylindrical billet during the process of pressing can be described by the following two-dimensional heat transfer equation[6,8]:

$$\frac{\partial\theta_3(l,y,\varphi)}{\partial\varphi} = \frac{\partial^2\theta_3(l,y,\varphi)}{\partial l^2} + \frac{1}{l}\frac{\partial\theta_3(l,y,\varphi)}{\partial l} + \beta^2\frac{\partial^2\theta_3(l,y,\varphi)}{\partial y^2} -$$

$$-\gamma^*\beta V_y(l,y,V_P)\frac{\partial\theta_3(l,y,\varphi)}{\partial y} - \gamma^* V_l(l,y,V_P)\frac{\partial\theta_3(l,y,\varphi)}{\partial l} + W^*(l,y,V_P); \tag{6.7}$$

$$l, y \in (0,1);\ \varphi_1^0 + \Delta_T \le \varphi \le \varphi_2^0$$

with boundary and initial conditions similar to Equation (6.2), Equation (6.3), Equation (6.5), and Equation (6.6):

$$\begin{cases} \dfrac{\partial\theta_3(1,y,\varphi)}{\partial l} = Bi\left(\theta_a^0 - \theta_3(1,y,\varphi)\right) + q(y,V_P); \quad \dfrac{\partial\theta_3(0,y,\varphi)}{\partial l} = 0; \\ \dfrac{\partial\theta_3(l,1,\varphi)}{\partial y} = Bi_1\left(\theta_{a1}^0 - \theta_3(l,1,\varphi)\right); \\ -\dfrac{\partial\theta_3(l,0,\varphi)}{\partial y} = Bi_0\left(\theta_{a0}^0 - \theta_3(l,0,\varphi)\right); \end{cases} \tag{6.8}$$

$$\theta_3(l,y,\varphi_1^0 + \Delta_T) = \theta_2(l,y,\varphi_1^0 + \Delta_T). \tag{6.9}$$

Here, in contrast to Equation (6.1), instead of internally generated heating power $W_d(\xi,l,y)u(\varphi)$, a volume density of heat sources $W^*(l,y,V_P)$ is used in Equation (6.7). The density of heat sources $W^*(l,y,V_P)$ is defined by energy of plastic deformation of metal during the direct extrusion with speed V_P.

The value of $W^*(l,y,V_P)$ can be calculated by expression[8–10]:

$$W^*(l,y,V_P) = LH(l,y,V_P)\,, \tag{6.10}$$

where L is intensity of the tangential stresses under plastic deformation and H is intensity of velocity of shear deformation during pressing.

The differential operator in the right side of Equation (6.1) is complemented in Equation (6.7) by components that take into account velocities V_l and V_y of the metal flow in radial and axial directions, respectively. Field of velocities is nonuniformly distributed within the billet volume during the pressing process.

Heat losses can be estimated by values Bi, Bi_1, and Bi_0 of Biot criterion in boundary conditions in Equation (6.7). The ambient temperatures θ_a^0, θ_{a1}^0, θ_{a0}^0 at lateral surface ($l = 1$, $y \in [0,1]$) and at butt-ends ($y = 1$, $l \in [0,1]$) and $y = 0$, $l \in [0,1]$) represent appropriate calculated values (Figure 4.37).

The density of heat flow $q(y,V_p)$ caused by contact friction is taken into account in the boundary condition in Equation (6.8).[8–10] Values V_p, V_l, V_y, and W^* in Equation (6.7) should be calculated as fractions of appropriate basic values V_δ and W_δ^*, where W_δ^* replaces P_{max} in formulas in Equation (1.28), Equation (1.29), Equation (1.33), and Equation (1.35). The factor β can be found from the correlation in Equation (4.76), and γ^* is defined by the equality

$$\gamma^* = \frac{XV_\delta}{a}\,,$$

where a is the factor of temperature conductivity.

The final product of technological complex "induction heater–hot working equipment" is a press product with length z that can be defined according to:

$$\frac{dz}{d\varphi} = kV_p; \ z(\varphi_1^0 + \Delta_T) = 0; \ z(\varphi_2^0) = z_{end} . \tag{6.11}$$

Here, k is elongation ratio[7] and the required value of $z(\varphi_2^0)$ is equal to z_{end} at the end of pressing stage $\varphi = \varphi_2^0$.

Metal heating processes impose two general requirements with regard to the temperature distribution during heating processes. The first requirement of the type in Equation (2.11) demands that the maximum value of the temperature within a heated workpiece should be not above a certain admissible value θ_{adm}. According to the second restraint in the form of Equation (2.13), the temperature differences within the whole volume of a heated workpiece should be restricted during heating in such a way that the maximum value of tensile thermal stresses would not exceed prescribed admissible value σ_{adm} (see Section 2.5.2). The restraints in Equation (2.11) and Equation (2.13) can be rewritten as follows:

$$\max_{l,y\in[0,1]} \theta_1(l,y,\varphi) \le \theta_{adm}; \ \max_{l,y\in[0,1]} \sigma(l,y,\varphi) \le \sigma_{adm}; \ 0 \le \varphi \le \varphi_1^0. \tag{6.12}$$

Temperature within the heated workpiece during extrusion is restricted by maximum admissible temperature in the die hole θ_{cr}:

$$\theta_3(l_k, 1, \varphi) \le \theta_{3cr}, \ \varphi_1^0 + \Delta_T \le \varphi \le \varphi_2^0 . \tag{6.13}$$

This requirement demands that the maximum temperature $\theta_3(l_k,1,\varphi)$ in the die deformation zone ($l = l_k$, $y = 1$, $0 < l_k < 1$) (see Figure 6.2) should not exceed admissible limit θ_{3cr}. Violation of these conditions could result in typical irreversible spoilage of product, i.e., crack development.[10]

Due to certain technical limitations that deal with power of press, instrument strength, temperature interval of metal plasticity, and other factors, it is possible to perform extrusion only in the case when initial temperature states belong to a certain region Ω:

$$\theta_3(l, y, \varphi_1^0 + \Delta_T) \in \Omega. \tag{6.14}$$

Heating power $u(\varphi)$ and ram speed $V_p(\varphi)$ can be considered as control inputs for heating and extrusion processes. The values of these control inputs are also restricted by certain admissible values $U_{\max}$ and $V_{p\max}$:

$$\begin{gathered} 0 \le u(\varphi) \le U_{\max}, \ 0 \le \varphi \le \varphi_1^0; \\ 0 \le V_p(\varphi) \le V_{p\max}, \ \varphi_1^0 + \Delta_T \le \varphi \le \varphi_2^0 . \end{gathered} \tag{6.15}$$

Here, maximum admissible value $V_{p\max}$ of ram speed is defined by deformed material properties, the system of the hydraulic press drive, and press power limits. It is assumed that the value of $V_{p\max}$ is known a priori.[11]

Equation (6.1) through Equation (6.11) and the restraints in Equation (6.12) through Equation (6.15) represent a mathematical model of the technological complex 'induction heating–direct extrusion" as controlled system with output controlled variable $z(\varphi)$. A unique feature of this production system deals with the fact that the separate operations are performed one after the other. This fact corresponds to boundary conditions in Equation (6.6) and Equation (6.9) and can be written in the form of the following equalities:

$$\theta_1(l, y, \varphi_1^0) = \theta_2(l, y, \varphi_1^0);\ \theta_3(l, y, \varphi_1^0 + \Delta_T) = \theta_2(l, y, \varphi_1^0 + \Delta_T). \qquad (6.16)$$

The expressions in Equation (6.16) are imperative to bind workpiece temperature fields at times $\varphi = \varphi_1^0$ and $\varphi = \varphi_1^0 + \Delta_T$ of transitions from one stage of technological cycle to another.

A similar description can be obtained with respect to more complicated nonlinear mathematical models of temperature fields within the heated billet during the heating, transportation, and extrusion stages.

6.2 GENERAL PROBLEM OF OPTIMIZATION OF A TECHNOLOGICAL COMPLEX

Let us consider the problem of optimization of a technological complex for induction heating of cylindrical billets and their subsequent hot forming using hydraulic presses. Under previously described conditions, the statement of the optimal control problem can be formulated for the technological complex "heater–hot working equipment" as follows.

It is necessary to select such controls by heat power $u^*(\varphi)$ $(0 < \varphi \le \varphi_1^0)$ and extrusion speed $V_p^*(\varphi)$ $(\varphi_1^0 + \Delta_T \le \varphi < \varphi_2^0)$ restricted by the preassigned set of constraints in Equation (6.12) through Equation (6.15) that provide a final press product of desired length $z = z_{end}$ and hold overall cost criteria I at extremum value. Time Δ_T should be considered here as a fixed transportation time and the system under control is described by Equation (6.1) through Equation (6.11).

Typically, system productivity or production cost[5,6] is considered as objective function I. If maximum productivity should be obtained, then a minimal time, φ_c, required for total cycle of processing a single billet can be considered as a cost function:

$$I_1 = \varphi_c \to \min\,. \qquad (6.17)$$

When minimum product cost is required, the following overall function can be used as a combined criterion of the optimal control problem:

$$I_2 = \sum_{i=1}^{s} C_i P_i + C_t \varphi_c \rightarrow \min . \tag{6.18}$$

This criterion (Equation 6.18) includes all cost components P_i, $i = \overline{1,s}$ with weighting coefficients C_i. Those coefficients make sense with respect to a relative costing (that is a function of control inputs) representing an importance of each of the overall costs. Increase of the heating time leads to expense growth, according to the criterion (Equation 6.17) with correspondent weight coefficients C_t. The value of P_i is primarily defined by energy expenses for heating and pressing, as well as by metal losses due to scale formation during the process of heating (see Section 2.9.2, Section 3.5, and Section 4.4.1).

The formulated problem provides freedom in choosing the billet temperature $\theta_2(l,y,\varphi_1^0 + \Delta_T)$ before the hot forming stage in a way that would satisfy requirements imposed by heating and hot working operations (with regard to transportation time). An appropriate temperature distribution, $\theta_{2\text{opt}}(l,y,\varphi_1^0 + \Delta_T)$, that provides minimal values of I_1 or I_2 should be found, as well as optimal control functions $u^*(\varphi)$ and $V_p{}^*(\varphi)$. The selection of $\theta_{2\text{opt}}(l,y,\varphi_1^0 + \Delta_T)$ out of set Ω of admissible temperature distribution allows varying technological parameters within process flow sheets in order to provide optimal operational modes of IHI and pressing equipment.

On the other hand, specification of process parameters allows one to perform a valid decomposition of the general problem to local optimization problems. These problems can be considered independently for induction heater and extrusion press. The proper choice of temperature distribution $\theta_{2\text{opt}}(l,y,\varphi_1^0 + \Delta_T)$, which does not lead to additional restraint on temperature profile at the inductor exit, seems to be complicated. However, an a priori set of requirements with respect to initial temperature (prior to extrusion) can be used to simplify the problem.[5] These requirements are based on experience and available information.

We shall limit our consideration to the simplest typical case. It is necessary to have a uniformly heated workpiece at the inductor exit. This means that, at the end of the heating stage, the temperature in any point of the billet should deviate not more than by prescribed value ε from desired temperature θ_2^* (see Section 2.7). Therefore, a family, Ω, of single-parameter functions in Equation (6.14) provides the unique uniform temperature distribution.

The single parameter θ_2^* determines the final temperature distribution $\theta_2(l,y,\varphi_1^0 + \Delta_T)$ within the heated billet at the end of the transfer stage. At the same time, this parameter defines (according to the correlation in Equation 6.9) the initial temperature distribution $\theta_3(l,y,\varphi_1^0 + \Delta_T)$ before hot working operation. Nevertheless, the value of θ_2^* is unknown a priori and should be defined when solving the optimal control problem.

In this case, in the course of searching for optimal heating and extrusion modes, the value of temperature θ_2^*, which is optimal with respect to overall cost function, should be found. Let us denote this value as θ_{2e}^*. The moments φ_1^0 and

φ_2^0 (when technological stages are over) will be determined as well in the course of searching for optimal modes. Parameters θ_{2e}^*, φ_1^0, and φ_2^0 represent the most important factors in technology formation.

Particular computation of optimal value of θ_2^* can be noticeably different in comparison with optimal control problems considered in previous chapters. It is important to underline that the value of θ_2^* was included in a set of initial data.

According to the condition in Equation (6.14), the range of admissible values of θ_2^* should satisfy the following inequality:

$$\theta_2^* \geq \theta_{2\min}^* \tag{6.19}$$

where $\theta_{2\min}^*$ is minimal admissible value of θ_2^* that is limited by strength of extrusion equipment.[5,6] The inequality in Equation (6.19) usually defines the admissible temperature interval of pressurized metal plasticity.

6.3 MAXIMUM PRODUCTIVITY PROBLEM FOR AN INDUSTRIAL COMPLEX "INDUCTION HEATER–EXTRUSION PRESS"

Let us consider the general problem (Equation 6.17) that provides maximum productivity of a technological complex "induction heater–hot working equipment" in the steady-state operation mode. For each fixed value of θ_2^*, the time-optimal modes of heater operation and extrusion press operation correspond to appropriate values of $\varphi_{1\min}^0(\theta_2^*)$ and $\varphi_{g\min}(\theta_2^*)$. Here, the value of $\varphi_{1\min}^0(\theta_2^*)$ represents the minimum time required for a billet heating up to the temperature $\theta_2^* \pm \varepsilon$. This also includes the billet transportation stage.

On the other hand, the value of $\varphi_{g\min}(\theta_2^*)$ is minimal possible time required for pressing the product of given length z_{end} under initial temperature $\theta_2^* \pm \varepsilon$, where $\varphi_{g\min} = (\varphi_2^0 - \varphi_1^0 - \Delta_T)_{\min}$.

For steady-state mode, the minimum duration of production cycle $\varphi_{c\min}(\theta_2{}^*)$ can be chosen as maximum between the values $\varphi_{1\min}(\theta_2{}^*)$ and $\varphi_{g\min}(\theta_2{}^*)$, keeping in mind the type of IHI and specific way of combined operations of IHI and press:

$$\varphi_{c\min}(\theta_2^*) = \max\left(\frac{1}{B_1}\varphi_{1\min}^0(\theta_2^*), \varphi_{g\min}(\theta_2^*) + \psi\Delta_T\right). \tag{6.20}$$

Here, B_1 is the number of the billets that can be simultaneously heated in the IHI. The condition $B_1 = 1$ is met for static heating. The expression $B_1 > 1$ holds true for progressive heating mode with step-by-step inline processing of billets through the induction heater (see Section 5.2). In this case, the time of extrusion, φ_g, is close to step duration that decreases in B_1 times compared to φ_1^0.

The value of $\psi \in [0,1]$ in Expression (6.20) is the factor that takes into account all possible variants of the correlation between the moment of unloading of the next billet and completion of the previous billet pressing. If a heated billet exits an induction heater after complete unloading of the press product, then $\psi = 1$ and IHI output rate should be close to the sum $\varphi_g + \Delta_T$. In this case, the idle time of press equipment during one cycle always turns out to be not less than billet transportation time.

If there is no interaction between these operations, then unloading of billets could be performed regardless of press operational mode and billets could be delivered to press with the rate equal to IHI output rate. In this case, the value of φ_g would be close to the value of φ_1^0/B_1 and $\psi = 0$. Intermediate variants take place under condition of $0 < \psi < 1$.

Obviously, the minimum value $\varphi_{c\min}(\theta_2^*)$ is the optimal duration of the production cycle φ_c^* that provides maximum possible productivity of the technological complex "induction heater–extrusion press." Therefore, according to Equation (6.20), it is possible to write:

$$\varphi_c^* = \min_{\theta_2^* \geq \theta_{2\min}^*} \varphi_{c\min}\left(\theta_2^*\right) = \min_{\theta_2^* \geq \theta_{2\min}^*} \left[\max\left(\frac{1}{B_1} \varphi_{1\min}^0\left(\theta_2^*\right), \varphi_{g\min}\left(\theta_2^*\right) + \psi\Delta_T \right) \right]. \tag{6.21}$$

If the functions $\varphi_{1\min}^0(\theta_2^*)$ and $\varphi_{g\min}(\theta_2^*)$ are known, the optimal temperature profile at the end of heating stage can be defined from Expression (6.21) as follows:

$$\theta_{2e}^* = \arg \min_{\theta_2^* \geq \theta_{2\min}^*} \left[\max\left(\frac{1}{B_1} \varphi_{1\min}^0\left(\theta_2^*\right), \varphi_{g\min}\left(\theta_2^*\right) + \psi\Delta_T \right) \right]. \tag{6.22}$$

This value corresponds to the optimal duration, φ_c^*, of the production cycle; therefore, the value of φ_c^* can be defined as well. When the value of θ_{2e}^* is found, it is possible to perform the valid decomposition of the joint maximum productivity optimization problem of induction heater and extrusion press.

Decomposition is reduced to statement of the maximum productivity problem for the cycle stage, which duration is longer, under the condition of $\theta_2^* = \theta_{2e}^*$. This means that, to reduce the value of φ_c^* (see Expression 6.21), the technological stage with a maximum duration should be optimized with respect to the time-optimal criterion. The technological stage that has the shortest duration should be optimized with respect to a cost function of the type in Equation (6.18).

The functions $\varphi_{1\min}^0(\theta_2^*)$ and $\varphi_{g\min}(\theta_2^*)$ should be defined as a result of solutions of separate local time-optimal control problems for heating process and billet extrusion. A sequence of optimization problems should be solved for fixed values θ_2^* that satisfy the condition in Equation (6.19). The first problem can be solved by methods described in previous chapters, but the second one represents an independent problem and requires special consideration. Section 6.5 discusses several possible ways of finding the required solution.[5]

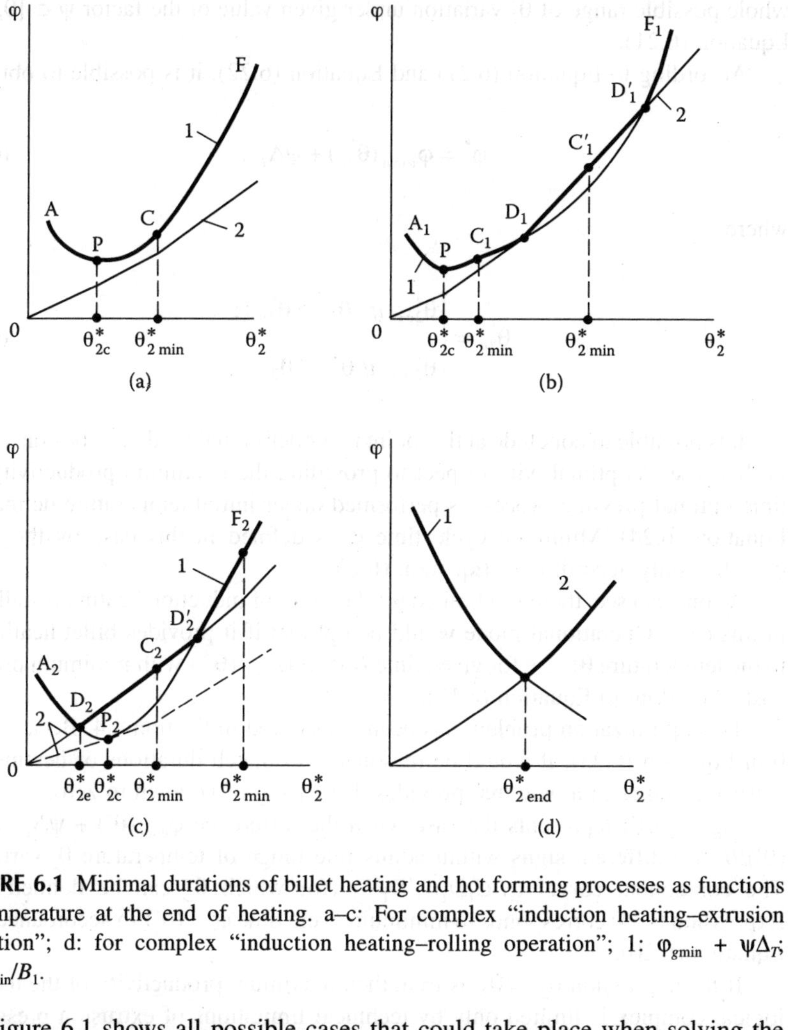

FIGURE 6.1 Minimal durations of billet heating and hot forming processes as functions of temperature at the end of heating. a–c: For complex "induction heating–extrusion operation"; d: for complex "induction heating–rolling operation"; 1: $\varphi_{gmin} + \psi\Delta_T$; 2: φ^0_{1min}/B_1.

Figure 6.1 shows all possible cases that could take place when solving the problem in Equation (6.21) and Equation (6.22) with respect to different variants of relations between minimal durations of billet heating and pressing processes under considered temperature range of θ^*_2 variation.

The value φ^0_{1min} increases monotonously with θ^*_2 growth and the curve $\varphi_{gmin}(\theta^*_2)$ has an evident extremum. Minimum of $\varphi_{gmin}(\theta^*_2)$ is reached at the temperature $\theta^*_2 = \theta^*_{2c}$. Growth of $\varphi_{gmin}(\theta^*_2)$ under $\theta^*_2 < \theta^*_{2c}$ occurs due to a reduction of maximal velocity of extrusion for the press hydraulic drive because pressing force increases with billet temperature reduction.[11] A similar effect under $\theta^*_2 > \theta^*_{2c}$ can be explained by forced reduction of pressing velocity under the restraint in Equation (6.13).[10] Figure 6.1a shows the case when the expression $\varphi_{gmin}(\theta^*_2) + \psi\Delta_T > \varphi^0_{1min}(\theta^*_2)/B_1$ is valid for all values $\theta^*_2 \geq \theta^*_{2min}$. The maximum productivity of complex is limited by only technical limitations of extrusion press for the whole possible range of θ_2^* variation under given value of the factor $\psi \in [0,1]$ in Equation (6.21).

According to Equation (6.21) and Equation (6.22), it is possible to obtain:

$$\varphi_c^* = \varphi_{g\min}(\theta_{2e}^*) + \psi\Delta_T\ , \tag{6.23}$$

where

$$\theta_{2e}^* = \begin{cases} \theta_{2c}^*, \ if \ \ \theta_{2c}^* \geq \theta_{2\min}^*; \\ \theta_{2\min}^*, if \ \theta_{2c}^* \leq \theta_{2\min}^*. \end{cases} \tag{6.24}$$

It is possible to conclude at this point that operational mode can be considered in this case as optimal with respect to providing the maximum productivity if a time-optimal pressing process is performed under initial temperature defined by Equation (6.24). Minimum cycle time φ_c^* is defined in this case by the value $\varphi_{g\min}(\theta_{2e}^*)$ only according to Equation (6.23).

As one can see, there is a limited productivity of induction heating installation in this case. Operational mode would be optimal if it provides billet heating up to the temperature $\theta_{2e}^* \pm \varepsilon$ for given time $B_1\varphi_c^* > \varphi_{1\min}^0\ (\theta_{2e}^*)$ with minimum product cost, according to Equation (6.18).

This optimization problem has been considered in Section 4.4. Because $\psi >$ 0 in Equation (6.23), it would be reasonable to match durations of heating and extrusion stages in a way that provides decreasing of ψ down to zero.

Figure 6.1b represents the case when the difference $\varphi_{g\min}(\theta_2^*) + \psi\Delta_T - \varphi_{1\min}^0(\theta_2^*)/B_1$ has different signs within admissible range of temperature θ_2^* variation. The component curve $A_1PC_1D_1C_1{}'D_1{}'F_1$ (for which θ_{2e}^* can be defined from Expression 6.24) corresponds to minimum cycle time $\varphi_{c\min}(\theta_2^*)$ in accordance with Equation (6.20).

If the expression $\theta_{2e}^* = \theta_{2c}^*$ is true, then maximum productivity of the technological complex is limited only by technical limitations of extrusion press, and the situation turns out to be the same as in Figure 6.1a. However, if the expression $\theta_{2e}^* = \theta_{2\min}^*$ is true under condition $\theta_{2c}^* < \theta_{2\min}^*$, then the value of φ_c^* is limited by press (the point C_1 in Figure 6.1b) or by heating installation (the point C_1') capabilities that depend on value of $\theta_{2\min}^*$. Therefore, the extrusion stage will be time optimal if the expression $\varphi_c^* = \varphi_{g\min}(\theta_{2\min}^*) + \psi\Delta_T$ is true; in the case of $\varphi_c^* = \varphi_{1\min}^0(\theta_{2\min}^*)/B_1$, the heating stage should be optimized with respect to time-optimal criterion.

At the same time, the operational mode of the other stage of the discussed technological complex should be optimal with respect to economical criterion (Equation 6.18) under specified values φ_c^* and $\theta_{2e}^* = \theta_{2\min}^*$.

If the value $\theta_{2\min}^*$ exceeds the level of θ_{2c}^* and corresponds to points D_1 or D_1' (Figure 6.1b), then the technological complex has maximum productivity under time-optimal control of operational mode for heating and extrusion stages. This means that maximum productivity of the technological complex can be provided only if both heater and hot working equipment operate with maximum productivity.

Figure 6.1c shows the case that differs from the previous one. Here, maximum press productivity under $\theta_2^* = \theta_{2c}^*$ cannot be reached because of insufficient output rate of the heating installation under the condition:

$$\varphi^0_{1\min}(\theta^*_{2c}) / B_1 > \varphi_{g\min}(\theta^*_{2c}) + \psi\Delta_T .$$

Component curve $A_2D_2C_2D_2'F_2$ corresponds to the value $\varphi^*_{c\min}(\theta^*_{2c})$ (see Expression 6.20) and defines the value of θ^*_{2e} in the point D_2 (in accordance with Expression 6.22). If the restraint (Equation 6.19) is not violated in this point, then the following equation can be written:

$$\varphi^0_{\min}(\theta^*_{2e}) / B_1 = \varphi_{g\min}(\theta^*_{2e}) + \psi\Delta_T \tag{6.25}$$

under time-optimal control of pressing and heating processes. Otherwise, the situation is similar to the one shown in Figure 6.1b.

In all cases, the value of θ^*_{2e} can be found according to Expression (6.22). At the same time, φ^*_c could be found using the algorithm in Equation (6.21), and optimal durations of the heating ($\varphi^0_{1\text{opt}}$) and pressing ($\varphi^0_{g\,\text{opt}}$) processes can be computed according to the following correlations:

$$\begin{aligned}
\varphi^0_{1\text{opt}} &= \begin{cases} \varphi^0_{1\min}(\theta^*_{2e}), & if\ \varphi^*_c = \dfrac{1}{B_1}\varphi^0_{1\min}(\theta^*_{2e}); \\[2ex] B_1\left[\varphi_{g\min}(\theta^*_{2e}) + \psi\Delta_T\right], & if\ \varphi^*_c = \varphi_{g\min}(\theta^*_{2e}) + \psi\Delta_T; \end{cases} \\
\varphi_{g\,\text{opt}} &= \begin{cases} \varphi_{g\min}(\theta^*_{2e}), & if\ \varphi^*_c = \varphi_{g\min}(\theta^*_{2e}) + \psi\Delta_T; \\[2ex] \dfrac{1}{B_1}\varphi^0_{1\min}(\theta^*_{2e}) - \psi\,\Delta_T, & if\ \varphi^*_c = \dfrac{1}{B_1}\varphi^0_{1\min}(\theta^*_{2e}). \end{cases}
\end{aligned} \tag{6.26}$$

The optimal control problem could be solved in a similar way with respect to maximal productivity of the technological complex that consists of another type of metal working operation — for example, for the technological complex "induction heating–rolling operation."

In case of optimizing the "induction heating–rolling" system, the dependency $\varphi^*_{g\min}(\theta^*_2)$ has a monotonous decreasing nature[12] and it is possible to obtain the condition shown in Figure 6.1d that is similar to the case considered previously in Figure 6.1c for the point D_2. If minimum production cost (Equation 6.18) is required for the technological complex "heating–extrusion," then the joint optimization problem could be solved by using the same technique as for maximum productivity problem.

Instead of dependencies $\varphi^0_{1\min}(\theta^*_2)$ and $\varphi_{g\min}(\theta^*_2)$ in Equation (6.21) and assuming that the values φ^0_1 and φ_g are given, it would be necessary to find separate values for heating ($I_{2H\min}(\theta^*_2)$) and for extrusion ($I_{2g\min}(\theta^*_2)$) that correspond to minimum expenses in the form of appropriate items of the sum (Equation 6.18). These values can be found by solving the appropriate local optimization problems under fixed values θ^*_2.

For each fixed value of θ^*_2, the minimal value of production cost $I_{2\min}(\theta^*_2)$ can be considered as the sum of expenses for heating and extrusion (according to Equation 6.18). Thus, the following minimization procedure with respect to θ^*_2 defines optimal value $I_2 = I^*_2$ in accordance with the algorithm of Equation (6.21):

$$I^*_2 = \min_{\theta^*_2 \geq \theta^*_{2\min}} I_{2\min}(\theta^*_2) = \min_{\theta^*_2 \geq \theta^*_{2\min}} \left[I_{2H\min}(\theta^*_2) + I_{2g\min}(\theta^*_2) \right]. \tag{6.27}$$

It should be pointed out here that, similarly to the case of solving the maximum productivity problem, it is possible to find optimal value θ^*_{2e} that represents an initial extrusion temperature as follows:

$$\theta^*_{2e} = \arg \min_{\theta^*_2 \geq \theta^*_{2\min}} \left[I_{2H\min}(\theta^*_2) + I_{2g\min}(\theta^*_2) \right]. \tag{6.28}$$

6.4 MULTIPARAMETER STATEMENT OF THE OPTIMIZATION PROBLEM FOR TECHNOLOGICAL COMPLEX "HEATING–HOT FORMING"

Solution of the problem that provides maximum productivity of the technological complex "induction heater–hot working equipment" (Section 6.3) is based on consideration of all possible initial temperature states of metal prior to hot forming in the class of one-parameter functions. The temperature θ^*_2 uniformly distributed within the billet with a priori fixed absolute inaccuracy ε can be considered as the single parameter.

Such an approach can be extended to more complex problems of combined optimization of the technological complex "heating–hot forming" when temperature profiles of treated billets are given in the parametric form as a function of two and more parameters. Heating accuracy ε can be considered as an example of these parameters. Optimal values ε_e and θ^*_{2e} should be determined in the course of searching optimal control functions. The value ε_e can be included in the set of technological parameters within process flow sheets that should be optimized.

Multiparameter optimization problems for technology of gradient heating prior to pressing are of great interest to industry. These problems deal with nonuniform initial temperature distribution along the length of a pressurized billet[10] that provides greater pressing velocities under restriction (Equation 6.13). Here, two- or multiparameter representation of temperature distribution will be required.

Let us assume that temperature $\theta_2(l,y,\varphi_1^0+\Delta_T)$ evolves along axial coordinate according to linear law and is uniformly distributed (with fixed accuracy $\pm\,\varepsilon$) in a radial direction. In this case, the final temperature θ_2^* in any billet cross-section and its drop along the length of heated billet can be considered as two unknown parameters.

In a general case, it is possible to go to indirect parametric definition of the region Ω of possible initial pressing temperatures on the parameter ensemble/set. Here, these parameters form the argument list in the a priori definition of optimal control input function.

As was mentioned earlier, optimal values of interval durations Δ_i, $i = 1, 2, \ldots, N$ uniquely specify time-optimal control of the billet heating process. Optimal control varies within each of Δ_i, according to certain a priori known law. Final temperature distribution at the end of heating and transfer stages can be represented under time-optimal control by expressions $\theta_1(l,y,\Delta)$ and $\theta_2(l,y,\Delta,\Delta_T)$, respectively, which can be obtained as functions of Δ — for example, in the form of Equation (2.28) and Equation (2.29); Equation (4.15) and Equation (4.16); Equation (4.34) through Equation (4.36); Equation (4.94) and Equation (4.95). Therefore, the final temperature state can be considered as a function of N parameters of control input. Now it will be necessary to consider the basic correlation (Equation 6.21) on the set of these parameters instead of θ_2^*:

$$\varphi_c^* = \min_{\Delta\in\Omega^*} \varphi_{c\,\min}(\Delta) = \min_{\Delta\in\Omega^*}\left[\max\left(\frac{1}{B_1}\varphi_{1\min}^0(\Delta),\ \varphi_{g\,\min}(\Delta)+\psi\Delta_T\right)\right]. \quad (6.29)$$

Here, Ω^* is such set of values Δ, on which the minimal initial pressing temperature within the billet volume turns out to be not less than the value $\theta_{2\min}^*$ in Equation (6.19):

$$\Omega^* = \left\{\Delta:\ \min_{l,y,\in[0,1]} \theta_2(l,y,\Delta,\Delta_T) \ge \theta_{2\min}^*\right\}. \quad (6.30)$$

The dependency $\varphi_{1\min}^0(\Delta)$ turns out to be preset in the simplest form, according to the criterion (Equation 2.31):

$$I_1(\Delta) = \varphi_{1\min}^0(\Delta) = \sum_{i=1}^{N} \Delta_i\ . \quad (6.31)$$

Therefore, only the function $\varphi_{g\min}(\Delta)$ should be found by solving the time-optimal problem for the pressing process with respect to an admissible set (Equation 6.30) of fixed values Δ. These values uniquely define initial temperature state $\theta_3(l,y,\varphi_1^0 + \Delta_T) \equiv \theta_2(l,y,\Delta,\Delta_T)$ of a pressurized billet. As a result, similarly to θ_{2e}^* in Equation (6.22), extremum values Δ_{ie}, $i = 1, 2, \ldots, N$ could be defined. Under these values, Expression (6.29) holds true for φ_c^*. Optimal temperature state θ_{20pt}^* $(l, y, \Delta_e, \Delta_T)$ corresponds to values Δ_{ie}, $i = 1, 2, \ldots, N$.

The described method of parametric assignment of the initial pressing temperatures set in general does not predestine the shape of spatial temperature distribution within the billet volume during the pressing process. Therefore, due to extension of the region Ω in Equation (6.14) (in comparison with the condition in Equation 6.19), the choice of temperature states can be performed more freely, and the value of combined cost function for the technological complex optimization can be improved.

At the same time, a multidimensional search procedure with respect to values Δ_{ie}, $i = 1, 2, \ldots, N$ that is performed according to algorithm (Equation 6.29) will be more complicated in comparison with the one-dimensional case (Equation 6.21).

When minimum product cost of the technological complex is required, a similar optimization problem can be formulated with respect to set of parameters $\Delta = (\Delta^{(1)},\Delta^{(2)})$ that specify optimal control of heating $(\Delta^{(1)})$ and pressing $(\Delta^{(2)})$ processes. Here, $\Delta^{(1)}$ and $\Delta^{(2)}$ could make another physical sense in comparison with parameter vector Δ in Expression (6.29). Because the value $\Delta^{(1)}$ uniquely specifies temperature distribution $\theta_2(l,y,\Delta^{(1)},\Delta_T)$, and, in the same time, criterion I_{2g} depends on this temperature distribution, $\Delta^{(2)}$ incorporates the components of vector $\Delta^{(1)}$. As a result, similarly to Equation (6.29), the expression for I_2^* can be obtained instead of Equation (6.27) in the following form:

$$I_2^* = \min_{\Delta^{(1)}\in\Omega_1^*,\Delta^{(2)}\in\Omega_2^*} I_2(\Delta^{(1)},\Delta^{(2)}) = \min_{\Delta^{(1)}\in\Omega_1^*,\Delta^{(2)}\in\Omega_2^*} \left[I_{2H}(\Delta^{(1)}) + I_{2g}(\Delta^{(2)}) \right]. \quad (6.32)$$

Here, Ω_1^* can be defined similarly to Equation (6.30), and Ω_2^* can be found according to restriction (Equation 6.13) as follows:

$$\begin{cases} \Omega_1^* = \left\{ \Delta^{(1)} : \min\limits_{l,y\in[0,1]} \theta_2(l,y,\Delta^{(1)},\Delta_T) \geq \theta_{2\min}^* \right\}; \\ \Omega_2^* = \left\{ \Delta^{(2)} : \theta_3(l_k,1,\Delta^{(2)},\varphi) \leq \theta_{3cr},\ \varphi_1^0 + \Delta_T \leq \varphi \leq \varphi_2^0 \right\}, \end{cases} \quad (6.33)$$

where $\theta_3(l_k,1,\Delta^{(2)},\varphi)$ represents the temperature in the die hole ($l = l_k$, $y = 1$) at any time φ during the pressing process for each fixed value of the vector $\Delta^{(2)}$.

6.5 COMBINED OPTIMIZATION OF HEATING AND PRESSING MODES FOR ALUMINUM ALLOY BILLETS

The described method has been used for analysis of productivity of the technological complex for heating of aluminum alloy D16 cylindrical billets and their subsequent hot forming on the horizontal hydraulic press. Length of the billet is 350 mm; applied frequency is 50 Hz; diameter of the press product (rods) is 22 mm; and diameter of the press container is 90 mm.[5,6] Technological requirement demands that required uniformity of temperature distribution within the billet should be provided prior to the hot forming operation.

At the first step, local problems of time-optimal control should be solved separately for heating and pressing processes. It is important to note that here the given set of permissible values of initial pressurized metal temperatures θ_2^* should be taken into account. At the next step, the minimum duration of production cycle should be chosen according to the condition in Equation (6.20).

6.5.1 TIME-OPTIMAL HEATING MODES

Dependency $\varphi_{1\min}^0(\theta_2^*)$ can be obtained with regard to models (Equation 6.1 through Equation 6.6) by using techniques described in Chapter 3 and Chapter 4. Restraint on maximum temperature at the level of 500° should be taken into account. Maximum value of heating power (that is uniformly distributed along the billet length) is equal to 60 kW/m².

Heat losses from the billet surface during heating and transfer stages (Δ_T = 30 s) can be estimated by values of $Bi_h = 0.015$ and $Bi_T = 0.030$, respectively, in boundary conditions in Equation (6.2) and Equation (6.5). It is assumed that the expression $\Delta_T > \Delta_{T2}^*$ (Section 4.2) holds true.

Calculations have been conducted for several fixed values of final heating temperatures θ_2^* in the range of 320 to 470° under required accuracy $\varepsilon = \varepsilon_{\min}^{(2)}$. Here, the lower value, $\theta_{2\min}^* = 320°$, in Equation (6.19) is in agreement with requirements to press instrument strength, and the upper value of 480° represents the maximum permissible temperature in the die hole, according to the condition in Equation (6.13).

Calculated values $\varepsilon_{\min}^{(2)}$ do not exceed 5°C and practically do not depend on the value θ_2^* in the considered temperature range. This fact allows obtaining the expression $\theta_2(l,y,\varphi_1^0 + \Delta_T) \cong \theta_2^* = const$ that holds true with acceptable accuracy. This expression can be used in order to define temperature distribution during the pressing process.

6.5.2 TIME-OPTIMAL PRESSING MODES

Dependency $\varphi_{g\min}(\theta_2^*)$ in Equation (6.21) can be defined after solving the local time-optimal problem for the pressing process. The fixed values of initial billet temperatures prior to hot forming operation should be considered within the range of 320 to 470°C. This temperature interval represents the whole possible range of final temperature (after heating and transfer stages) variation.

The billet temperature during the pressing process can be described by Equation (6.7) through Equation (6.9). The local time-optimal control problem can be formulated as follows. It is necessary to select such control of pressing speed $V_{P\,opt}$ (φ) restricted by a preassigned set of constraints (Equation 6.13 through Equation 6.15) that provides the required value, z_{end}, of final press product (see Equation 6.11) for minimum possible time under given initial temperature distribution within the billet $\theta_3(l,y,\varphi_1^0 + \Delta_T) \equiv \theta_2^* = const.$

According to the Equation (6.11), the optimal program $V_{P\,opt}$ (φ) can be defined by the choice of maximum admissible value of pressing speed at each time instant. Pressing speed is limited only by conditions in Equation (6.13) and Equation (6.15). Thus, at the initial stage, the function $V_{P\,opt}$ (φ) should be held at its admissible limit, $V_{P\max}$ (φ), up to the moment, $\varphi = \varphi_{\lim}$, when the condition $\theta_3(l_k,1,\varphi_{\lim}) = \theta_{3cr}$ becomes true in Equation (6.13). During the next step, until the end of the pressing process, the optimal control $V_{Popt}(\varphi) = V_{Pcr}$ should hold the temperature $\theta_3(l_k,1,\varphi_{\lim})$ at the level of θ_{3cr} (isothermal extrusion mode).[10]

As a result, the following algorithm of time-optimal control of the extrusion process can be obtained:

$$V_{P\,opt}(\varphi) = \begin{cases} V_{P\max}(\varphi), \; \varphi_1^0 + \Delta_T < \varphi < \varphi_{\lim}; \\ V_{Pcr}(\varphi) : \theta_3(l_k, 1, \varphi)\big|_{V_P(\varphi)=V_{Pcr}(\varphi)} \equiv \theta_{3cr}, \varphi_{\lim} < \varphi \le \varphi_2^0. \end{cases} \tag{6.34}$$

The value of $V_{P\max}$ in Equation (6.15) is equal to the value of function $V_{P\max}$ (φ) that evolves in time according to relatively complex law. In particular, for an accumulator hydraulic drive with throttle adjustment of extrusion speed, the following equation of press velocity characteristic,[11]

$$V_{P\max}(\varphi) = \sqrt{\frac{P_a - P_P(\varphi)}{u^*_{\min}}}, \tag{6.35}$$

directly links the value of $V_{P\max}(\varphi)$ with the minimum value $u^*_{\min}$ of the hydraulic resistance factor of the throttle valve. In Expression (6.35), P_a represents constant pressure in an accumulator and P_P is extrusion pressure that depends on values of L and H in Equation (6.10).

Using a simplified approach, it is possible to calculate $V_{P\,opt}$ (φ) assuming that average value $V_{P\max}(\varphi)$ is constant during the interval $(0,\varphi_{\lim})$.

If the control algorithm is given in the form of Equation (6.34), then the total time, $\varphi_{g\min}$, required for extrusion can be defined as a root of the equation $z(\varphi_2^0, V_{P\,opt}(\varphi)) = z_{end}$. This equation could be obtained by integration of Equation (6.11) under $V_P = V_{Popt}$ (φ).

With regard to the optimal control algorithm (Equation 6.34), the time, $\varphi_{\lim}$, and the optimal control function, $V_{Pcr}(\varphi)$, during the interval of isothermal pressing can be found by using analytical expressions for the temperature distribution within pressurized metal $\theta_3(l,y,\varphi)$ as function of the value V_P. These expressions can be obtained as approximate solution of Equation (6.7) through Equation (6.10).[13]

6.5.2.1 Temperature Distribution within Pressurized Metal

Computation of temperature distribution $\theta_3(l,y,\varphi)$ evolving over the extrusion process represents a problem that cannot be solved easily. One of the difficulties deals with the necessity to solve previously a highly complicated problem of mechanics of continua in order to define velocity field of metal flow and spatial distribution of plastic deformation energy.[7–10]

In the considered case, simple analytical approximations for sought functions $V_y(l,y,V_P)$, $V_l(l,y,V_P)$, and $W^*(l,y,V_P)$ can be used in Equation (6.7) under conditions of axially symmetric deformation. According to the well-known hypothesis of spherical sections, metal flow velocities V_c remain constant along concentric spheres that have the common center in the point O_1, and velocities have the direction to this point within the pressed part of the plastic zone (PPPZ) (Figure 6.2).[7] Based on the hypothesis of spherical sections, the following equations for velocity field within figure ABCD in Figure 6.2 can be obtained[5,13]:

$$V_l = -V_P \frac{l\, tg\alpha}{v\sqrt{v}\cos^2\alpha}; \; V_y = V_P \frac{l_k + \dfrac{1-y}{\beta} tg\alpha}{v\sqrt{v}\cos^2\alpha}; \; v = l^2 tg^2\alpha + \left(\frac{1-y}{\beta} tg\alpha + l_k\right)^2. \tag{6.36}$$

Here, $l_k = X_k/X$, where X_k and X are radius of the press product (rod) and radius of the pressurized billet, respectively; α is the angle that takes into account existence of the elastic zone CBF during the extrusion stage. It is possible to assume that the value of α will be equal to $\pi/3$.[7] The factor β in Equation (6.7) can be defined by correlation in Equation (4.76).

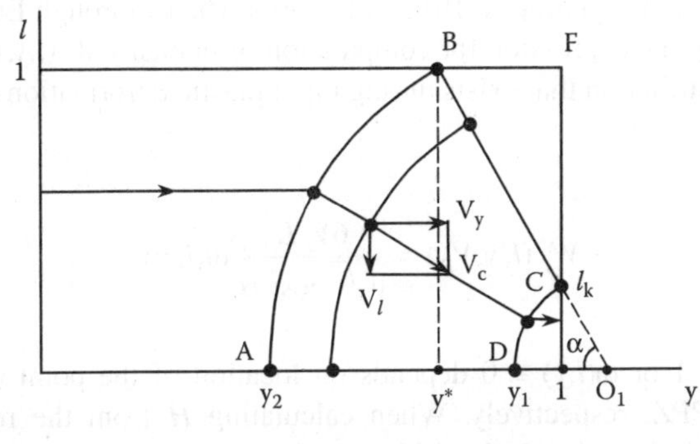

FIGURE 6.2 Definition of metal flow velocities according to hypothesis of spherical cross-sections.

Within elastic zone CBF, there will be no movement of metal flow:

$$V_l = V_y \equiv 0 \, . \qquad (6.37)$$

Outside this zone and outside the pressed part of the plastic zone, the following conditions will be valid:

$$V_y \equiv V_P ; \; V_l = 0 \, . \qquad (6.38)$$

Neglecting the radial component of flow velocity, the following expressions can be obtained for metal flow velocities to the left and to the right of PPPZ:

$$V_y = \frac{V_P}{l_k^2} ; \; V_l = 0 \, . \qquad (6.39)$$

Velocities V_y and V_l of metal flow within the whole pressurized billet volume can be described approximately by Equation (6.36) through Equation (6.39).

Intensity H of velocity of shear deformation during pressing can be calculated by the formula in Equation (6.10) after substituting the known relation between the value of H and components of deformation velocity tensor in the cylindrical coordinate system. The following expression for H can be obtained in relative units[13]:

$$H = \sqrt{2\left[\left(\frac{\partial V_l}{\partial l}\right)^2 + \beta^2\left(\frac{\partial V_y}{\partial y}\right)^2 + \frac{V_l^2}{l^2} + \frac{1}{2}\left(\frac{\partial V_y}{\partial l} + \beta\frac{\partial V_l}{\partial y}\right)^2\right]} \qquad (6.40)$$

Correlations in Equation (6.10) and Equation (6.36) through Equation (6.40) allow obtaining the expression for computation of intensity $W^*(l,y,V_P)$ of internal heat source distribution that exists during billet plastic deformation in the process of extrusion[13]:

$$W^*(l,y,V_P) = \frac{6V_P L}{\nu\sqrt{\nu}\cos^2\alpha}\,\omega(l,y) \, . \qquad (6.41)$$

Here, $\omega(l,y) \equiv 1$ or $\omega(l,y) \equiv 0$ depends on location of the point (l,y) inside or outside the PPPZ, respectively. When calculating H from the relationship in Equation (6.40), the value L should be considered as a part of the basic value

$$L_b = \frac{W_b^* X}{V_b}.$$

In Expression (6.41), it is reasonable to use linear approximation of the function $L(\theta)$[10] as follows:

$$L(\theta) = T_0 - K_\theta \theta \,, \tag{6.42}$$

where T_0 corresponds to value of T under basic temperature, and K_θ is the constant factor.

Let us assume that heat flow density of contact friction $q(y,V_P)$ in the boundary condition in Equation (6.8) is proportional to velocity $V_y(l,y)$ at the billet surface $l = 1$.[14] In this case, the following correlation can be obtained, taking into account Expression (6.37) and Expression (6.38):

$$q(y,V_P) = \begin{cases} K_q V_P, \;\; 0 \le y \le y^*; \\ 0, \;\; y^* < y \le 1; \; y^* = 1 - \dfrac{\beta}{tg\alpha}(1 - l_k). \end{cases} \tag{6.43}$$

Here, K_q = const is coefficient of proportionality.

Analytical solution of heat transfer problems in Equation (6.7) through Equation (6.10) with respect to temperature distribution $\theta_3(l,y,\varphi)$ will be greatly complicated due to highly pronounced nonuniformity of spatial distribution of metal flow velocities. This is so even in the case when simplified Expression (6.37) through Expression (6.43) represent explicit functions V_l, V_y, W^*, and q from their arguments.

A simplified approach provided in Rapoport[13] allows one to solve the problem for initial and steady-state stages of extrusion with constant speed $V_P = const.$ The numerical model that describes temperature distribution $\theta_3(l,y,\varphi)$ during a variety of hot forming processes[15] can be successfully applied to optimization of different modes of a hot forming operation. This numerical model has been obtained in the course of the problem (Equation 6.7 through Equation 6.10) solution by using the finite-difference method under conditions in Equation (6.37) through Equation (6.43).

6.5.2.2 Optimal Program of Extrusion Speed Variation

In the considered case, the constant averaged value of extrusion speed is equal to $V_{P\max} \cong 5$ m/min during the interval $(\varphi_1^0 + \Delta_T, \varphi_{\lim})$. The correlation $\theta_{2c}^* < \theta_{2\min}^*$ (Figure 6.1) will be valid for each time point that belongs to the interval $(\varphi_1^0 + \Delta_T, \varphi_{\lim})$. The value $\varphi_{\lim}$ should be defined as the moment when maximum permissible temperature $\theta_{3cr} = \theta_3(l_k, 1, \varphi_{\lim})$ in the die hole is reached. Under extrusion with constant speed of 5 m/min, the temperature $\theta_3(l_k,1,\varphi_{\lim})$ is assumed to be equal to 480°. This temperature can be calculated using the analytical model described earlier.

A stepwise approximation of extrusion speed $V_{Pcr}(\varphi)$ during the isothermal stage of the process can be found by selection of extrusion speed $V_P^{(j)} = const$ on each jth interval of its constancy $(\varphi_j, \varphi_{j+1})$, $j = \overline{1,M}$. Lengths of these intervals, $\varphi_{j+1} - \varphi_j$, are relatively small for fixed a priori values φ_j and φ_{j+1}. Extrusion speed $V_P^{(j)} = const$ should provide holding of the temperature $\theta_3(l_k,1,\varphi)$ at the level of 480° with required accuracy for all times $\varphi = \varphi_j$ and $\varphi = \varphi_{j+1}$.

Computation of $\theta_3(l_k,1,\varphi)$ for all $\varphi \in (\varphi_j,\varphi_{j+1})$ was conducted on the basis of the proposed analytical model under $V_P = V_P^{(j)} = const$ using the special computational algorithm. This algorithm allows one to simplify sharply the definition of initial temperature distribution $\theta_3(l,y,\varphi_j)$ using initial temperature distribution at a previous interval with pressing speed equal to $V_P^{(j-1)}$.[16]

The values $V_P^{(j)}$ can be found consecutively for all $j = \overline{1,M}$ as roots of equation $\theta_3(l_k,1,\varphi_j,V_P^{(j)}) = \theta_{3cr}$. Here, the function $\theta_3(l_k,1,\varphi_j,V_P^{(j)})$ can be defined by analytical expressions for temperature in the die hole as functions of $V_P^{(j)}$ at the fixed a priori times $\varphi = \varphi_j$.

All computations for the temperature field of a pressurized billet were conducted for the following initial values of parameters of models in Equation (6.7) through Equation (6.15) and Equation (6.36) through Equation (6.43)[13]:

$$\gamma^* = 1.8;\ \beta = 0.26;\ Bi = Bi_0 = Bi_1 = 0.15;\ \theta_a = \theta_a^0 = \theta_{a0}^0 = \theta_{a1}^0 = 0;$$

$$l_k = 0.25;\ \alpha = \pi/3;\ K_\theta = 0.064;\ K_q = 0.49.$$

Comparison with experimental data exhibits acceptable accuracy of temperature computation using analytical expressions for extrusion under constant speed.[16]

6.5.3 Computational Results

Figure 6.3 shows optimal extrusion speed vs. time according to the algorithm (Equation 6.34) for initial temperature of the billet $\theta_2^* = 450$°C.[5,6] Figure 6.4 presents curves $\varphi_{1\min}^0(\theta_2^*)/B_1$ and $\varphi_{g\min}(\theta_2^*) + \psi\Delta_T$ that have been calculated according to the previously described method using absolute units ($\tau_{1\min}^0(t_2^*)/B_1$ and $\tau_{g\min}(t_2^*) + \psi\Delta_T$, respectively). The value of $\psi\Delta_T$ is equal to $\Delta_T = 30$ s and $B_1 = 3$.

The component curve $C_2D_2'F_2$ corresponding to the variant in Figure 6.1c represents the maximum complex productivity as a function of temperature t_2^* that determines the final temperature within the heated billet at the end of the transfer stage. As can be seen in Figure 6.4, optimal temperature value t_{2e}^* coincides in this case with minimal admissible value $t_{2\min}^* = 320$°C that is limited by strength of extrusion equipment. As one can see, there is a limited productivity of induction heating installation in this case; therefore, the maximum productivity of the technological complex is limited by heating installation capabilities. The

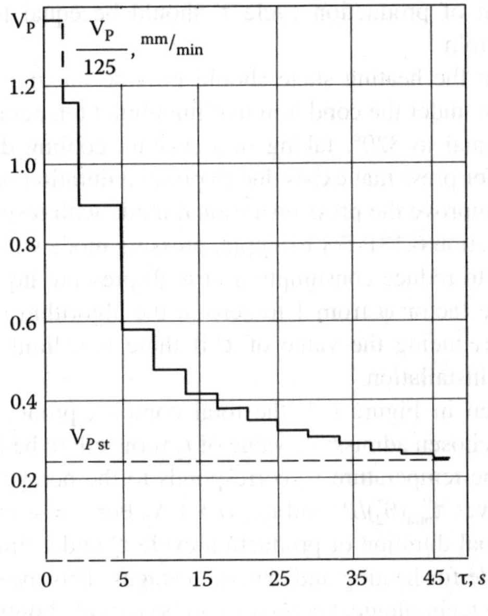

FIGURE 6.3 Optimal pressing velocity vs. time.

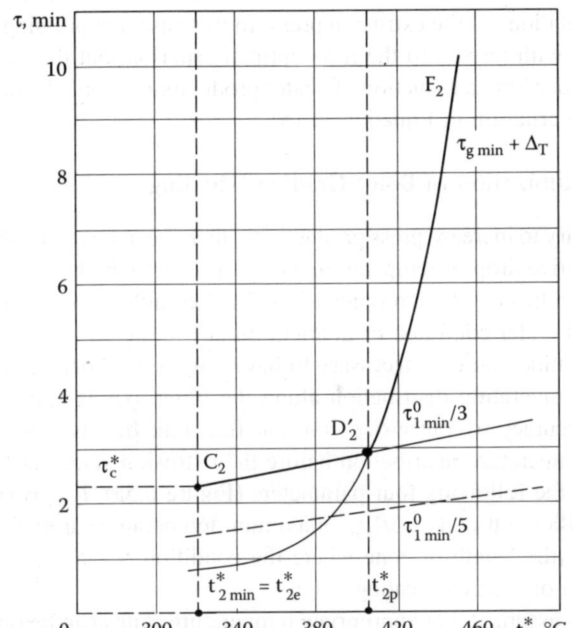

FIGURE 6.4 Minimal durations of billet heating and pressing processes as functions of temperature at the end of heating.

minimum duration of production cycle τ_c^* should be equal to IHI output rate $\tau_{1\min}^0/B_1$, i.e., 2.25 min.

Control during the heating stage should provide maximum productivity of heating installation under the condition that final billet temperature prior to press feeding will be equal to 320°, taking into account cooling during the transfer stage. Extra time for press that exists due to optimal duration of production cycle could be used to improve the press operational mode with respect to an economical criterion (Equation 6.18). For example, pressing mode with reduced velocity can be performed to reduce consumption of high-pressure liquid.[11]

Decreasing the factor ψ from 1 to zero in the algorithm in Equation (6.20) does not lead to reducing the value of τ_c^* if there is a limited productivity of induction heating installation.

As can be seen in Figure 6.4, the total complex productivity will be not optimal while the chosen admissible value of t_2^* turns out to be less than the value of $t_{2p}^* = 408°C$. The temperature t_{2p}^* corresponds to the point D_2' that represents the crossing of curves $\tau_{\min}^0(\theta_2^*)/B_1$ and $\tau_{g\min}(t_2^*) + \Delta_T$. For $t_2^* = 408°C$, the production complex has optimal duration of production cycle τ_c^* under time-optimal control of operational mode for heating and extrusion stages. This means that maximum productivity of the technological complex can be provided only if heater and hot working equipment operate with maximum productivity.

If the expression $t_2^* \geq 408°C$ is true, then the value of τ_c^* is limited only by technical limitations of the extrusion press. In this case, the pressing stage should be optimized with respect to the time-optimal criterion, but the heating process performed under forced reduction of heater productivity should be optimized with respect to the criterion in Equation (6.18).

6.5.3.1 Optimization of Billet Gradient Heating

If it is necessary to increase press productivity, then it can be reasonable to provide an initial positive drop of temperature along the billet length (in the direction of extrusion).[10] In this case, the problem of optimal complex control and IHI design can be solved under condition of gradient billet heating.[5,17]

Let us assume that it is necessary to have a uniformly heated billet along its radius, and temperature distribution along the billet axis is approximated with acceptable accuracy by piecewise constant function. In this case, the set Ω of admissible temperature distribution before the extrusion stage can be defined as a function of the following four parameters (Figure 6.5a): $t_{2\max}^*$ (maximum temperature of billet butt-ends); $\Delta t_{2\max}^*$ (maximum temperature drop along the billet length; $\Delta y_{\max}$ (the length of zone where the condition $t_2^* = t_{2\max}^*$ holds true); and Δy (the length of gradient heating zone).

Here, it is assumed that appropriate temperature states can be obtained in the process of heating with relatively small absolute inaccuracy ε, which could be neglected when calculating temperature distribution during the extrusion stage. Optimal values $t_{2\max e}^*$, $\Delta t_{2\max e}^*$, $\Delta y_{\max e}$, and Δy_e of specified parameters correspond to maximum complex productivity. These values can be found according to the

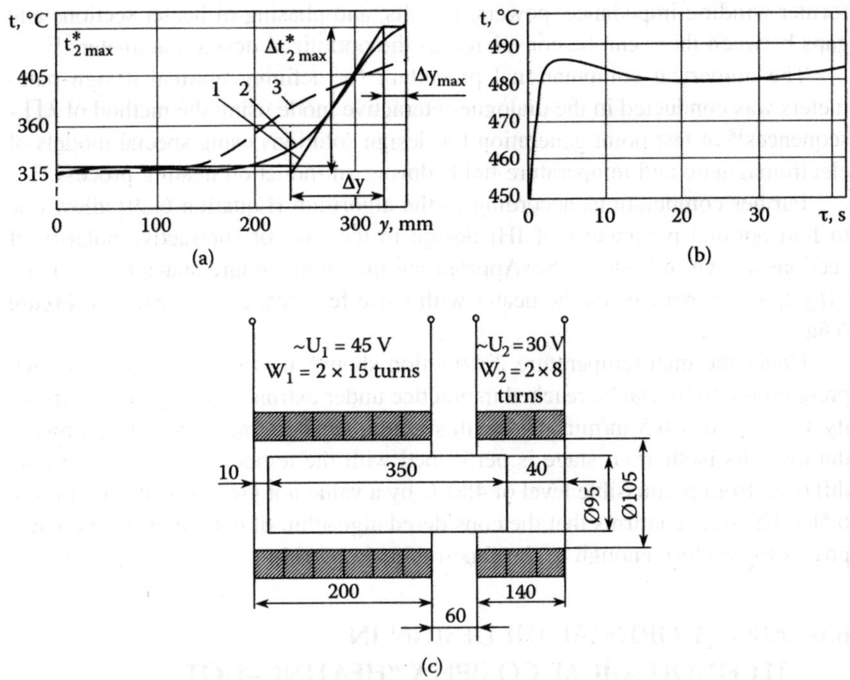

FIGURE 6.5 Optimization of design and operational modes of production complex "induction heating–extrusion operation" with gradient heating. a: Temperature distribution along the billet length prior to extrusion (1: required distribution; 2: real temperature distribution without time of transportation; 3: real temperature distribution with time of transportation). b: Temperature in the die hole vs. time during extrusion process under $V_P \cong 0.5$ m/min. c: Heater design.

previously described general method using the numerical model for pressurized metal temperature field.[15] In the course of computational procedure, it is also possible to find optimal temperature distribution $t_{2opt}(l,y,\varphi_1^0 + \Delta_T)$ under the optimal values of process parameters.

Some results of calculations for aluminum alloy billet gradient heating and subsequent pressing operation are presented in Figure 6.5 under initial data of the case shown in Figure 6.4.

The temperature distribution $t_2(l,y,\varphi_1^0 + \Delta_T)$ at the end of heating and transfer stages under a certain set of parameters t^*_{2max}, Δt^*_{2max}, Δy_{max}, and Δy can be considered as a requirement to IHI operation. At the same time, the required final temperature distribution within the heated billet should be provided with accuracy ε.

If we limit our consideration to typical single-stage heating mode under constant inductor voltage and two-section heater design suitable for gradient heating (Figure 6.5c), then the problem can be reduced to the problem of optimal inductor design that provides the minimal value of ε. Coil overhang, autotransformer winding impedance, powers, lengths, and phasing of heater sections and gaps between them can be considered as the optimized design parameters.[17]

The numerical computational procedure for defining optimal design parameters was conducted in the dialogue/interactive mode using the method of $\lambda\Pi_\tau$-sequences[18] of test point generation for design solutions using special models of electromagnetic and temperature fields during an induction heating process.[17]

Further computations according to the algorithm (Equation 6.29) allow one to find optimal parameters of IHI design in the case of subtractive polarity of sections shown in Figure 6.5c. Appropriate final temperature states $t_1(l,y,\tau_1)$ and $t_2(l,y,\tau_1 + \Delta_T)$, realized in the heater with these features, are presented in Figure 6.5a.

Under the final temperature distribution shown in Figure 6.5a, the maximal press productivity can be reached in practice under extrusion with constant velocity $V_P \cong const = 0.5$ m/min. Under this mode, during almost the whole process duration, its isothermal stage is performed with the temperature in the die hole differing from permissible level of 480°C by a value not greater than 5°C (Figure 6.5b). This fact confirms that the considered algorithm of extrusion stage control proves to be close enough to the time-optimal algorithm.

6.6 ABOUT OPTIMAL IHI DESIGN IN TECHNOLOGICAL COMPLEX "HEATING–HOT FORMING"

The best effect of technological process optimization is reached in the cases when design solutions and technological instructions (charts) are chosen on the stage of technology development, and while designing the appropriate part of the whole production complex. These technological and design solutions provide the extreme value of optimized cost function while equipment operates in optimal modes.[1–6]

Solutions of similar problems allow one to implement elements of the optimal IHI design in production complexes for metal hot forming. In this regard, the different approaches could be offered, including methods proposed in the present book; some of them represent direct solutions of particular optimal design problems.

1. It was shown earlier that the optimization problem solution for the technological complex "heating — hot forming" allows one to define the optimal values of parameters of initial temperature distribution prior to hot working operations and to develop appropriate optimal control algorithms for IHI and hot forming equipment. This conclusion is valid under the condition that final temperature distribution within heated workpiece at the end of controlled heating process can be chosen freely. The parameters of final temperature distribution represent the key characteristics of heating modes and subsequent hot working processes, and they can be considered as the basis for development of normative technological instruction (charts).

The general method, examples of statement and solution of similar problem for maximum productivity of the complex "IHI–hot forming equipment" are discussed in Sections 6.3 through 6.5 under the condition that the permissible region for temperature distributions at the end of heating is presented in parametric form.

2. The solution of the combined optimization problem for processes of heating and subsequent hot forming could lead to the optimal design solutions for the production complex. The proper choice of operating characteristics of heating installations and hot working equipment would provide extreme values of overall cost function for the whole technological line.

In particular, when optimizing the billet output cycle time φ_c in the system "IHI — hot forming equipment," the complex productivity could be increased, according to the algorithm in Equation (6.21) in the case shown in Figure 6.1c and Figure 6.4, due to the increase of heating installation power and due to the larger number B_1 of simultaneously heated billets (see dotted lines, displacing point D_2 or D_2').

The same problem could be solved in the case when the condition $\theta^*_{2\min} > \theta^*_{2c}$ is true, which means that the stronger press instrument allows less value of minimum admissible temperature $\theta^*_{2\min}$ with regard to constraint in Equation (6.19) (see Figure 6.1 through Figure 6.4).

Another way of optimal IHI design respective to the best approximation to required initial temperature distribution before hot forming is described in Section 6.5. An example of the maximum productivity problem solution for the complex "IHI — hot forming equipment" with preliminary gradient billet heating before pressing is also provided in this section.

3. Definition of spatial-distributed control inputs and optimal inductor length in the problem of continuous heating mode optimization (Chapter 5) give answers to the main questions of continuous heater optimal design.

4. Based on the results of the solution of induction heating optimization problem, the surfaces in the space of IHI design parameters could be built. Within the limits of these surfaces the fixed values of considered cost function I remain constant (equiscalar surfaces of quality I = const). The set of these surfaces, built for different values I = const, allows to define in this space the regions of attainable quality according to chosen performance indexes. The further choice of particular parameter values within specified regions leads to partial realization of the optimal IHI design solutions.

As an example, equiscalar curves $\varphi^0_{1\min} = const$ are shown in Figure 6.6 for $\varepsilon = \varepsilon^{(1)}_{\min}$, $\varepsilon = \varepsilon^{(2)}_{\min}$ and $\varepsilon_0 = \varepsilon^{(i)}_{\min}$ = const, i = 1,2. These curves have been built using the solution of several time-optimal problems (see Section 3.1 through Section 3.4, and Section 3.6) with respect to the basic model (Equation (1.27) through Equation (1.35)) in the plane of dimensionless parameters ξ and Bi under θ_0 = const [6].

Each point on these curves corresponds to the computed value of minimum heating process time $\varphi^0_{\min}$ (ξ', Bi') or to maximum heating accuracy $\varepsilon^{(i)}_{\min}$ (ξ', Bi')

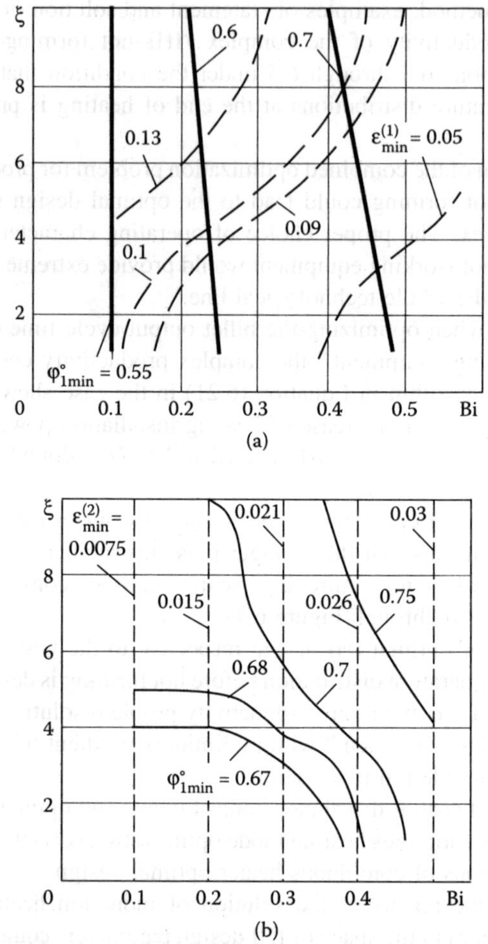

FIGURE 6.6 Results of time-optimal heating process under single-stage (a) and two-stage (b) control ($\theta_0 = -1$, $\Gamma = 1$; $\theta^* = 0$; $U_{max} = 1$).

for corresponding to these point values $\xi = \xi'$ and $Bi = Bi'$ under chosen value $\theta_0 = \text{const}$.

The set of these equiscalar curves allows to find admissible (according to given cost function, i.e., with respect to process time φ_1^0 and heating accuracy ε) intervals of parameters ξ and Bi variation. These intervals determine possible ranges of values that define the basic characteristics of IHI design solutions, for example, possible ranges of supplying current frequency (possible variations of ξ) and parameters of inductor thermal insulation (possible values of Bi) for IHI with given maximum heating power (under $\theta_0 = \text{const}$).

Building of corresponding *equiscalar surfaces* in the space of parameters ξ, Bi, θ_0 and their sections for ξ = const and Bi = const allows one to include the power of heating installation (the value θ_0) into the set of design parameters.

5. If we supplement unknown parameters Δ_i, $i = \overline{1, N}$ of sought optimal control algorithms (see Section 3.7) by parameters Δ_i, $i = \overline{N+1, N_1}$, characterizing IHI design solutions, then the set of all these values Δ_i, $i = \overline{1, N_1}$ could be found as a result of the combined optimization problem solution for optimal IHI control and design with regard to chosen cost criteria.

A representative example was presented earlier in Section 4.5.3, where the billet overhang in static induction installation can be considered as additional sought-for parameter.

A similar problem of combined optimization could be reduced to the statement of mathematical programming problem of the type in Equation (3.38), but it should be solved with respect to "extended" vector $\Delta = (\Delta_i)$, $i = \overline{1, N_1}$. In this case, the main feature consists in the fact that for typical economic criteria of combined optimization (in contrast to local criteria for optimal control problems), the global minimum of overall cost function is reached, as a rule, on the optimal design solution. Consequently, equality of the type in Equation (3.42) becomes true on the set of values Δ_i, $i = \overline{N+1, N_1}$. As was shown in Section 3.7, this equality violates assumptions, for which main properties of final temperature distribution within heated billet (see Equation (3.39) and Equation (3.40)) remain valid.

In that case it is not possible to use method proposed in the present text for parametric optimization of induction heating processes, because it is based on these properties. To find the solution of considered problem, it is necessary to apply the additional procedure of searching (by known methods) for extreme value of overall cost function within the set of optimized design solutions. This procedure should provide (as an intermediate step) computation of optimal control algorithms according to the general techniques of Chapter 3 through Chapter 5 on all tested elements of the design solutions set.

Let us illustrate the obtained conclusions by the example, when the power P_u of the heating installation presents the single sought parameter of the design solution. Let us also consider only the typical situation, when maximum IHI throughput corresponds to maximum of obtained economical profit. Then the total cost of the heating process implementation can be accepted as the overall cost function for optimal IHI design and control. The total cost is proportional to the duration of the heating stage and capital expenditures (proportional to P_u^2):

$$I = k_1 \tau_{1\min}\left(P_u\right) + k_2 P_u^2 \to \min_{P_u} . \tag{6.44}$$

Here τ_{1min} is the minimum time required for the heating process, that is defined by parameter vector Δ^0 of time-optimal control inputs (for instance, in the form of sum in Equation (2.31)).

The problem is reduced to searching for optimal heating power $P_{u\,opt}$ and parameters Δ^0 providing minimum value of criterion in Equation (6.44) under given cost factors K_1 and K_2.

For each fixed value P_u, the corresponding "own" solution Δ^0 (P_u) of properly set time-optimal problem exists. This solution is found using methods described earlier in Chapter 3 through Chapter 5.

The subsequent solutions for different values of P_u allows us to find monotonously decreasing dependency $\tau_{1min}(P_u)$. When increasing the second component in Equation (6.44), the cost function in Equation (6.44) reaches its global minimum at the point $P_u = P_{u\,opt}$ (Figure 6.7) which is defined by correlation

$$P_{u\,opt} = \arg\min_{P_u} \left[k_1 \tau_{1\min}\left(P_u\right) + k_2 P_u^2 \right]. \tag{6.45}$$

The special case takes place when given cost criteria have no extreme points on the set of considered design solutions. Under existing real-life constraints, the minimum of cost function is always reached at one of boundaries of possible design parameters range.

This information usually allows us to find optimal values of design parameters, and then it remains only to solve the optimal control problem without considering "extended" set of sought-for unknowns.

Such a situation appears, in particular, when searching for optimal values of supplying current frequency (parameter ξ) and *Bi* criterion (defining the quality of inductor thermal refractory) in the time-optimal problem with heating power

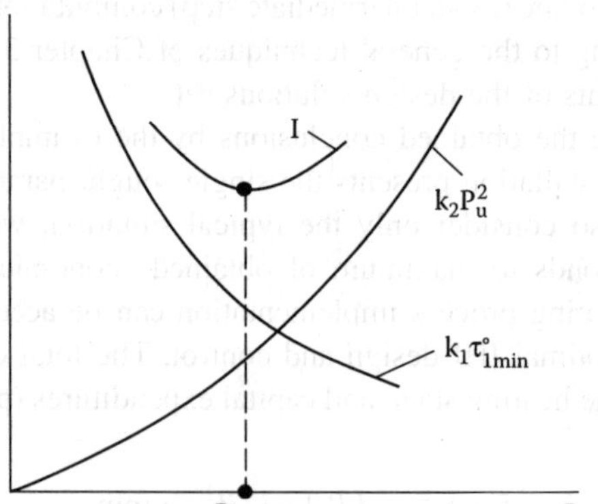

FIGURE 6.7 Value of economic cost I criterion as a function of heat power.

control. Optimal values of parameter ξ and *Bi* criterion comply often with their minimal admissible values.

REFERENCES

1. Butkovskij, A.G., Malyj, S.A., and Andreev, Yu.N., *Optimal Control of Metals Heating,* Metallurgy, Moscow, 1972.
2. Butkovskij, A.G., Malyj, S.A., and Andreev, Yu.N., *Control of Metals Heating*, Metallurgy, Moscow, 1981.
3. Malyj, S.A., *Energy Conserving Heating of Metals*, Metallurgy, Moscow, 1967.
4. Andreev, Yu.N., *Optimal Design of Heating Installations*, Mashinostroenie, Moscow, 1983.
5. Rapoport, E.Ya., Optimization of heating and extrusion modes in technological complex "heater–press," *Physica i himija obrabotki materialov*, 3, 66, 1985.
6. Rapoport, E.Ya., *Optimization of Induction Heating of Metals*, Metallurgy, Moscow, 1993.
7. Perlin, I.L. and Gajtbarg, L.H., *Theory of Metals Extrusion*, Metallurgy, Moscow, 1975.
8. Gun, G.Ya., Poluhin, P.I., and Ganelin, D.Yu., Mathematical modeling of axially symmetrical steady-state processes of metals extrusion, *Izvestija vuzov, Chernaja metallurgija,* 5, 82, 1976.
9. Gun, G.Ya., *Basics of Metals Treatment by Pressure*, Metallurgy, Moscow, 1980.
10. Gun, G.Ya. et al., *Extrusion of Aluminum Alloys (Mathematical Modeling and Optimization)*, Metallurgy, Moscow, 1974.
11. Scholobov, V.V. and Zverev, G.I., *Metals Extrusion*, Metallurgy, Moscow, 1971.
12. Khenzel, A.A., *Optimization of Energy Consumption in Extrusion Processes*, Metallurgy, Moscow, 1985.
13. Rapoport, E.Ya., Mathematical modelling of metals temperature fields during extrusion process, *Physica i Khimija obrabotki materialov,* Moscow, 1, 29, 1980.
14. Gun, G.Ya. et al., *Plastic Deformation of Metals*, Metallurgy, Moscow, 1968
15. Zimin, L.S., Rapoport, E.Ya., and Kondrashov, S.V., Determining temperature field of aluminum alloy cylindrical billet during axially symmetrical extrusion, *Algoritmy i programmy,* 1. 64, 1986.
16. Rapoport, E.Ya., Theory and algorithms optimal control of metals induction heating prior to pressure processing, D.D. thesis, Institut stali i splavov, Moscow, 1983.
17. Zimin L.S. et al., Reducing energy usage efficiency in Volga region, Saratov State University, Saratov, 122, 1990.
18. Sobol, I.M. and Statnikov, R.B., *Selection of Optimal Parameters in Multi-Criteria Problems,* Nauka, Moscow, 1981.

Conclusion

The main goal of this book is to provide a detailed systematic description of basic theory and practical applications of new methods for solving engineering optimization problems. The authors show the advantages of new, highly-effective approaches for the optimization of induction heating processes prior to metal working. These new optimization approaches deliver a essential advantages over the presently used classical approaches. Novel optimization techniques can be applied not only to induction heating applications but also to the optimization of a much wider range of technological processes.

The described technique proves to be efficient not only for a wide range of IHP optimization problems and a large variety of specific mathematical models, but for various cost functions, control input types, special requirements, and restraints of practical technology as well.

The method is based on the fundamental properties of temperature distributions of induction heating processes, which were ascertained and described here by authors; these properties remain constant for all of the stated problems.

In a particular application, the use of these properties allows the development of an appropriate set of equations to sought-for parameters of the optimal process taking into consideration specifics of the particular physical properties of the technological process.

The authors hope that readers will be able to apply the described optimization technique to similar problems and for the problems that remain beyond the bounds of this monograph.

Rigorous proof of the alternance method is not given in this book because it requires the authors to be fluent in advanced mathematics and have special knowledge in the area that is usually far beyond area of interest of majority of users of induction heating technology. We would suggest to those who are interested in the rigorous study of a particular topic discussed in this book to see the listings given in the reference sections of each chapter.

The authors will deeply appreciate the readers' comments regarding novel, alternate optimization techniques that were described here in the application of the optimization of induction metal heating processes.

Index

D

E

U

Related Other Titles

Kakac, Sadik
Heat Exchangers: Selection, Rating, and Thermal Design, Second Edition (0902)
ISBN: 0849309026 Date of Publication: March 2002
520 pages

Rudnev, Valery I.
Handbook of Induction Heating (DK2027)
ISBN: 0824708482 Date of Publication: December 2002
796 pages

Deshmukh, Yeshvant V.
Industrial Heating: Principles, Techniques, Materials, Applications and Design (DK340X)
ISBN: 0849334055 Date of Publication: May 2005
800 pages